Y0-AAN-302

# The Structure and Interpretation of the Standard Model

# Philosophy and Foundations of Physics
Series Editors: Dennis Dieks and Miklos Redei

*In this series:*

Vol. 1: The Ontology of Spacetime
Edited by Dennis Dieks

Vol. 2: The Structure and Interpretation of the Standard Model
By Gordon McCabe

# The Structure and Interpretation of the Standard Model

By

Gordon McCabe

*Dorchester, UK*

ELSEVIER

Amsterdam – Boston – Heidelberg – London – New York – Oxford – Paris
San Diego – San Francisco – Singapore – Sydney – Tokyo

Elsevier
Radarweg 29, PO Box 211, 1000 AE Amsterdam, The Netherlands
The Boulevard, Langford Lane, Kidlington, Oxford OX5 1GB, UK

First edition 2007



**Library of Congress Cataloging-in-Publication Data**
A catalog record for this book is available from the Library of Congress

**British Library Cataloguing in Publication Data**
A catalogue record for this book is available from the British Library

ISBN-13: 978-0-444-53112-4
Series ISSN: 1871-1774

For information on all Elsevier publications
visit our website at books.elsevier.com

Printed and bound in The Netherlands

07 08 09 10 11 10 9 8 7 6 5 4 3 2 1

*To my parents*

# Preface

As a philosophy of physics student in the 1990s, I approached particle physics with a desire to understand the types of thing which exist at the most fundamental physical level. Already converted to structural realism, I sought to find the mathematical structures used to represent elementary particles. What I found, in the physics textbooks, were some structures, but also a turgid and confusing array of calculational recipes; in the mathematical texts, meanwhile, I found structures such as fibre bundles and gauge connections, but often presented in frustrating abstraction from the realities of theoretical particle physics. Although Andrzej Derdzinski's 1992 monograph, *Geometry of the Standard Model of Elementary Particles*, expounded the structure of the first-quantized standard model, Derdzinski's work was difficult for a non-mathematician to grasp, and given that it promoted an unusual approach, it was often difficult to see the relationship between the structures in Derdzinski's book, and the things which particle physicists write about. The current work has therefore grown from my own exasperated failure, in the 1990s, to understand the mathematical structure of particle physics.

It is the purpose of this book to duly plug the hole which exists in the literature, and to provide an introduction to the mathematical structure of particle physics which is not dependent upon any particular approach, and which is accessible to those who have some competency in mathematics, but do not number themselves amongst the ranks of mathematicians or physicists. Whilst the mathematical material in this book outweighs the philosophical by some margin, the presentation is motivated by structural realism, and the mathematical material is framed by the foundational discussions.

I would like to take this opportunity to extend my gratitude to Dennis Dieks, for his patient and intelligent editorial skills; to the anonymous referees, for their constructive criticism; and to Andrzej Derdzinski, Karl Hofmann, John C. Baez, and Shlomo Sternberg, for their assistance on various technical matters. I would also like to thank Michael Proudfoot, John Preston, and David Oderberg in the Philosophy department at the University of Reading, for their assistance in the early stages of my academic work.

Gordon McCabe
Dorchester, Dorset
October 2006

# Contents

Chapter 1

# Introduction

The purpose of this book is to provide a mathematically rigorous and philosophically informed exposition of the 'standard model' of particle physics. The standard model is the currently accepted, empirically adequate[1] model of the particles and interactions in our universe. All the elementary particles in our universe, and all the non-gravitational interactions — the strong nuclear force, the weak nuclear force, and the electromagnetic force — are collected together and, in the case of the weak and electromagnetic forces, unified in the standard model.[2]

The standard model is an application of relativistic quantum theory, and in relativistic quantum theory the two basic types of thing which are represented to exist are matter fields and gauge force fields. The forces which matter fields exert on each other are mediated by gauge fields; a matter field interacts locally with a gauge field, which then interacts locally elsewhere with other matter fields.[3]

Relativistic quantum theory is obtained by applying quantization procedures to classical relativistic particle mechanics and classical relativistic field theory. The quantization of relativistic particle mechanics can be broken down into first-quantization and second-quantization, and as a consequence, there is a first-quantized version of the standard model, and a second-quantized version. It is possible in first quantization to represent interacting fields in a tractable mathematical manner, and the first-quantized approach is empirically adequate to the extent that it enables one to accurately represent many of the structural features of the physical world. Second-quantization, quantum field theory proper, is required to generate quantitatively accurate predictions, but quantum field theory proper is incapable of directly representing interacting fields.[4] The main focus of this text is upon the first-quantized

[1] A theory which is 'empirically adequate' is one which explains, predicts and organises the experimental and observational data.

[2] Note, however, that because the standard model fails to unify the strong force with the electroweak force, and because it contains numerous free parameters whose values must be set by observation and experiment, the standard model is not considered to be the final word on particle physics.

[3] This statement requires a couple of qualifications: firstly, in principle, matter fields can interact directly with each other under the Yukawa interaction (see Section 4.1); and secondly, the gauge bosons of some gauge fields (a notion introduced below) themselves possess charge, which entails that such gauge bosons can exert forces on each other. Gauge fields are therefore not always necessary as mediators between matter fields, and they can also play a non-mediating role.

[4] It is impossible to provide a mathematically rigorous representation of an interacting quantum field in Fock space, as demonstrated by Haag's theorem, and there is no known alternative to Fock space,

standard model, this being the arena in which interacting systems can be treated in a tractable manner, and the arena in which the mathematically interesting structures arise. Whilst this book contains an account of second quantization, and its possible implications for the status of the first-quantized theory (see Section 2.11), there is no intention to provide a detailed interpretation of second-quantized quantum field theory. Only a general introduction will be provided to the various numerical recipes and algorithms which constitute the second-quantized standard model. Accordingly, the discussion of quantum electrodynamics (the second-quantized theory of particles interacting via the electromagnetic force) is confined to Section 2.11, and there is no treatment of quantum chromodynamics (the second-quantized theory of particles interacting via the strong force). The interested reader should consult a text such as Nachtmann (1990), Huang (1992) or Ticciati (1999) for a detailed account of these topics.

The first-quantized standard model will be interpreted according to a philosophical doctrine referred to as 'structuralism.'[5] Structuralism holds that an axiomatic mathematical structure is the specification of a concept, and, in mathematical physics at least, the domain of a true theory is an instance of such a mathematical structure.[6] An axiomatic mathematical structure consists of a set, or collection of sets ('base' sets), equipped with properties, relationships, operations and distinguished elements, which are collectively required to satisfy a set of conditions, called the axioms of the structure. A base set equipped with such a structure is said to be a structured set.

The significance of structuralism is that it provides a *semantic* conception of mathematical physics. Structuralism stands in opposition to the *syntactical* view of theories, which is based upon the notion that an axiomatic mathematical system is merely an uninterpreted calculus, in which well-formed formulae are strings of symbols composed according to 'formal' syntactical rules, and theorems are deduced by the 'formal,' syntactical manipulation of these well-formed formulae. Under the syntactical view, a physical theory is then a partially interpreted calculus, in which the non-logical symbols are divided into those which are observational and those which are theoretical, and the observational terms and predicates have meaning because they are linked by 'correspondence rules' to objects and properties of the empirical world, but the theoretical terms and predicates are left uninterpreted (Torretti, 1999, pp. 410–412; Giere, 2000, pp. 515–520).

Two versions of structuralism will be of relevance in this text. Firstly, there is structural *realism*, which asserts that there is a physical domain which exists beyond

hence, as Teller puts it, "there appears to be no known consistent mathematical formalism within which interacting quantum field theory can be expressed" (1995, p. 115). Fock space and Haag's theorem will be covered in Section 2.11.

[5] This notion was originally advocated by, amongst others, Patrick Suppes (1969), Joseph Sneed (1971), and Frederick Suppe (1989).

[6] Whilst the definition of structuralism is most often expressed in terms of the set-theoretical, Bourbaki notion of a species of mathematical structure, one can reformulate the definition in terms of other approaches to the foundations of mathematics, such as mathematical category theory.

the empirical phenomena, and that, in mathematical physics at least, this objective physical domain is an instance of a mathematical structure. Secondly, there is structural *empiricism*, which holds that empirical phenomena are organised, explained and predicted by embedding the empirical data into mathematical structures, but those mathematical structures do not characterise anything which lies beyond the empirical phenomena.

It is reasonable to suppose that there are some theories of mathematical physics which are susceptible to a realist interpretation, and other theories which are susceptible to an empiricist interpretation. For example, one could reasonably propose that whilst general relativity provides a characterisation of the objective ontology of space–time, quantum theory provides only an empiricist calculus for particle phenomena. If so, then structural realism would be applicable to the former, and structural empiricism would be applicable to the latter. If quantum theory admits only an empiricist interpretation, then none of the structures to be found within it capture the ontology of a world which lies beyond the phenomena. Whilst not dismissing such a possibility, the interpretation proposed here, and elaborated in Section 2.11, will apply structural realism to the standard model of particle physics.

Ladyman (1998) has distinguished between two versions of structural realism: the ontic version and the epistemic version. Epistemic structural realism holds that mathematical physics provides knowledge of the structure which the physical world possesses beyond the empirical phenomena, but accepts that there is more to the world beyond the structure that it possesses. In contrast, the ontic version holds that the structure of the physical world is the only thing which exists. It is the ontic version of structural realism which provides the rationale for the mathematical exposition in this book.

Many authors take structural realism to merely assert that the physical world possesses a *second-order* structure. Different sets, with different first-order properties and relationships, can possess the same second-order structure. Hence, the assertion that a mathematical structure has a physical instance, does not pick out any particular set amongst those which have the first-order properties and relationships possessing this structure (see Votsis, 2003). Such an approach to structural realism is inspired by Bertrand Russell, who argued in *The Analysis of Matter* (Russell, 1927) that, at most, what we can *know* is the second-order structure of the world which lies beyond the empirical phenomena, not the first-order physical properties and relationships which possess that structure.

Whilst we might not be able to know more than the second-order structure of the physical world beyond the empirical data, this does not entail that the world beyond the empirical data isn't a specific set, whose elements are equipped with specific, first-order physical properties and relationships, which exist beyond the empirical first-order properties and relationships, and which satisfy the conditions of a certain second-order structure. Structural realism can be taken to assert that the physical world is a specific structured set, not merely a second-order structure. In particular, ontic structural realism can be taken to assert that the physical world is a structured set, and nothing more than a structured set.

However, structural realism still faces a number of problems, of which the biggest is arguably the so-called Newman problem. As Pooley puts it, "it is a theorem of set theory and second-order logic that any consistent proposition to the effect that a certain set of properties and relations exist, no matter what structural constraints are placed upon this set, will be true of any domain, provided that the domain has the right cardinality" (Pooley, 2005). This suggests, then, that structural realism is trivial as an account of the reality beyond the phenomena; if a physical domain is a set of a specific cardinality, then what lies beyond the observable properties and relations of that domain will possess any structure of that cardinality. The empirical properties and relationships, and the structural conditions they satisfy, might indeed pick out a specific structure, but this is merely the structure of the empirical phenomena. Structural realism, then, faces the danger of either being trivial, or collapsing into structural empiricism. For structural realism to work, it must be possible to define a specific physical structure beyond the empirical structure. In terms of the standard model of particle physics, a proposal will be made below to identify the structure which lies beyond the empirical phenomena. First, however, we need to understand a little more of the mathematical specifics of the standard model.

Given that the standard model is an application of relativistic quantum theory, and given that the two basic types of thing which are represented to exist in relativistic quantum theory are matter fields and gauge force fields, a structural realist interpretation of the standard model must analyse the mathematical structures used to represent matter fields and gauge fields. In the first-quantized theory, a matter field can be represented by a cross-section of a vector bundle, and a gauge force field can be represented by a connection upon a principal fibre bundle. For the uninitiated, Appendix C precisely defines the concepts of fibre bundle, vector bundle, and principal fibre bundle, but a brief digression to introduce these concepts may be helpful now:

A fibre bundle is something which enables one to attach a structured set to each point of an underlying mathematical space, called the base of the bundle. In the case of interest to us, the base space will be a manifold. The structured set assigned to a point is called the fibre over that point, and the collection of all the fibres is the fibre bundle. Each fibre is isomorphic to something called the typical fibre of the fibre bundle, hence a fibre bundle assigns isomorphic sets to the various different points of the base space. If each fibre is a vector space, then one has a vector bundle; if each fibre is a group, then one has a group bundle; and so on. A cross-section of a fibre bundle is something which picks out an element from each one of the fibres. Thus, a cross-section of a vector bundle, by which a matter field is represented, picks out an element from each one of the vector spaces assigned to the points of the base manifold. Now, a principal fibre bundle, without prematurely delving into its subtleties, is a fibre bundle in which there is a group, called the structure group, which acts as a group of transformations upon each of the fibres. A connection $\nabla$ upon a fibre bundle, by which a gauge field is represented, enables one to define the notion of path-dependent parallelism between the different fibres. Thus, equipped with a connection, one can determine whether elements in fibres over two distinct points are parallel to each other along a particular path joining those points. Chapter 2 will provide a full exposition of matter fields, and Chapter 3 will provide an exposition of gauge fields and connections.

The fact that, in the first-quantized theory, a matter field can be represented by a cross-section of a vector bundle, and a gauge force field can be represented by a connection upon a principal fibre bundle, is rather curious because the matter fields are obtained by quantizing the point-like objects of classical relativistic particle mechanics, whilst, at first sight, the gauge fields have undergone no quantization at all. If one treats the first-quantized matter fields as classical fields, and if one treats those matter fields as interacting with classical gauge fields, then there is no inconsistency. However, on both counts, such a treatment may be misleading:

On the first count, given that the matter fields in the first-quantized theory are the upshot of quantizing classical particles, they are provisionally interpretable as wave-functions, i.e., they are provisionally interpretable as vectors in a quantum state space, coding probabilistic information.[7] One of the outputs from the first quantized theory is a state space for each type of elementary particle, which becomes a so-called 'one-particle subspace' of the multiple-particle state space ('Fock space'), used in the second-quantized theory. The vector bundle cross-sections which represent a matter field in the first-quantized theory, are vectors from the one-particle subspace of the second-quantized theory. As we will see in Section 2.11, there are creation and annihilation operators defined upon the Fock spaces of the second-quantized theory, and these create or annihilate particle states from the first-quantized theory.

On the second count, the connections which represent a gauge field can be shown, under a type of symmetry breaking called a 'choice of gauge,' to correspond to cross-sections of a vector bundle direct sum[8] (Derdzinski, 1992, p. 91). The cross-sections of the individual direct summands are vectors from the one-particle subspaces of particles called 'interaction carriers,' or 'gauge bosons.' Hence, neither the matter fields nor the gauge fields of the first-quantized theory can be unambiguously treated as classical fields.

Given that the two basic types of thing which exist in relativistic quantum theory are matter fields and gauge fields, one must countenance the possibility that particles are merely derivative notions, perhaps merely particular states of an underlying quantum field. In particular, it has been suggested that particles are merely 'excitation modes' of quantum fields. Given that the state of an object is the set of all that

[7] See Section 2.11 for reasons why this provisional interpretation cannot be maintained.

[8] The direct sum $\bigoplus_i V_i$ of a collection of vector spaces $\{V_i\}$ is the vector space formed by taking the sums of the vectors in the constituent spaces. The direct sum contains all the constituent vector spaces as linearly independent subspaces. A direct sum of a collection of vector bundles over the same underlying space possesses, over each point, a fibre consisting of the direct sum of the vector spaces over that point in each one of the constituent vector bundles.

object's properties, if a particle is merely a state of an underlying quantum field, it follows that a particle is not an object in its own right, but merely a property, or a collection of properties, possessed by an object. Whilst the question, 'What is a particle?', addresses the metaphysical problem of whether a particle is an object or a property, one can pose the secondary metaphysical question, 'What is an *elementary* particle?' To some extent, the answer to this question can be separated from the answer to the first question. If particles are objects in their own right, then one can define an elementary particle to be a particle which is not composed of other particles. Alternatively, if a particle is taken to be merely an excitation mode or special type of state of an underlying quantum field, then one can speak of a matter field or gauge field as being elementary if the excitation modes or particle states ('particles') of that field cannot be decomposed into the excitation modes or particle states ('particles') of other fields. Note that in both cases here the question, 'What is an elementary particle?', is answered in terms of the notions of whole and part; i.e., the answer is expressed in mereological terms. In Section 4.13 we shall explore a metaphysically distinct answer in terms of extrinsic properties and intrinsic properties.

The basic interpretational doctrine in this text can be stated as follows: *Each type of matter field or gauge field in first-quantized relativistic quantum theory, is the type of thing which is represented by a cross-section of such-and-such a bundle over space–time, satisfying such-and-such conditions. In the guise of one-particle states, these are, more or less, the types of thing which are created or annihilated in second-quantized relativistic quantum theory, but these things are there re-cast as properties of a quantum field, rather than things in their own right.*

In terms of structural realism, the structures which represent matter fields, gauge fields, and their associated particles, are fibre bundle cross-sections, and these fibre bundle cross-sections can be treated as the structures which lie beyond the empirical phenomena. The cross-sections represent physical particles, but they lie beyond properties such as energy, momentum and charge, which I shall take to be empirical properties (even if they are not directly observable), and which are merely derivable from the fibre bundle cross-sections. In terms of the standard model of particle physics at least, this provides a potential response to the Newman problem.

To make this introductory discussion a little more concrete, let us provide a brief roster of the elementary particles known to exist. All particles, including elementary particles, are divided into fermions and bosons according to the value they possess of a property called 'intrinsic spin.' If a particle possesses a non-integral value of intrinsic spin, it is referred to as a fermion, whilst if it possesses an integral value, it is referred to as a boson. The particles of the elementary matter fields are fermions and the interaction carriers of the gauge force fields are bosons. The elementary fermions come in two types: leptons and quarks. Whilst quarks interact via both the strong and electroweak forces, leptons interact via the electroweak force only. There are six types of lepton and six types of quark. The six leptons consist of the electron and electron-neutrino ($e$, $\nu_e$), the muon and muon-neutrino ($\mu$, $\nu_\mu$), and the tauon and tauon-neutrino $(\tau, \nu_\tau)$. The six quarks consist of the up-quark and down-quark $(u, d)$, the charm-quark and strange-quark $(c, s)$, and the top-quark and bottom-quark $(t, b)$. The six leptons have six anti-leptons, $(e^+, \bar{\nu}_e)$, $(\mu^+, \bar{\nu}_\mu)$, $(\tau^+, \bar{\nu}_\tau)$, and the six quarks have six anti-quarks $(\bar{u}, \bar{d})$, $(\bar{c}, \bar{s})$, $(\bar{t}, \bar{b})$. The historical process which led to the discovery of all these elementary particles is charted in Table 1.1.[9]

The elementary fermions are partitioned into three generations, of which the first generation, $(e, \nu_e, u, d)$, and its anti-particles, is responsible for most of the macroscopic phenomena we observe. Quarks are bound together into composite systems by the strong force, a short-range force acting over distances of the order $10^{-13}$ cm. Over these distance scales, the strong force binds triples of up-quarks and down-quarks together to form protons and neutrons. The residual effects of the strong inter-quark forces are responsible for binding protons and neutrons together in an atomic nucleus. In particular, the strong force is able to overcome the electrical repulsion between neighbouring protons in the nucleus. The electromagnetic forces between atomic nuclei and electrons then lead to the formation of atoms and molecules (Manin, 1988, p. 3; Sternberg, 1994, p. 156; Derdzinski, 1992, p. 6).

The application of ontic structural realism to particle physics has the consequence that each type of matter field or gauge field, and their associated particle types listed above, is considered to be an instance of some species of mathematical structure. In this vein, a typical definition of a particle is proffered by Huang, who states that "a particle, be it 'fundamental' or composite, is defined as a state of a quantum field that transforms under elements of the Poincaré group according to a definite irreducible representation" (1992, p. 2) (and note here the use of the idea that a particle is merely a type of state of an underlying quantum field). In particular, one frequently finds in the literature the assertion that an elementary particle 'is' an irreducible, unitary representation of the local space–time symmetry group (e.g. Sternberg, 1994, p. 149; Streater, 1988, p. 144). As such, these assertions constitute expressions of structuralism. Whilst this text endorses such a structuralist approach, it proposes an alternative definition of what an elementary particle is in the first-quantized standard model. Whilst the *state space* of a free elementary particle in our universe is indeed an irreducible, unitary representation of the universal cover of the restricted Poincaré group (see Section 2.1), the structure of a particle state space does not itself represent the structure of the particle. The structure of the state space associated with an elementary particle does not tell us what type of thing the particle is. There are different levels of structure here, and, to echo Bueno (2001), it is important to establish which levels of structure the structural realist should be realist about. As stated above, this text treats the matter fields, gauge fields, and associated particles in the first-quantized standard model, as the types of thing which are represented by cross-sections of such-and-such bundles over space–time, satisfying such-and-such conditions. A particle state space can then be constructed from such cross-sections, but once constructed,

[9] This is a modified version of the table in Rajasekaran (1997).

Table 1.1
Theoretical and experimental history of the standard model

| | Theory | | Experiment |
|---|---|---|---|
| | | 1897 | Thomson's discovery of electron $e^-$. |
| 1918 | Weyl's first gauge concept. | | |
| | | 1919 | Rutherford's discovery of proton $p$. |
| | | 1922 | Confirmation that photon is elementary (Compton). |
| 1928 | Dirac's prediction of anti-particles. | | |
| 1929 | Weyl's gauge theory of electromagnetism. | | |
| | | 1932 | Anderson discovers positron.<br>Evidence for neutron (Chadwick). |
| 1934 | Fermi's theory of weak interactions. | | |
| 1935 | Yukawa's prediction of the meson. | | |
| | | 1947 | Discovery of $\pi$-meson and $\mu$-lepton. |
| 1954 | Yang–Mills/Utiyama gauge field theory. | | |
| 1956 | Lee and Yang predict non-conservation of parity in weak interactions. | 1956 | Detection of neutrino (Reines and Cowan).<br>Wu *et al.* discover parity violation. |
| 1958 | V–A theory of weak interactions. | | |
| 1961 | Weak neutral-currents predicted (Glashow). | | |
| 1964 | Higgs mechanism.<br>Quarks and strong force (Gell-Mann; Zweig).<br>Coloured quarks and gluons (Greenberg; Han and Nambu). | | |
| 1967 | Electroweak unification (Weinberg; Salam; Glashow). | | |
| 1971 | Renormalizability of gauge theories with Spontaneous Symmetry Breaking ('t Hooft). | | |
| 1973 | Quantum Chromodynamics Lagrangian (Fritzsch, Gell-Mann and Leutwyler). | 1973 | Weak neutral-currents detected. |
| | | 1974 | Evidence of c-quark from the $J/\psi$ resonance. |
| | | 1975 | Evidence of $\tau$-lepton. |
| | | 1977 | Evidence of b-quark from the $\Upsilon$ resonance. |
| | | 1979 | Evidence for the gluon in $e^+e^- \to 3$ jet. |
| | | 1983 | $W^{\pm}$, $Z$ bosons discovered. |
| | | 1994 | Evidence for the t-quark. |

the state space is only determined up to unitary isomorphism within an equivalence class; from the viewpoint of state-space calculations, the choice of a particular member from the unitary equivalence class is a matter of mere convenience, but from the viewpoint of ontological considerations, the choice is vitally important.

There is one point which needs to be emphasised here: a question such as, 'What type of thing is an elementary particle?', cannot be answered without a final, definitive theory, and neither first-quantized relativistic quantum theory, nor the second quantized theory, deserve this status. Consequently, one can only state that an elementary particle is *represented to be* a such-and-such mathematical object in the standard model. If, however, a final, definitive theory were to be discovered, then ontic structural realism would enable one to assert that an elementary particle is nothing more than a such-and-such mathematical object.

Let us change tack slightly to consider another purpose of this text. There is a natural extension of ontic structural realism which proposes that an entire physical universe is an instance of a mathematical structure. Those expressions of structural realism which state that 'the' physical universe is an instance of a mathematical structure, tacitly assume that our physical universe is the only physical universe. If one removes this assumption, then structural realism can be taken as the two-fold claim that (i) our physical universe is an instance of a mathematical structure, and (ii), other physical universes, if they exist, are either different instances of the same mathematical structure, or instances of different mathematical structures. Given that mathematical structures are arranged in tree-like hierarchies, other physical universes may be instances of mathematical structures which are sibling to the structure possessed by our universe. In other words, the mathematical structures possessed by other physical universes may all share a common parent structure, from which they are derived by virtue of satisfying additional conditions. This would enable us to infer the mathematical structure of other physical universes by first generalising from the mathematical structure of our own, and then classifying all the possible specialisations of the common, generic structure.

This point can be understood from the perspective of mathematical logic, in which a theory is a set of sentences in some language, which is closed under logical implication. A 'model' for such a set of sentences is an interpretation of the language in which those sentences are expressed, which renders each sentence as true. In this sense, each theory in mathematical physics has a class of models associated with it. As Earman puts it, "a practitioner of mathematical physics is concerned with a certain mathematical structure and an associated set $\mathfrak{M}$ of models with this structure. The... laws $L$ of physics pick out a distinguished sub-class of models $\mathfrak{M}_L := \mathrm{Mod}(L) \subset \mathfrak{M}$, the models satisfying the laws $L$ (or in more colorful, if misleading, language, the models that "obey" the laws $L$)" (2002, p. 4).

The existence of numerous free parameters in the standard model, whose values are set by experiment and observation, rather than theoretically derived, has prompted the suggestion that there are collections of universes which realise all possible value-combinations for these parameters. At first sight, such universe collections are different in type from those obtained by varying a mathematical structure, or by taking all the models of a fixed mathematical structure. However, this appearance is deceptive because the values chosen for the free parameters of a theory correspond to a choice of structure at various points in the theory. For example, the free parameters of the standard model of particle physics include the coupling constants of the strong and electromagnetic forces, the Weinberg angle, the masses of the elementary quarks and leptons, and the values of the four parameters in the Cabibbo–Kobayashi–Maskawa matrix, which specifies the 'mixing' of the $\{d, s, b\}$ quark flavours in weak force interactions. In terms of structure, the value chosen for the coupling constant of a gauge field with gauge group $G$ corresponds to a choice of metric in the Lie algebra $\mathfrak{g}$; the Weinberg angle corresponds to a choice of metric in the Lie algebra of the electroweak force (see Section 4.3); the values chosen for the masses of the elementary quarks and leptons correspond to the choice of a finite family of irreducible unitary representations of the local space–time symmetry group, from a continuous infinity of alternatives on offer; and the choice of a specific Cabibbo–Kobayashi–Maskawa matrix corresponds to the selection of a specific orthogonal decomposition $\sigma_{d'} \oplus \sigma_{s'} \oplus \sigma_{b'}$ of the fibre bundle which represents a generalisation of the $\{d, s, b\}$ quark flavours (see Section 4.8).

Hence, it is the aim of this book not only to explain how the particle world of our universe is an instance of certain mathematical structures, but to also emphasise how the mathematical structures of our particle world are special cases of more general mathematical structures, from which one may be able to infer the nature of the particle world in other universes. To this end, Section 2.6 considers free particles in other universes, Section 4.10 considers gauge fields in other universes, and Sections 5.1 and 5.2 consider the possible standard model gauge groups and representations in other universes. It is structural realism which provides the rationale for this material.

The text assumes that the reader knows nothing about the mathematics of particle physics or the standard model. All the terms of discourse are explained wherever practical, and the book is as self-contained as possible. However, the reader is assumed to have some experience of mathematical techniques and terminology, and a general knowledge of the mathematical landscape. Particular attention is given to explaining the different approaches to particle physics and the standard model which can be found within the literature: Chapter 2 explains, at length, the relationship between the configuration space approach and the momentum space approach to the representation of free matter fields; Chapter 3 explains the relationship between the principal fibre bundle approach and the interaction bundle approach to the representation of gauge fields; Chapter 5 expounds and clarifies the standard model gauge groups and their representations; Chapter 6 unravels the relationship between the gauge group approach and the vector bundle approach to the standard model.

The book can be read as a comprehensive reference on the first-quantized fibre bundle formulation of the standard model and the group theoretical representations of elementary particles. For those already familiar with the standard model, the mathematical generality of the text, the links made between the different approaches, and the interpretational/philosophical asides, will provide some new perspectives upon the subject. Whilst a number of philosophical questions are discussed at length, the purpose of the book is not to fully develop all of the philosophical questions raised; rather, the intention is, metaphorically, to excavate a network of hitherto unexplored tunnels and caverns, whose rich seams can be mined by other researchers. Given that many philosophers of physics may have been deterred from investigation of the standard model because, as Torretti puts it, they are "sickened by untidy math" (1999, p. 396), one purpose of this book is to demonstrate that the standard model can be understood with the use of very elegant mathematics.

Chapter 2

# Matter Fields

This chapter is concerned with the representation of *free* matter fields in the first-quantized theory. Free matter fields are those which are idealised to be free from interaction with gauge force fields. There are convincing arguments that the notion of a particle in relativistic quantum theory is derivative from the notion of a field, with particles perhaps corresponding to a subset of field states. There are also arguments, to which we will turn in Section 2.11, which claim to show that there are no states in the first-quantized theory which possess the characteristics one would expect of particles. Despite these arguments, we shall fall into the, no doubt, shameful habit of using the terms 'matter field' and 'particle' interchangeably in this chapter, and throughout much of the book. This is done only with the tacit understanding that particles, if such a notion can be found at all within relativistic quantum theory, are derivative concepts. Repeatedly making this explicit would give the prose a rather forced, didactic feel, and given that this book is not primarily concerned with the notions of field and particle, it is hoped that the reader will excuse this equivocation, and that the inclusion of the arguments in Section 2.11 will prevent the reader from drawing any unwarranted conclusions.

## 2.1. The local space–time symmetry group

This chapter is particularly concerned with the representation of free *elementary* particles, i.e., particles which are not represented to be composed of other particles. To specify the free elementary particles capable of existing in a universe, i.e., the free elementary 'particle ontology,' one specifies the *projective*, unitary, irreducible representations of a 'local' space–time symmetry group. However, because free particles are physical idealisations, one works backwards from the space–time symmetry group of *interacting* systems, and one bestows this symmetry group on free particles to ensure that when the idealisation is removed, and interactions are included, the space–time symmetry group is correct. Hence, the symmetry group for free particles is not necessarily the largest local space–time symmetry group permitted by the structure of space–time, but possibly a subgroup determined by the space–time symmetry of interacting particles. In turn, the space–time symmetry group of interacting particles is determined by which gauge fields exist, and the way in which matter fields couple to those gauge fields.

To understand what a projective, unitary, irreducible representation is, some explanation of the mathematical concepts may be helpful here. A linear representation of a group assigns to each group element a one-to-one linear transformation of a vector space $V$. Given that the linear transformations of a vector space are the structure-preserving transformations of a vector space, each group element is mapped to a one-to-one, structure-preserving transformation of a vector space. An irreducible linear representation is one for which there is no closed subspace, apart from the zero subspace and the entire vector space, which is invariant under the action of the represented group elements. In other words, one cannot restrict the representation to a smaller closed subspace. An inner product on a vector space enables one to define the length of vectors and the angles between vectors; if a representation space is equipped with an inner product, then the linear transformations which leave this inner product between vectors unchanged are referred to as orthogonal transformations (in the case of a real vector space), or unitary transformations (in the case of a complex vector space). A unitary group representation is a representation which assigns a unitary transformation of a complex vector space to each group element. A projective representation of a group assigns to each group element a one-to-one, structure preserving transformation of the set of one-dimensional subspaces $\mathscr{P}(V)$ in a vector space $V$. In the case of interest in this text, the vector space will be a Hilbert space $\mathscr{H}$, a special type of vector space equipped with an inner product, and the projective space in this case is considered to inherit its own product from the inner product on the Hilbert space. Defining a symmetry of $\mathscr{P}(\mathscr{H})$ to be a one-to-one mapping which preserves this product, each such symmetry can be implemented by a unitary operator $U$ on the Hilbert space,[1] but the unitary operator is only unique up to a complex number of unit modulus $z = e^{i\theta}$, a so-called 'phase factor.' A projective unitary representation is a projective representation implemented by unitary operators, which are only determined up to a phase factor. One can choose a unitary operator from each such equivalence class, but doing so does not generally define a unitary representation. Instead, such operators will satisfy the equation

$$U(g_1 \circ g_2) = \omega(g_1, g_2)U(g_1)U(g_2),$$

with $\omega(g_1, g_2)$ a complex number of unit modulus. If one can judiciously choose elements from each equivalence class so that $\omega = 1$ everywhere, then the projective unitary representation is said to be unitarizable as an ordinary representation.

The 'pure' states of a quantum system, the states which provide a maximal description, can be represented by the one-dimensional subspaces of a Hilbert space. However, these one-dimensional subspaces only specify the state of a quantum system with respect to a particular reference frame. Under a change of reference frame, the state of the system must also change, hence a change of reference frame corresponds to a transformation of the set of one-dimensional subspaces. Each element of

[1] In general, the operator can be either unitary or anti-unitary.

the local space–time symmetry group can be treated as a change of reference frame (this is referred to as the 'passive' sense of such a symmetry), hence the requirement that an elementary system correspond to a projective, unitary, irreducible representation of the local symmetry group.[2]

To understand what the local symmetry group of a space–time is, one needs to begin by understanding that the large-scale structure of a space–time is represented by a pseudo-Riemannian manifold $(\mathcal{M}, g)$. At each point $x$ of a manifold $\mathcal{M}$ the set of all possible tangents to the curves which pass through that point form a special vector space, called the tangent vector space $T_x\mathcal{M}$. A metric tensor field $g$ on the manifold assigns an inner product $\langle\ ,\ \rangle$ to the tangent vector space at each point of the manifold, and this is considered to provide the manifold with a geometrical structure. The metric $g$ has a signature $(p, q)$ determined by the number $p$ of orthogonal unit vectors which have a positive inner product $\langle v, v\rangle > 0$, and the number $q$ of orthogonal unit vectors which have a negative inner product $\langle w, w\rangle < 0$. The dimension $n$ of the manifold $\mathcal{M}$, and the signature $(p, q)$ of the metric $g$, determine the largest possible local symmetry group of the space–time. The automorphism group[3] of $T_x\mathcal{M}$, treated as an affine space,[4] defines the largest possible local symmetry group of such a space–time, the semi-direct product $O(p, q) \ltimes \mathbb{R}^{p,q}$. The semi-orthogonal group $O(p, q)$ is the group of linear transformations which preserve a signature $(p, q)$ inner product on a real vector space. $\mathbb{R}^{p,q}$ is $(p + q)$-dimensional Euclidean space, equipped with an inner product of signature $(p, q)$, and treated as a group of translations in this context. A semi-direct product is a special type of product of two groups in which the first factor acts as a group of transformations upon the second factor in a particular way (defined precisely in Section 2.4).

If there is no reason to restrict to a subgroup of this, then one specifies the possible free elementary particles in such a universe by specifying the projective, unitary, irreducible representations of $O(p, q) \ltimes \mathbb{R}^{p,q}$. As a consequence of $O(p, q) \ltimes \mathbb{R}^{p,q}$ being a non-compact Lie group,[5] these representations are infinite-dimensional.

In the case of our universe, the dimension is $n = 4$, and the signature is $(p, q) = (3, 1)$, indicating three spatial dimensions and one time dimension. An $n$-dimensional pseudo-Riemannian manifold such as this, with a signature of $(n-1, 1)$, is said to be a Lorentzian manifold. Each tangent vector space of a 4-dimensional Lorentzian manifold is isomorphic to Minkowski space–time, hence the automorphism group of such

[2] An *irreducible* projective representation is one for which there is no non-trivial closed subspace $W \subset \mathscr{H}$ such that $\mathscr{P}(W)$ is invariant under the group representation.

[3] The automorphisms of a structured set are the one-to-one maps of the set onto itself which preserve the structure possessed by that set.

[4] An affine space is a set which is acted upon transitively and effectively by the additive group structure of a vector space, the so-called 'translation space.' Given the vector space $T_x\mathcal{M}$, it can be rendered as an affine space if $T_x\mathcal{M}$ is treated as a space with no preferred origin, upon which $T_x\mathcal{M}$ itself defines the transitive and effective action.

[5] See Appendix A for a definition of compactness, and Appendix B for a definition of a Lie group.

a tangent vector space is the Poincaré group, $O(3,1) \ltimes \mathbb{R}^{3,1}$, the largest possible symmetry group of Minkowski space–time.[6] In the case of our universe, particle interactions involving the weak force violate space-reflection symmetry ('parity') and time reversal symmetry, hence the Lorentzian manifold appears to be equipped, at the very least, with a local time orientation and a local space orientation.[7] This entails that time reversal operations, parity transformations, and combinations thereof, are not considered to be local space–time symmetries. Hence, the local space–time symmetry group of our universe appears to be a subgroup of the Poincaré group, called the *restricted* Poincaré group, $SO_0(3,1) \ltimes \mathbb{R}^{3,1}$. The projective, unitary, irreducible representations of the restricted Poincaré group are in bijective correspondence with the ordinary, unitary, irreducible representations of its universal covering group,[8] $SL(2,\mathbb{C}) \ltimes \mathbb{R}^{3,1}$. In particular, every projective unitary representation of $SL(2,\mathbb{C}) \ltimes \mathbb{R}^{3,1}$ is unitarizable, hence the projective unitary representations of $SO_0(3,1) \ltimes \mathbb{R}^{3,1}$ can be lifted to projective unitary representations of $SL(2,\mathbb{C}) \ltimes \mathbb{R}^{3,1}$ with the aid of the covering map, and then unitarized. Thus, one specifies the free elementary particle ontology of our universe by specifying the ordinary, irreducible, unitary representations of $SL(2,\mathbb{C}) \ltimes \mathbb{R}^{3,1}$.

Let us examine the selection of the restricted Poincaré group in a little more detail. The largest possible local symmetry group of our space–time, the Poincaré group $O(3,1) \ltimes \mathbb{R}^{3,1}$, is the group of *diffeomorphic*[9] isometries of Minkowski space–time $\mathcal{M}$, and a semi-direct product of the Lorentz group $O(3,1)$ with the translation group $\mathbb{R}^{3,1}$. In contrast, the Lorentz group is the group of *linear* isometries of Minkowski space–time. The group of linear isometries doesn't include the translation group $\mathbb{R}^{3,1}$, because such translations don't leave the zero vector fixed. The Poincaré group is a disconnected group which possesses four components; one component contains the isometry which reverses the direction of time, another component contains the isometry which performs a spatial reflection, another component contains the isometry which both reverses the direction of time and performs a spatial reflection, whilst the identity component $SO_0(3,1) \ltimes \mathbb{R}^{3,1}$ preserves both the direction of time and spatial parity. The identity component $SO_0(3,1) \ltimes \mathbb{R}^{3,1}$, the restricted Poincaré group, is also referred to as the 'proper isochronous' Poincaré group, and is often denoted as $\mathscr{P}_+^\uparrow$ in the physics literature. Similarly, the identity component of the Lorentz group, $SO_0(3,1)$, is variously referred to as the *restricted* Lorentz group, or the 'proper isochronous' Lorentz group, and is often denoted as $\mathscr{L}_+^\uparrow$ in physics literature.

---

[6] Minkowski space–time and the tangent vector space of an arbitrary Lorentzian manifold are both treated here as affine spaces, hence their mutual symmetry group includes the translation group $\mathbb{R}^{3,1}$.

[7] These orientations could, quite conceivably, be defined by local physical fields on space–time. There is no intention here to imply that these orientations are properties which space–time possesses independently of the physical fields on space–time. Whilst space–time itself possesses the property of local orientability, the choice of an orientation may be supplied by local physical fields.

[8] See Appendix B for a definition of the universal covering group.

[9] Diffeomorphisms are the isomorphisms of manifolds.

In 1956 it was discovered that interactions involving the weak nuclear force violate parity symmetry, and, in 1964, a single weak interaction process, the decay of the $K^0$-meson, was discovered to violate time reversal symmetry. Thus, the physical evidence seems to indicate that our universe possesses, at the very least, a separate local time orientation and local space orientation. As a consequence, the local symmetry group of our space–time is the restricted Poincaré group $SO_0(3, 1) \ltimes \mathbb{R}^{3,1}$, the group of local symmetries which separately preserve time and space orientation. However, Geroch and Horowitz argue to the contrary that "the strongest conclusion to be drawn... using the presently observed symmetry violations in elementary particle physics, is that our spacetime must be total orientable. One cannot conclude from this, for example, that our spacetime must be separately time- and space-orientable" (1979, p. 229). If our universe merely possessed a 'total' space–time orientation, the local space–time symmetry group would not be the restricted Poincaré group, but $SO(3, 1) \ltimes \mathbb{R}^{3,1}$.

Some further explanation may help to make this distinction clear. Whilst a Lorentzian manifold equipped with a separate time orientation and space orientation must also possess a space–time orientation (a 'total' orientation), the converse is not true. The presence of a space–time orientation does not entail the presence of a separate time orientation and space orientation, and the absence of a space–time orientation does not entail the separate absence of a time orientation and space orientation. Letting $T$ denote the operation of time reversal, and letting $P$ denote the operation of parity reversal, one says that, whilst $P$ and $T$ entails $PT$, the converse is not true. Sternberg states that "the current belief is that the correct [local space–time symmetry] group has two components corresponding to simultaneous space and time inversion, but that this transformation must be accompanied by reversal of all electric charges" (Sternberg, 1994, p. 150). The comments of Sternberg and Geroch–Horowitz seem to stem from the *CPT* theorem of second-quantized quantum field theory, which states that a combination of time reversal, parity transformation, and charge conjugation ($C$) provide a physical symmetry. If the local symmetry group of space–time were $SO(3, 1) \ltimes \mathbb{R}^{3,1}$, then $PT$ would be a physical symmetry, but it is known that such is not the case in weak interactions. The *CPT* theorem is consistent with the notion that neither $P$, nor $T$, nor $PT$ are physical symmetries, and is consistent with the belief that the restricted Poincaré group $SO_0(3, 1) \ltimes \mathbb{R}^{3,1}$ is the local space–time symmetry group in our universe.

As stated at the outset, because free particles are physical idealisations, one works backwards from the space–time symmetry group of interacting systems, and one bestows this symmetry group on free particles to ensure that when the idealisation is removed, and interactions are included, the space–time symmetry group is correct. As we will subsequently explain at length, in the first-quantized theory, interacting particles are represented by the cross-sections of bundles constructed from the tensor products of free-particle bundles with so-called interaction bundles, hence every free particle has a corresponding interacting particle, and the space–time symmetry group of a free particle is inherited by an interacting counterpart. Because weak interactions violate parity symmetry, and because weak interactions themselves are a consequence of a world in which the unified electroweak force has undergone spontaneous symmetry breaking (see Section 4.4), the choice of $SO_0(3,1) \ltimes \mathbb{R}^{3,1}$, rather than $SO(3,1) \ltimes \mathbb{R}^{3,1}$, as the local space–time symmetry group for free particles, is not determined by space–time structure alone, but by the space–time symmetry of interacting particles after the unified electroweak force has been broken. Local space–time structure alone does not determine the permitted elementary particles; rather, it is the combination of local space–time structure and the coupling of matter fields to the gauge force fields present, which determine the permitted elementary particles. The experimentally observed parity violations are a consequence of the fact that right-handed leptons and quarks do not couple at all to the $W$ boson of the weak force, and the coupling strength of leptons and quarks to the $Z$ boson of the weak force depends upon whether they are left-handed or right-handed.[10] The interaction carriers of the strong force, so-called 'gluons,' couple in the same way to left-handed and right-handed fermions, hence the strong force preserves parity symmetry.

The physically relevant irreducible unitary representations of $SL(2,\mathbb{C}) \ltimes \mathbb{R}^{3,1}$ are parameterized by one continuous parameter, the mass $m$, and one discrete parameter, the spin $s$. One can present these representations in either the (energy–)momentum representation (the Wigner representation), or the configuration representation. Whilst the tangent vector space $T_x\mathcal{M}$ at an arbitrary point $x$ of a 4-dimensional Lorentzian manifold is isomorphic to Minkowski space–time, the cotangent vector space[11] $T_x^*\mathcal{M}$ can be treated as Minkowski energy–momentum space. In the Wigner approach, free particles of mass $m$ and spin $s$ correspond to vector bundles $E_{m,s}^{\pm}$ over mass hyperboloids and light cones $\mathscr{V}_m^{\pm}$ in Minkowski energy–momentum space $T_x^*\mathcal{M} \cong \mathbb{R}^{3,1}$. Let us introduce the following indefinite metric on $T_x^*\mathcal{M}$:

$$\langle p, p\rangle = -p_0 p_0 + p_1 p_1 + p_2 p_2 + p_3 p_3.$$

For any $m \in \mathbb{R}_+$ the mass hyperboloid $\mathscr{V}_m$ is the set $\mathscr{V}_m = \{p \in \mathbb{R}^{3,1}\colon \langle p, p\rangle = -m^2 < 0\}$. Each mass hyperboloid consists of two connected components, the forward mass hyperboloid $\mathscr{V}_m^+ = \{p \in \mathbb{R}^{3,1}\colon \langle p, p\rangle = -m^2 < 0,\ p_0 > 0\}$ and the backward mass hyperboloid $\mathscr{V}_m^- = \{p \in \mathbb{R}^{3,1}\colon \langle p, p\rangle = -m^2 < 0,\ p_0 < 0\}$. The forward light cone is $\mathscr{V}_0^+ = \{p \in \mathbb{R}^{3,1}\colon \langle p, p\rangle = m^2 = 0,\ p_0 > 0\}$ and the backward light cone is $\mathscr{V}_0^- = \{p \in \mathbb{R}^{3,1}\colon \langle p, p\rangle = m^2 = 0,\ p_0 < 0\}$.

[10] These facts will be recast in the final chapter in the language of interacting-particle bundles, and the irreducible representations of the standard model gauge group $SU(3) \times SU(2) \times U(1)$. For example, whilst a right-handed electron corresponds to cross-sections of the bundle $\chi^{-2} \otimes \sigma_R$, a left-handed electron corresponds to cross-sections of the bundle $\tau \otimes \chi^1 \otimes \sigma_L$. In loose terms, the $W$ boson corresponds to $\tau$, the $Z$ boson corresponds to $\chi$, and the strength of coupling to the $Z$ boson corresponds to the superscript over $\chi$, the so-called 'weak hypercharge.'

[11] The cotangent vector space $T_x^*\mathcal{M}$ is said to be the *dual* vector space. I.e., it is the vector space containing all the linear maps $f : T_x\mathcal{M} \to \mathbb{R}$.

It is the Hilbert spaces $\mathscr{H}^{\pm}_{m,s}$ of square-integrable[12] cross-sections of these vector bundles $E^{\pm}_{m,s}$ which provide the physically relevant irreducible unitary representations of $SL(2,\mathbb{C}) \ltimes \mathbb{R}^{3,1}$.

The irreducible unitary representation of $SL(2,\mathbb{C}) \ltimes \mathbb{R}^{3,1}$ upon the space $\Gamma_{L^2}(E^{+}_{m,s})$ of square-integrable cross-sections of $E^{+}_{m,s}$, for a particle of mass $m$ and spin $s$, is unique up to unitary equivalence. The anti-particle of mass $m$ and spin $s$ is represented by the conjugate representation[13] on the space $\Gamma_{L^2}(E^{-}_{m,s})$ of square-integrable cross-sections of the vector bundle $E^{-}_{m,s}$. In other words, if the particle is represented by the Hilbert space $\mathscr{H}$, then the anti-particle is represented by the conjugate Hilbert space $\bar{\mathscr{H}}$. The two representations are related by an anti-unitary transformation. A particle is represented by the cross-sections over a forward mass hyperboloid or light cone, and an anti-particle is represented by the cross-sections over a backward mass hyperboloid or light cone. If the local symmetry group included time reversal transformations, each forward mass hyperboloid and light cone would lie within the same orbit as its backward counterpart, and there would be no distinction between particle and anti-particle. Particles would possess charge as a property, but the state space would decompose into a positive charge subspace and a negative charge subspace, between which superpositions would be possible (Schwartz, 1968, p. 75).

In relativistic quantum theory there are so-called superselection rules for electric charge, baryonic charge, and leptonic charge.[14] A superselection rule for charge prohibits the physical superposition of particle states with different charges. The self-adjoint operator[15] representing the charge is said to be a superselection operator, and the Hilbert space decomposes into an orthogonal direct sum $\bigoplus_f \mathscr{H}_f$, where each direct summand is an eigenspace of the charge operator. The direct summands of such a decomposition are called superselection 'sectors.' The presence of a superselection rule entails that not all self-adjoint operators on the Hilbert space represent physical observables. Rather, a physical observable corresponds to a self-adjoint operator $A$

---

[12] 'Square-integrable' simply means that the square of the modulus of the cross-section, when integrated over the base space, is finite. After taking a set of equivalence classes of square-integrable cross-sections, one obtains a Hilbert space, a special type of vector space which is equipped with an inner product, and which is complete as a metric space.

[13] Given any complex vector space $V$, not only can one define a dual vector space $V^*$, one can define a conjugate vector space $\bar{V}$. The conjugate space consists of all the anti-linear mappings from $V^*$ into the set of complex numbers. There is a canonical bijective anti-linear mapping from $V$ into $\bar{V}$ called the complex conjugation, which, for any $v \in V$ and $\phi \in V^*$ maps $v$ to $\bar{v}$ defined by $\bar{v}(\phi) = \overline{\phi(v)}$. Equivalently, one obtains $\bar{V}$ from $V$ by replacing the action of the imaginary number $i$ by its complex conjugate $-i$. Given a representation on a complex vector space $V$, there is a corresponding representation on $\bar{V}$, called the conjugate representation.

[14] The baryonic charge, or 'baryon number' $B$, is $+1$ for baryons, $-1$ for anti-baryons, and 0 for all other particle types (see Section 4.12 for the definition of a baryon). Similarly, the leptonic charge, or 'lepton number' $L$, is $+1$ for leptons, $-1$ for anti-leptons, and 0 for all other particle types.

[15] The adjoint $A^*$ of an operator $A$ on a vector space $V$ equipped with an inner product is such that $\langle A^* v, w\rangle = \langle v, Aw\rangle$ for all $v, w \in V$. A self-adjoint operator $A$ is such that $A = A^*$.

which leaves each sector invariant. I.e., if $\psi_f \in \mathscr{H}_f$, then $A\psi_f \in \mathscr{H}_f$. This entails that $\langle \psi_f, A\psi_{f'} \rangle = 0$ if $f \neq f'$, and this, in turn, means that a linear combination of states from different sectors is statistically equivalent to a mixture of those states. In this context, statistical equivalence means that the two states generate the same expectation values on all *physical* observables.

In relativistic quantum theory, the most general one-particle Hilbert space for a particle can be decomposed as the direct sum (Bogolubov *et al.*, 1990, p. 247)

$$\mathscr{H} = \bigoplus_{q,b,l \in \mathbb{Z}} \mathscr{H}(q, b, l).$$

Each summand here possesses a unitary irreducible representation of the Poincaré group, but the entire algebra of observables is represented on the direct sum Hilbert space, and within this algebra of observables there are self-adjoint operators which represent electric charge, baryonic charge, and leptonic charge, which commute with all the other operators representing physical observables. These three operators possess integer-valued spectra, and each superselection sector corresponds to a different combination $(q, b, l)$ of these eigenvalues.

Universes which differ in their local symmetry group, differ in the mathematical structure their particle world is an instance of. However, even with the local symmetry group fixed, the projective, unitary, irreducible representations of this group only determine the set of *possible* free particles. In our universe, only a finite number of elementary free particle types, of specific mass and spin, have been selected from the infinite number of possible free elementary particle types. Thus, in terms of describing the particle world in our universe at least, there is a type of 'surplus structure' in the standard model. However, this is not the type of surplus structure which involves a degree of redundancy in the mathematical representation, or the use of multiple structural elements to describe physically equivalent objects. Hence, this type of surplus structure is quite distinct from the notion of 'gauge equivalence' (see Section 3.4). One could infer that the discrepancy between actual and possible particle types in the standard model is either (i) an indication of the incomplete nature of the standard model, or (ii) an indication that the masses and spins of the actual particle types in a universe is a matter of contingency. The second option enables one to suggest that other universes may utilise the extra structure. Under this approach, universes with the same local symmetry group, and the same set of possible free particles, can be further sub-classified by the particular collection of actual free particles instantiated.

We conclude this section by noting that the free-particle Hilbert spaces obtained in first-quantization are equipped with irreducible representations of (the universal covering of) the restricted Poincaré group, rather than representations of the group of all diffeomorphisms in $\mathbb{R}^4$. The Poincaré group, in the so-called 'passive' sense, provides the transformations between inertial (non-acceleratory) reference frames, whilst the diffeomorphism group of $\mathbb{R}^4$, in the passive sense, provides the transformations between *any* reference frames. The Poincaré group will not change an inertial reference frame into a non-inertial reference frame, or vice versa. Given a coordinate chart which corresponds to an inertial reference frame $\sigma$, each $g \in SL(2, \mathbb{C}) \ltimes \mathbb{R}^{3,1}$ maps $\sigma$ to another inertial reference frame $g\sigma$. For each type of free particle, the group element $g$ is represented by a unitary linear operator $U_g$ on a Hilbert space. If $v$ is the state of a system as observed from the inertial reference frame $\sigma$, then $w = U_g v$ will be the state of the system as observed from the inertial reference frame $g\sigma$. This is the change of observed particle state under a change of inertial reference frame. (In a general Lorentzian space–time, the Poincaré group, in the passive sense, provides the coordinate transforms between the *local* inertial reference frames which have a shared domain).

A particle state space equipped with an irreducible unitary representation of (the universal covering of) the restricted Poincaré group does not possess unitary linear operators that represent transformations between inertial and non-inertial reference frames. Hence, the automorphisms of such a particle state space cannot represent the change of observed particle state under transformations between inertial and non-inertial reference frames.

## 2.2. The configuration space approach

In the configuration representation, each irreducible unitary representation is constructed from a space of mass-$m$ solutions, of either positive or negative energy, to a linear differential equation over Minkowski space–time $\mathcal{M}$. The Hilbert space of a unitary irreducible representation in the configuration representation is provided by the completion of a space of solutions which can be Fourier-transformed into square-integrable objects in Minkowski energy–momentum space.

Whilst the Wigner approach deals directly with the irreducible unitary representations of $SL(2, \mathbb{C}) \ltimes \mathbb{R}^{3,1}$, the configuration space approach requires two steps to arrive at such a representation. The first step is to introduce, for each possible spin $s$, a reducible, mass-independent representation of $SL(2, \mathbb{C}) \ltimes \mathbb{R}^{3,1}$ upon an infinite-dimensional space. For each spin $s$, there is a finite-dimensional vector space $V_s$, such that the mass-independent representation space can be either a set of vector-valued functions $\mathcal{F}(\mathcal{M}, V_s)$, or, more generally, a space of cross-sections $\Gamma(\eta)$ of a vector bundle $\eta$ over $\mathcal{M}$ with typical fibre $V_s$. $\mathcal{F}(\mathcal{M}, V_s)$ coincides with $\Gamma(\eta)$ in the special case where $\eta$ is the trivial bundle $\mathcal{M} \times V_s$, hence one can speak in terms of vector bundle cross-sections $\Gamma(\eta)$ in full generality.

The representations of $SL(2, \mathbb{C}) \ltimes \mathbb{R}^{3,1}$ upon such $\Gamma(\eta)$ can be defined by a combination of the finite-dimensional irreducible representations of $SL(2, \mathbb{C})$, and the action of $SL(2, \mathbb{C}) \ltimes \mathbb{R}^{3,1}$ upon the base space $\mathcal{M}$. The complex, finite-dimensional, irreducible representations of $SL(2, \mathbb{C})$ are indexed by the set of all ordered pairs $(s_1, s_2)$ (Bleecker, 1981, p. 77), with

$$(s_1, s_2) \in \frac{1}{2}\mathbb{Z}_+ \times \frac{1}{2}\mathbb{Z}_+.$$

In other words, the complex, finite-dimensional, irreducible representations of $SL(2, \mathbb{C})$ form a family $\mathscr{D}^{s_1,s_2}$, where $s_1$ and $s_2$ run independently over the set $\{0, 1/2, 1, 3/2, 2, \ldots\}$. The number $s_1 + s_2$ is called the spin of such a representation.

Letting $\psi(x)$ denote an element of $\Gamma(\eta)$, the representation of $SL(2, \mathbb{C}) \ltimes \mathbb{R}^{3,1}$ upon $\Gamma(\eta)$ is defined as

$$\psi(x) \to \psi'(x) = \mathscr{D}^{s_1,s_2}(A) \cdot \psi\big(\Lambda^{-1}(x - a)\big),$$

where it is understood that $A \in SL(2, \mathbb{C})$, $a \in \mathbb{R}^{3,1}$, $\Lambda$ is shorthand for $\Lambda(A)$, and $\Lambda$ is the covering homomorphism $\Lambda : SL(2, \mathbb{C}) \to SO_0(3, 1)$.[16]

These reducible, mass-independent representations do not correspond to single particle species. Each space of vector-valued functions, or each space of vector bundle cross-sections, represents many different particle species. To obtain the mass-$m$, spin-$s$ irreducible unitary representations of $SL(2, \mathbb{C}) \ltimes \mathbb{R}^{3,1}$ in the configuration representation, the second step is to introduce a linear differential equation, such as the Dirac equation or Klein–Gordon equation, which contains mass as a parameter. These differential equations are imposed upon the vector-valued functions or cross-sections in the reducible, mass-independent, spin-$s$ representation. The state space for each individual particle species is constructed from the functions or cross-sections for a particular value of mass.

To understand the relationship with the Wigner approach, we need to look at the Fourier transform in more detail. The Fourier transform is well-defined for any smooth cross-section or function which is 'rapidly decreasing' at spatial infinity. The space of such objects is typically denoted as $\mathscr{S}$. This space contains the space of smooth functions or cross-sections of compact support $\mathscr{D}$ as a dense subspace. The space of continuous linear functionals defined upon the space of rapidly decreasing cross-sections/functions is referred to as the space $\mathscr{S}'$ of 'tempered' distributions. There is a continuous linear injection of $\mathscr{S}'$ onto a subspace of the space of distributions $\mathscr{D}'$, and within the space of distributions it is the tempered distributions $\mathscr{S}'$ which can be said to have Fourier transforms (Schwartz, 1968, p. 58).

With a slight abuse of the mathematical convention, I will hereafter refer to rapidly decreasing cross-sections/functions themselves as tempered. Thus, the Fourier transform is well-defined for any tempered cross-section $\psi \in \mathscr{S}(\eta)$, or any tempered $V_s$-valued function, $f(x) \in \mathscr{S}(\mathcal{M}, V_s)$. The mass-$m$, tempered solutions of the linear differential equations correspond, under a Fourier transform, to vector valued functions over $\mathscr{V}_m$, or the cross-sections of vector bundles over $\mathscr{V}_m$. The Fourier transform $\hat{f}(p)$ of a mass-$m$ tempered function $f(x)$ defined on Minkowski space–time $\mathcal{M}$, is a function defined throughout Minkowski energy–momentum space $T^*_x\mathcal{M}$, but with support upon the mass hyperboloid or light cone $\mathscr{V}_m$. The same is true of the Fourier transform $\hat{\psi}(p)$ of a mass-$m$ tempered cross-section $\psi(x)$. The mass-$m$

[16] Appendix D provides an alternative perspective upon this representation in the special case where the vector bundle is the product bundle $\mathcal{M} \times \mathbb{C}^n$, and the space of cross-sections is the space of functions $\mathcal{F}(\mathcal{M}, \mathbb{C}^n)$.

tempered solutions can be split into positive-energy and negative-energy solutions. Under Fourier transform, the positive-energy mass-$m$ solutions have support on the 'forward-mass' hyperboloid or light cone $\mathcal{V}_m^+$, whilst under Fourier transform, the negative-energy mass-$m$ solutions have support on the 'backward-mass' hyperboloid or light cone $\mathcal{V}_m^-$.

Given a function or cross-section on Minkowski configuration space $\mathcal{M}$, and letting $\psi(x)$ denote either case here, the Fourier transform $\hat{\psi}(p)$ on Minkowski energy–momentum space $T_x^*\mathcal{M}$ is defined to be

$$\hat{\psi}(p) = \int e^{-i\langle p,x\rangle}\psi(x)\,d^4x,$$

or

$$\hat{\psi}(p_0,\mathbf{p}) = \int e^{-i(\mathbf{p}\cdot\mathbf{x}-p_0t)}\psi(x)\,d^4x.$$

Note that the indefinite Minkowski space inner product, with the following sign convention,

$$\langle p,x\rangle = -p_0t + p_1x_1 + p_2x_2 + p_3x_3,$$

is used to define the Fourier transform here. The inverse Fourier transform is defined to be

$$\psi(x) = \int e^{i\langle p,x\rangle}\hat{\psi}(p)\,d^4p,$$

or

$$\psi(\mathbf{x},t) = \int e^{i(\mathbf{p}\cdot\mathbf{x}-p_0t)}\hat{\psi}(p)\,d^4p.$$

Given that the positive energy solutions $\psi^+(x)$ for mass $m$ are those which are the inverse Fourier transform of functions with support upon the forward mass-$m$ hyperboloid, and from the fact that $p_0(\mathbf{p}) = \omega(\mathbf{p}) = +(m^2 + \|\mathbf{p}\|^2)^{1/2}$ on a forward mass hyperboloid, it follows from the expression for an inverse Fourier transform that a positive energy solution can be expressed as

$$\begin{aligned}\psi^+(\mathbf{x},t) &= \frac{1}{(2\pi)^{3/2}}\int_{T_x^*\mathcal{M}} e^{i(\mathbf{p}\cdot\mathbf{x}-p_0t)}a(p)\theta(p_0)\delta\big(m^2 - p^2\big)\,d^4p\\ &= \frac{1}{(2\pi)^{3/2}}\int_{\mathcal{V}_m^+} e^{i(\mathbf{p}\cdot\mathbf{x}-\omega(\mathbf{p})t)}a(p)\,d^3\mathbf{p}/2\omega(\mathbf{p}),\end{aligned}$$

where $a(p)$ has support upon the forward mass hyperboloid $\mathcal{V}_m^+$, a subset of measure zero, but $a(p)\theta(p_0)\delta(m^2 - p^2)$ is a tempered distribution in $\mathcal{S}'(T_x^*\mathcal{M})$, hence its inverse Fourier transform is well-defined.

In the natural measure $d^4p$ upon Minkowski energy–momentum space $T_x^*\mathcal{M}$, each forward mass hyperboloid or light cone $\mathcal{V}_m^+$ is a subset of measure zero. Given that the

Fourier transform of a tempered, mass-$m$, positive-energy solution on $\mathcal{M}$ is a function with support upon a subset $\mathscr{V}_m^+$, the inverse Fourier transform of this function with respect to $d^4p$ would equal zero. The integral of a function over a set of measure zero, equals zero. Hence, one takes the inverse Fourier transform of the tempered distribution $a(p)\theta(p_0)\delta(m^2 - p^2)$ instead. $\theta(p_0)$ is the Heaviside function, defined to be $+1$ when $p_0 > 0$, and $0$ when $p_0 < 0$. Needless to say, $\delta(m^2 - p^2)$ is the dirac delta function.

Given that the negative energy solutions $\psi^-(x)$ for mass $m$ are those which are the inverse Fourier transform of functions with support upon the backward mass-$m$ hyperboloid, and from the fact that on a backward mass hyperboloid, the energy component is negative, with $p_0(\mathbf{p}) = -\omega(\mathbf{p}) = -(m^2 + \|\mathbf{p}\|^2)^{1/2}$, it follows that a negative energy solution can be expressed as

$$\begin{aligned}\psi^-(\mathbf{x}, t) &= \frac{1}{(2\pi)^{3/2}} \int_{T_x^*\mathcal{M}} e^{i(\mathbf{p}\cdot\mathbf{x} - p_0 t)} c(p)\theta(-p_0)\delta\big(m^2 - p^2\big)\, d^4p \\ &= \frac{1}{(2\pi)^{3/2}} \int_{\mathscr{V}_m^-} e^{i(\mathbf{p}\cdot\mathbf{x} + \omega(\mathbf{p})t)} c(p)\, d^3\mathbf{p}/2\omega(\mathbf{p}),\end{aligned}$$

where $c(p)$ has support upon the backward mass hyperboloid $\mathscr{V}_m^-$, a subset of measure zero, but $c(p)\theta(-p_0)\delta(m^2 - p^2)$ is a tempered distribution in $\mathscr{S}'(T_x^*\mathcal{M})$.

For the sake of simplicity, let us concentrate upon the case of $V_s$-valued functions. Supposing that there is a linear differential equation which contains mass as a parameter, and which can be imposed upon the $V_s$-valued functions, one can find subspaces $\mathcal{F}_m^+(\mathcal{M}, V_s)$ which are composed of mass-$m$, positive energy solutions to this differential equation. Furthermore, one can find subspaces $\mathscr{S}_m^+(\mathcal{M}, V_s)$ of tempered functions which are mass-$m$, positive energy solutions to this differential equation. For each such space, there is a yet further subspace consisting of functions with square-integrable Fourier transforms on $\mathscr{V}_m^+$. (Note that the square-integrability is defined with respect to a measure on $\mathscr{V}_m^+$, not a measure on Minkowski space–time $\mathcal{M}$.) The completion of this topological vector space of $V_s$-valued functions on $\mathcal{M}$ takes one into the space $\mathscr{D}'(\mathcal{M}, V_s)$ of $V_s$-valued distributions on $\mathcal{M}$.[17] The completion

$$\mathscr{H}_{m,s}^+ \subset \mathscr{D}'(\mathcal{M}, V_s),$$

is a Hilbert space which provides the configuration representation for a particle of mass $m$ and spin $s$ (Dautray and Lions, 1988, Chapter IX, §1, 6f; Schwartz, 1968, p. 69).[18] This Hilbert subspace of $\mathscr{D}'(\mathcal{M}, V_s)$ is equipped with an irreducible unitary representation of $SL(2, \mathbb{C}) \ltimes \mathbb{R}^{3,1}$, and is unitarily isomorphic to the Hilbert space for a mass $m$, spin $s$ particle in the Wigner representation.

[17] Private communication with Veeravalli Varadarajan.

[18] Note that one has the continuous linear injections $\mathscr{S} \subset \mathscr{S}' \subset \mathscr{D}'$.

One can also find subspaces $\mathcal{F}_m^-(\mathcal{M}, V_s)$ which are composed of mass-$m$, *negative* energy solutions to the relevant differential equation, and further subspaces $\mathcal{S}_m^-(\mathcal{M}, V_s)$ of tempered, mass-$m$, negative energy solutions. The further subspace of functions with square-integrable Fourier transforms on $\mathcal{V}_m^-$, once completed, is unitarily isomorphic to the Hilbert space for a mass $m$, spin $s$ *anti*-particle in the Wigner representation.

In terms of the Wigner representation, 'first quantization' is the process of obtaining a Hilbert space of cross-sections of a vector bundle over $\mathcal{V}_m^{\pm}$. In terms of the configuration representation, first quantization is the two-step process of obtaining a vector bundle/function space over $\mathcal{M}$, and then identifying a space of mass-$m$ solutions.

There are two mathematical directions one can go after first quantization. Firstly, one can treat the Hilbert space obtained as the 'one-particle' state space, and one can use this Hilbert space to construct a Fock space. This is the process of 'second quantization,' explained in Section 2.11. One defines creation and annihilation operators upon the Fock space, and thence one defines scattering operators. One can use the scattering operators to calculate the transition amplitudes between the incoming and outgoing free states of a system involved in a collision or decay process. Calculation of these transition amplitudes requires the so-called 'regularization' and 'renormalization' of perturbation series, but these calculations do enable one to obtain empirically adequate predictions. Nevertheless, a Fock space is a space of states for a free system. In the configuration representation, the space of 1-particle states is a linear vector space precisely because it is a space of solutions to the *linear* differential equation for a free system.

Although one could use either the Wigner representation or the configuration representation, second quantization conventionally uses a Wigner representation for the one-particle Hilbert spaces. This, however, is merely a matter of calculational convenience, and whilst the Wigner approach is empirically identical with the configuration approach, this text makes the interpretational proposal that, in the first-quantized quantum theory, a matter field, or its associated particles, are cross-sections of a fibre bundle over *space–time*. It is space–time which is the ontological arena in which other types of object can be considered to physically exist, hence it is the configuration representation which is assigned ontological priority.

Rather than second-quantizing the free fields, and then attempting to deal with interacting second quantized fields, one can introduce interacting fields at the level of first-quantization. This approach typically uses the configuration representation, and will occupy us for a large portion of the book. In the fibre bundle approach to first-quantization, where a mass $m$, spin $s$ particle can be represented by the mass-$m$ cross-sections of a spin-$s$ bundle $\eta$, this bundle $\eta$ can, following Derdzinski (1992), be referred to as a *free-particle bundle*. One can associate a vector bundle $\delta$ with a gauge field, which can, again following Derdzinski, be referred to as an *interaction bundle*. One can take the free-particle bundle $\eta$, and with an interaction bundle $\delta$, one can construct an *interacting-particle bundle* $\alpha$. The mass-$m$ cross-sections of this bundle represent the particle in the presence of the gauge field. This is the route of the first-quantized interacting theory. The first-quantized interacting theory is not itself empirically adequate. Moreover, it is not possible to subject the first-quantized interacting theory to second-quantization because the state space of an interacting system is not a linear vector space; in the configuration representation, the space of states for an interacting 1-particle system is non-linear (see Section 4.1). Nonetheless, the first-quantized interacting theory, being mathematically well-defined, offers a far better approach to the foundations of the standard model than the morass of computational recipes offered by the second-quantized theory. We shall therefore expound the first-quantized interacting theory in this text.

## 2.3. Curved space–time

It is assumed in Section 2.1 that representations of the *local* space–time symmetry group of interacting systems are an adequate means of determining the elementary particle ontology. Let us try to understand what locality means in this context.

The strong equivalence principle of general relativity holds that Minkowski space–time, and its symmetries, are valid in a neighbourhood about any point of space–time. I.e., the strong equivalence principle holds that the global symmetry group of Minkowski space–time is the local symmetry group of a general space–time. One can choose a neighbourhood $U$ about any point in a general space–time, which is sufficiently small that the gravitational field within the neighbourhood is uniform to some agreed degree of approximation (Torretti, 1983, p. 136). Such neighbourhoods provide the domains of 'local Lorentz charts.' A chart in a 4-dimensional manifold provides a diffeomorphic map $\phi : U \to \mathbb{R}^4$. If $\mathbb{R}^4$ is equipped with the Minkowski metric, a local Lorentz chart provides a map which is almost isometric, to some agreed degree of approximation (Torretti, 1983, p. 147). Unless the gravitational field is very strong, one can treat each elementary particle as 'living in' the domain of a local Lorentz chart within a general space–time $(\mathcal{M}, g)$. Consequently, the fibre bundles used in the standard model are usually assumed to be fibre bundles over Minkowski space–time. This is done with the understanding that the base space of such bundles represents the domain of an arbitrary local Lorentz chart, rather than the whole of space–time. Hence, with the exception of the regions where the gravitational field is very strong, the elementary particles which exist in a general Lorentzian space–time still transform under (a subgroup of) the global symmetry group of Minkowski space–time, the Poincaré group.

With the exception of regions where the gravitational field is very strong, a fully realistic representation of each individual elementary particle would begin with a Lorentzian manifold $(\mathcal{M}, g)$ which represents the entire universe, and would then proceed to identify a local Lorentz chart which the particle 'lives in.' In terms of the first-quantized quantum theory, the particle would then be represented by the cross-sections and connections of bundles over this local Lorentz chart. As far as practical physics is concerned, this would be an act of representational *largesse*, but it is, nevertheless, correct from an ontological perspective.

Because gravity is geometrized in general relativity, it is consistent to speak of *free* elementary particles in a gravitational field. Newtonian gravity represents the motion of a freely-falling particle in the vicinity of a massive body to be acceleratory motion, the result of a gravitational field exerting a force upon the particle. In contrast, general relativity represents a freely-falling particle in the vicinity of a massive body to follow a geodesic of a curved space–time geometry. A geodesic is the generalisation of straight line, constant speed motion to an arbitrary curved geometry; a straight line traversed at a constant speed is simply a geodesic in the special case of a flat Euclidean geometry. Given any curve $\gamma$ in a pseudo-Riemannian manifold, the covariant derivative[19] $\nabla$ of the geometry enables one to define the acceleration vector of the curve as $\nabla_{\dot{\gamma}}\dot{\gamma}$, where $\dot{\gamma}$ is the tangent vector to the curve. A geodesic $\gamma$ is a curve of zero-acceleration in the sense that $\nabla_{\dot{\gamma}}\dot{\gamma} = 0$. Because general relativity represents a freely-falling particle to follow a geodesic of the space–time geometry, general relativity represents a freely-falling particle to undergo non-acceleratory motion. A free particle is a particle which is free from acceleratory influences, hence a free particle is free from the influence of *non-gravitational* forces.

Where the gravitational field is very strong (i.e., where the space–time curvature is very large), the following applies: it is no longer valid to assume that the gravitational field is locally uniform; it is not valid to assume that elementary particles transform under the global symmetry group of Minkowski space–time, or a subgroup thereof; and elementary particles are represented by fibre bundles over general, curved space–times. Bundles over curved space–time are employed with the understanding that the base space of such bundles represents a subset of space–time, rather than the whole universe. These considerations weaken the assumption that the representations of the Poincaré group are an adequate means of determining the elementary particle ontology in a universe. In a general curved space–time, there might well be no isometry group at all, hence the possible elementary particles in such a space–time region cannot be classified by the irreducible unitary representations of that region's space–time symmetry group. If the elementary particle ontology is not determined in all regions of space–time by the irreducible unitary representations of the local symmetry group of interacting systems in the regions of weak gravitational field, then a classification scheme based upon representations of space–time symmetry groups will be incomplete.

Given the absence, in general, of a symmetry group for a curved space–time, practitioners of quantum field theory in curved space–time take the linear field equations associated with particles of mass $m$ and spin $s$ in the Minkowski space–time 'con-

[19] See Section 3.3 for a definition of the covariant derivative.

figuration representation,' and generalise them to curved space–time by replacing partial derivatives with covariant derivatives. The solutions of these equations can be considered to represent first-quantized free particles of mass $m$ and spin $s$ in curved space–time.[20] Whilst the solutions of these linear equations correspond to unitary irreducible representations of the space–time symmetry group in the case of Minkowski space–time, no such correspondence exists for the generalised equations. However, given a 'canonical' formulation of the classical field in curved space–time, a one-particle Hilbert space can be constructed from it.

In the lexicon of physics, a canonical formulation is one which uses the techniques of Hamiltonian mechanics. This typically involves introducing a configuration space, a phase space, a Hamiltonian, dynamical evolution equations generated from the Hamiltonian, and possibly some constraint equations. The configuration space $\mathscr{C}$ in this context is the set of all possible *spatial* configurations of the system. (Note that in this text we have used the phrase 'configuration space' in a different sense, to distinguish representations in Minkowski *space–time* from those in energy–momentum space.) A canonical approach necessarily involves splitting space–time into a one-parameter family of spacelike hypersurfaces, and studying the time evolution of the system over those hypersurfaces. Given a configuration space $\mathscr{C}$, the phase space $\Gamma$ is then typically the cotangent bundle $T^*\mathscr{C}$. In the canonical formulation of a classical free field, the phase space is an infinite-dimensional linear space, consisting of pairs $(\phi, \pi)$ of fields on a spacelike hypersurface $\Sigma$. This linear phase space can be equipped with a non-degenerate bilinear form $\Omega : \Gamma \times \Gamma \rightarrow \mathbb{R}$, and a complex structure $J$. Using the complex structure one obtains the complexified phase space, which can then be equipped with a sesquilinear inner product related to $\Omega$. Taking the completion with respect to the sesquilinear inner product then obtains the one-particle Hilbert space (Saunders, 1991, p. 92; Halvorson, 2001, p. 128). This approach is part of the 'geometric quantization' programme, and generalises more readily to quantum field theory in curved space–time.

However, the choice of the complex structure $J$ corresponds to the choice of a positive-energy subspace and a negative-energy subspace in the space of solutions, and for systems in curved space–time, such a decomposition is often non-unique. For this reason, most formulations of quantum field theory in curved space–time assume a 'stationary' space–time, in which there are global timelike Killing vector fields to single out such a decomposition. Wald claims that "in a general curved spacetime where no timelike Killing field is present, it should not be surprising that there is no natural, universally applicable procedure for defining particles and that, in a given spacetime, different constructions... will define different notions of

[20] However, for $s > 1$, there are reasons for thinking the solutions to these equations do not satisfy physical criteria. For example, such equations do not have a well-posed initial-value formulation (Wald, 1984, p. 375).

particles" (1984, p. 416). These are issues which we shall address further in Section 2.11.

## 2.4. The Wigner approach

Let us now consider the Wigner approach in detail. The details of this approach will be of relevance in much of what follows, particularly in the discussion of parity in Section 2.5. Moreover, the nature of the construction enables one to understand how particle representations can be constructed from a symmetry group itself, rather than being supplied *a priori*.

The Wigner approach uses the method of induced group representation,[21] applied to semi-direct product groups (Sternberg, 1994, pp. 135–139). Given a semi-direct product $G = H \ltimes N$, the method of induced representation can obtain, up to unitary equivalence, all the irreducible, strongly continuous,[22] unitary representations of the group $G$. Given that the local symmetry group of space–time is a semi-direct product $G = SL(2, \mathbb{C}) \ltimes \mathbb{R}^{3,1}$, the method of induced representation enables us to obtain, up to unitary equivalence, all the irreducible, strongly continuous, unitary representations of $G = SL(2, \mathbb{C}) \ltimes \mathbb{R}^{3,1}$. In fact, the method of induced representation enables us to classify all such representations of $G = SL(2, \mathbb{C}) \ltimes \mathbb{R}^{3,1}$, and to provide an explicit construction of one member from each unitary equivalence class (Sternberg, 1994, pp. 143–150; Emch, 1984, p. 503).

Given any group $H$, a vector space $N$, and a representation $\tau$ of $H$ on $N$, the semi-direct product $G = H \ltimes N$ is the Cartesian product of $H$ and $N$, equipped with the following group product operation:

$$(h_1, n_1) \cdot (h_2, n_2) = \big(h_1 \cdot h_2, n_1 + \tau(h_1)n_2\big).$$

$N$ is a normal Abelian subgroup of the semi-direct product, and the representation of $H$ on $N$ coincides with the conjugation action of $H$ on $N$.[23]

To obtain the irreducible representations of such a semi-direct product $G = H \ltimes N$, one needs to introduce (i) a dual group $\hat{N}$, (ii) an action of $H \ltimes N$ upon $\hat{N}$, (iii) the isotropy group[24] $L_j$ of each $H \ltimes N$ orbit in $\hat{N}$, and (iv) the irreducible representations of the isotropy groups on vector spaces $V_k$. For each orbit in $\hat{N}$, and each irreducible isotropy group representation on a vector space $V_k$, one obtains a vector bundle $E$ with typical fibre $V_k$ and base space $G/L_j$ (see Appendix E).

---

[21] See Appendix E.

[22] This simply means continuous with respect to a particular topology on the space of operators, called the strong topology.

[23] I.e. each $h \in H$ acts on $N$ by mapping $n \in N$ to $hnh^{-1}$.

[24] The isotropy group of a point under a group action is the subgroup which leaves that point fixed. It is also variously referred to as the 'little group' or 'stabilizer' of a point.

Given an Abelian group $N$, the dual group $\hat{N}$ is the set of continuous homomorphisms from $N$ into $U(1) = \mathbb{T}$.[25] $\hat{N}$ is also the set of characters on $N$ from the irreducible representations of $N$.[26]

In the case where $G = SL(2, \mathbb{C}) \ltimes \mathbb{R}^{3,1}$, we have $H = SL(2, \mathbb{C})$ and $N = \mathbb{R}^{3,1}$. In this case, $\hat{N} \cong N$, where Minkowski energy–momentum space $T_x^*\mathcal{M} \cong \mathbb{R}^{3,1}$ can be treated as the dual space $\hat{N}$. Each continuous homomorphism $p : \mathbb{R}^{3,1} \to U(1)$ can be mapped to a point $p = (p_0, p_1, p_2, p_3)$ in Minkowski energy–momentum space so that

$$p(x) = \exp\big(i\langle p, x\rangle\big) = \exp\big(i(-p_0x_0 + p_1x_1 + p_2x_2 + p_3x_3)\big).$$

The isomorphism between $\hat{N}$ and $N$ entails that the orbits of the $G$-action on $\hat{N}$ also correspond to the orbits of the $G$-action on $N = \mathbb{R}^{3,1}$. Given that the Abelian group $N = \mathbb{R}^{3,1}$ leaves the points of $N$ fixed under the conjugation action, and given that the conjugation action of $H = SL(2, \mathbb{C})$ on $N = \mathbb{R}^{3,1}$ is given by the representation $\tau : SL(2, \mathbb{C}) \to SO_0(3, 1)$ on $\mathbb{R}^{3,1}$, it follows that the orbits of the $G$-action on $N$ are simply the orbits of the Lorentz group action.

The Lorentz group is a group of linear isometries which preserve the value of the quadratic form $Q(p) = \langle p, p\rangle$. Hence, the orbits of the $G$-action on $\hat{N}$ are hypersurfaces in Minkowski energy–momentum space which have constant values for the quadratic form; the orbits of the $G$-action correspond to either mass hyperboloids or light cones. One constructs vector bundles over these hypersurfaces in Minkowski energy–momentum space. For each distinct orbit of $G = SL(2, \mathbb{C}) \ltimes \mathbb{R}^{3,1}$, there is an isotropy group $H_j \subset SL(2, \mathbb{C})$ of the $H$-action, and a corresponding isotropy group $L_j = H_j \ltimes \mathbb{R}^{3,1}$ of the $G$-action. For each distinct orbit of $SL(2, \mathbb{C}) \ltimes \mathbb{R}^{3,1}$, and for each distinct irreducible representation of the isotropy group $H_j \subset SL(2, \mathbb{C})$, one constructs a distinct vector bundle.

The orbits of the $G$-action in Minkowski energy–momentum space come in three types[27]:

- Spacelike hypersurfaces: For any $m \in \mathbb{R}_+$ the mass hyperboloid $\mathcal{V}_m$ is the set $\mathcal{V}_m = \{p \in \mathbb{R}^{3,1}\colon \langle p, p\rangle = -m^2 < 0\}$. Each mass hyperboloid consists of two connected components, the forward mass hyperboloid $\mathcal{V}_m^+ = \{p \in \mathbb{R}^{3,1}\colon \langle p, p\rangle = -m^2 < 0,\ p_0 > 0\}$ and the backward mass hyperboloid $\mathcal{V}_m^- = \{p \in \mathbb{R}^{3,1}\colon \langle p, p\rangle = -m^2 < 0,\ p_0 < 0\}$. Each component of each mass hyperboloid is an orbit of the $G$-action.

[25] The unitary group $U(1)$ can be defined as the set of complex numbers $e^{i\theta}$ of unit modulus. It is isomorphic as a group to the circle group $S^1 \cong \mathbb{T}$, where the circle is thought of as a one-dimensional torus $\mathbb{T}$.

[26] The character of a representation assigns to each group element the trace of the operator representing that group element.

[27] Given the isomorphism between $\mathbb{R}^{3,1}$ and $T_x^*\mathcal{M}$, we shall hereafter omit the dual space notation.

- Null hypersurfaces: The so-called forward light cone, $\mathscr{V}_0^+ = \{p \in \mathbb{R}^{3,1}: \langle p, p \rangle = m^2 = 0,\ p_0 > 0\}$, and backward light cone $\mathscr{V}_0^- = \{p \in \mathbb{R}^{3,1}: \langle p, p \rangle = m^2 = 0,\ p_0 < 0\}$.
- Timelike hypersurfaces: For any $n \in \mathbb{R}_+$, connected hyperboloids $\{p \in \mathbb{R}^{3,1}: \langle p, p \rangle = n > 0\}$.

Only the first two cases are considered to be physically relevant. The isotropy group of the $G$-action upon a mass hyperboloid $\mathscr{V}_m^+$ is different from the isotropy group of the $G$-action upon a light cone, hence we shall treat these two cases separately.

In the case where the orbit is a forward mass hyperboloid $\mathscr{V}_m^+$, the isotropy group $H_j \subset SL(2, \mathbb{C})$ of the $H$-action on an arbitrary point $\chi_j \in \mathscr{V}_m^+$, is $SU(2)$. The irreducible representations of $SU(2)$ are parameterized by $s \in \frac{1}{2}\mathbb{Z}_+$. As one possible realisation of these representations, let $V_s$, for all $s \in \frac{1}{2}\mathbb{Z}_+$, denote the space of homogeneous polynomials of degree $s$ on $\mathbb{C}^2$ (Sternberg, 1994, p. 181). I.e., $V_s$ is the space of functions

$$p(z_1, z_2) = \sum_{k=0}^{2s} a_k z_1^{2s-k} z_2^k.$$

$V_{\frac{1}{2}} = \mathbb{C}^2$, the standard representation of $SU(2)$ on $\mathbb{C}^2$, and this determines a representation of $SU(2)$ upon each of the function spaces.

Given that the isotropy group of the $H$-action is $SU(2)$, the isotropy group of the $G$-action is $SU(2) \ltimes \mathbb{R}^{3,1}$. Each irreducible representation of $SU(2) \ltimes \mathbb{R}^{3,1}$ generates an irreducible representation of $SL(2, \mathbb{C}) \ltimes \mathbb{R}^{3,1}$ by the method of induced representation.

For a particle of mass $m > 0$ and arbitrary spin $s$, the Wigner construction uses the spin-$s$ representation of $SU(2)$ on $V_s$ to obtain a vector bundle $E_{m,s}^+$ over $\mathscr{V}_m^+$, with typical fibre $V_s$, and an irreducible unitary representation of $SL(2, \mathbb{C}) \ltimes \mathbb{R}^{3,1}$ upon the space $\Gamma_{L^2}(E_{m,s}^+)$ of square-integrable cross-sections of the vector bundle $E_{m,s}^+$. This representation is unique up to unitary equivalence. The anti-particle is represented by the conjugate representation on the space $\Gamma_{L^2}(E_{m,s}^-)$ of square-integrable cross-sections of the vector bundle $E_{m,s}^-$ over the backward mass hyperboloid $\mathscr{V}_m^-$.

In the case where the $G$-orbit is the forward light cone $\mathscr{V}_0^+$, the isotropy group $H_j \subset SL(2, \mathbb{C})$ of the $H$-action is the double cover of $E(2) = SO(2) \ltimes \mathbb{R}^2$, the group of 'motions' of the Euclidean plane. The double cover can be denoted as $\tilde{E}(2)$ on the understanding that, on this occasion, the tilde does not indicate the universal cover.

Given that the isotropy group of the $H$-action is $\tilde{E}(2)$, the isotropy group of the $G$-action is $\tilde{E}(2) \ltimes \mathbb{R}^{3,1}$. Each irreducible representation of $\tilde{E}(2) \ltimes \mathbb{R}^{3,1}$ generates an irreducible representation of $G = SL(2, \mathbb{C}) \ltimes \mathbb{R}^{3,1}$ by the method of induced representation. However, in contrast with the $m^2 > 0$ case, not all of the irreducible representations of $\tilde{E}(2)$ are considered to be physically relevant.

As a semi-direct product, the irreducible representations of $\tilde{E}(2)$ can themselves be obtained by the method of induced representation. The orbits of $\tilde{E}(2)$ acting on $\mathbb{R}^2$

consist of concentric circles, and the point at the origin. Consider first the irreducible representations corresponding to the circles. The method of induced representation provides a vector bundle over each circle, and a representation of $\tilde{E}(2)$ on the space of cross-sections of each such vector bundle. Given that each such representation of the isotropy group $\tilde{E}(2)$ is infinite-dimensional, one obtains a vector bundle over the light cone $\mathscr{V}_0^+$ which has an infinite-dimensional fibre. These representations are not physically relevant.

The irreducible representations of $\tilde{E}(2)$ corresponding to the origin in $\mathbb{R}^2$ are the physically relevant representations. The isotropy group of the $\tilde{E}(2)$-action at the origin is the double cover of $SO(2)$. The irreducible representations of the double cover of $SO(2)$ are 1-dimensional, and parameterized by the half-integers $s \in \frac{1}{2}\mathbb{Z}$. Hence, for each such representation, one has a vector bundle over a single point, with typical fibre isomorphic to $\mathbb{C}^1$, and a representation of $\tilde{E}(2)$ upon the cross-sections of such a vector bundle. The cross-sections of a vector bundle over a single point coincide with the typical fibre, hence each representation of $\tilde{E}(2)$ on the cross-sections of such a bundle is a representation of $\tilde{E}(2)$ on a vector space isomorphic to $\mathbb{C}^1$. This can be extended to an irreducible representation of $\tilde{E}(2) \ltimes \mathbb{R}^{3,1}$.

For each such representation of the isotropy group $\tilde{E}(2) \ltimes \mathbb{R}^{3,1}$, parameterized by $s \in \frac{1}{2}\mathbb{Z}$, there is a vector bundle $E^+_{0,s}$ over the forward light cone $\mathscr{V}_0^+$, with typical fibre isomorphic to $\mathbb{C}^1$. The space of *square-integrable* cross-sections of the vector bundle $E^+_{0,s}$ is Wigner's (unitary) irreducible representation of the local space–time symmetry group $SL(2,\mathbb{C}) \ltimes \mathbb{R}^{3,1}$ for a particle of mass 0 and spin $s$. In the case of a mass zero particle, the discrete parameter $s$ is often alternatively referred to as the 'helicity' or the 'polarization' of the particle, and the spin is treated as the absolute value $|s|$. The anti-particle is represented by the conjugate representation that uses the backward light cone $\mathscr{V}_0^-$.

Note that if the local symmetry group of space–time were enlarged to include dilatations (i.e., scale factor transformations), then all the forward mass hyperbolae, for example, would be within the same orbit, and, consequently, the method of induced group representation would not use mass as one of the parameters defining elementary particle type. Whilst mass would still be a property of individual elementary particles, it would be a variable property (Schwartz, 1968, p. 75). In fact, given that light cones are invariant under dilatations, massless particles such as the photon include the dilatations amongst their symmetries. The free Maxwell equations are invariant under dilatations, and, in fact, their generalisation, the free Yang–Mills equations, are also invariant under the Weyl group, the 11-dimensional group product of the Poincaré group with the group of scale factors. Actually, massless particles have an even larger symmetry group, the 15-dimensional conformal group, $SO(2,4)$, or its covering group $SU(2,2)$. This is the symmetry group of the conformal compactification of Minkowski space–time. The latter is a space which preserves the causal structure of Minkowski space–time, but without the specific geometry or the non-compact spatial topology of Minkowski space–time. The free Maxwell equations, the

Weyl equation, and the free Yang–Mills equations are invariant under the conformal group (Baez, 2005). Note, however, that neither scale transformations nor conformal transformations[28] are isometries of a pseudo-Riemannian manifold. If either type of transformation were considered to be a general space–time symmetry, then space–time geometry would correspond, respectively, to either a homothetic or conformal equivalence class of metrics.

## 2.5. Parity

To recap, whilst the largest possible local symmetry group of a universe with three spatial dimensions and one temporal dimension is the full Poincaré group $O(3,1) \ltimes \mathbb{R}^{3,1}$, the way in which matter fields couple to gauge fields appears to equip our universe with a local time orientation and a local space orientation, hence the local symmetry group is the restricted Poincaré group $SO_0(3,1) \ltimes \mathbb{R}^{3,1}$. This, however, appears to be a contingent fact about our universe. There are other possible universes, of the same dimension and geometrical signature to our own, which possess a larger local symmetry group. For example, in a universe in which left-handed and right-handed matter fields couple impartially to gauge fields, and in which space reflections are therefore local space–time symmetries, the local symmetry group would be the isochronous Poincaré group $O^{\uparrow}(3,1) \ltimes \mathbb{R}^{3,1}$. (Also called the orthochronous Poincaré group.) This group consists of both the identity component of the Poincaré group, and the component which contains the operation of parity reversal, $\mathscr{P} : (x_0, x_1, x_2, x_3) \mapsto (x_0, -x_1, -x_2, -x_3)$. The possible free particles in such a universe are specified by the irreducible unitary representations of some double cover $H \ltimes \mathbb{R}^{3,1}$ of the isochronous Poincaré group, where $H$ is a $\mathbb{Z}_2$-extension[29] of $SL(2,\mathbb{C})$ to a group which double covers $O^{\uparrow}(3,1)$ (Sternberg, 1994, pp. 150–161).

In a universe in which space reflections $\mathscr{P}$, time reversals $\mathscr{T}$, and combinations thereof $\mathscr{P}\mathscr{T}$, are all local space–time symmetries, the local symmetry group is the entire Poincaré group $O(3,1) \ltimes \mathbb{R}^{3,1}$. The possible free particles in such a universe are specified by the irreducible unitary representations of some double cover $K \ltimes \mathbb{R}^{3,1}$ of the entire Poincaré group, where $K$ is a $\mathbb{Z}_2 \times \mathbb{Z}_2$-extension of $SL(2,\mathbb{C})$ to a group which double covers $O(3,1)$. There are eight such double covers of the Poincaré group (Sternberg, 1994, pp. 160–161). For each different choice of double cover $K$, one has a different family of unitary irreducible representations, hence for each different choice of double cover $K$, one has a different free-particle ontology.

One can treat $\{I, \mathscr{P}, \mathscr{T}, \mathscr{P}\mathscr{T}\}$ as the group $\mathbb{Z}_2 \times \mathbb{Z}_2$ (Sternberg, 1994, p. 160). In any double cover of $O(3,1)$, let $\mathscr{L}_{\mathscr{P}}$ denote any one of the two elements which

[28] A conformal transformation is effectively a locally variable scale transformation (see O'Neill, 1983, p. 92).

[29] A $\mathbb{Z}_2$-extension of $SL(2,\mathbb{C})$ is a group $H$ which contains $\mathbb{Z}_2$ as a normal subgroup, and which is such that $H/\mathbb{Z}_2 \cong SL(2,\mathbb{C})$.

cover $\mathscr{P}$, let $\mathscr{L}_{\mathscr{T}}$ denote any one of the two elements which cover $\mathscr{T}$, and let $\mathscr{L}_{\mathscr{P}}\mathscr{L}_{\mathscr{T}}$ denote any one of the two elements which cover $\mathscr{P}\mathscr{T}$. Each one of the eight possible double covers of $O(3,1)$ is specified by a triple of values for the following three variables (Berg *et al.*, 2001, p. 1021),

$$\mathscr{L}_{\mathscr{P}}^2 = a, \qquad \mathscr{L}_{\mathscr{T}}^2 = b, \qquad (\mathscr{L}_{\mathscr{P}}\mathscr{L}_{\mathscr{T}})^2 = c,$$

where $a, b, c \in \{1, -1\} = \mathbb{Z}_2$.

Two of the eight possible double covers are the well-known groups $Pin(3,1)$ and $Pin(1,3)$, closely related with Clifford algebras (see Section 2.8). In terms of the finite groups which cover $\{I, \mathscr{P}, \mathscr{T}, \mathscr{P}\mathscr{T}\}$ within a double cover of $O(3,1)$, one of the eight double covers uses $\mathbb{Z}_2 \times \mathbb{Z}_2 \times \mathbb{Z}_2$, three use $\mathbb{Z}_2 \times \mathbb{Z}_4$, three use $D_4$, the dihedral group of order eight, and one uses the quaternionic group $G_2$.

Let us consider the case where space reflections are included in the local space–time symmetry group from the viewpoint of the Wigner approach.[30] To find the irreducible unitary representations of $H \ltimes \mathbb{R}^{3,1}$, one studies the orbits and isotropy groups of the $H \ltimes \mathbb{R}^{3,1}$-action on $\mathbb{R}^{3,1}$. The physically relevant orbits of $H \ltimes \mathbb{R}^{3,1}$ are the same as the physically relevant orbits of $SL(2,\mathbb{C}) \ltimes \mathbb{R}^{3,1}$, namely the mass hyperboloids and light cones $\mathscr{V}_m$. However, the isotropy groups of these orbits change under the action of the enlarged group. For $m^2 > 0$ the isotropy group of a mass hyperboloid is an enlargement of $SU(2)$, and for $m = 0$ the isotropy group is an enlargement of $\tilde{E}(2)$. However, there is more than one choice for an enlargement of $SL(2,\mathbb{C})$ that covers $O^{\uparrow}(3,1)$, and as a consequence, there is more than one choice for an enlargement of the isotropy group of a mass hyperboloid or light cone.

Consider first the case of $m^2 > 0$. Let $\overline{SL}(2,\mathbb{C})$ denote an enlargement of $SL(2,\mathbb{C})$ that covers $O^{\uparrow}(3,1)$, and let $\overline{SU}(2)$ denote the consequent enlargement of $SU(2)$. The parity transformation $\mathscr{P} \in O^{\uparrow}(3,1)$ is covered by a pair of elements $\pm\mathscr{L} \in \overline{SL}(2,\mathbb{C})$. In the case of $m^2 > 0$, the parity transformation $\mathscr{P}$ leaves the representative point $(m,0,0,0)$ in a mass hyperboloid fixed, hence $\pm\mathscr{L}$ belong to the enlarged isotropy group $\overline{SU}(2)$ of a mass hyperboloid.

Let $I$ denote the identity element in $\overline{SU}(2)$. Bosonic representations of $\overline{SU}(2)$ (i.e., spin $s \in \mathbb{Z}$) map both $\pm I$ to $Id$, where $Id$ denotes the identity transformation on the representation space. In contrast, fermionic representations of $\overline{SU}(2)$ map $I$ to $Id$ and $-I$ to $-Id$.

The different choices of enlargement $\overline{SU}(2)$ for the isotropy group correspond to whether (a) $\mathscr{L}^2 = I$ or (b) $\mathscr{L}^2 = -I$. In both cases, the finite-dimensional, complex, irreducible representations of $\overline{SU}(2)$ are now parameterized not only by spin $s$, but also by parity $\epsilon = \pm 1$ (see Table 2.1).

In the case of enlargement (a), an irreducible representation $\pi$ must be such that

$$\pi\left(\mathscr{L}^2\right) = \pi(\mathscr{L})\pi(\mathscr{L}) = \pi(I) = Id,$$

[30] The treatment here closely follows Sternberg, 1994, pp. 153–154.

Table 2.1
Enlarged isotropy group representations for $m^2 > 0$

| Spin type | Parity $\epsilon$ | $\overline{SU}(2)$ |
|---|---|---|
| Bosonic | $\pi(\mathscr{L}) = Id\ (\epsilon = +1)$ | $\mathscr{L}^2 = I$ |
| Bosonic | $\pi(\mathscr{L}) = -Id\ (\epsilon = -1)$ | $\mathscr{L}^2 = I$ |
| Fermionic | $\pi(\mathscr{L}) = Id\ (\epsilon = +1)$ | $\mathscr{L}^2 = I$ |
| Fermionic | $\pi(\mathscr{L}) = -Id\ (\epsilon = -1)$ | $\mathscr{L}^2 = I$ |
| Bosonic | $\pi(\mathscr{L}) = Id\ (\epsilon = +1)$ | $\mathscr{L}^2 = -I$ |
| Bosonic | $\pi(\mathscr{L}) = -Id\ (\epsilon = -1)$ | $\mathscr{L}^2 = -I$ |
| Fermionic | $\pi(\mathscr{L}) = i \cdot Id\ (\epsilon = +1)$ | $\mathscr{L}^2 = -I$ |
| Fermionic | $\pi(\mathscr{L}) = -i \cdot Id\ (\epsilon = -1)$ | $\mathscr{L}^2 = -I$ |

hence either $\pi(\mathscr{L}) = Id$ or $\pi(\mathscr{L}) = -Id$. This is true for both bosonic and fermionic representations. Depending upon whether $\pi(\mathscr{L}) = \pm Id$, the representation is said to have $\epsilon = \pm 1$ parity.

In the case of enlargement (b), one needs to distinguish between bosonic and fermionic representations. A bosonic representation of a double cover $\overline{SU}(2)$ is a representation which has been lifted from a representation of $O^{\uparrow}(3)$. In the case of enlargement (b), a bosonic representation sends $\mathscr{P} \in O^{\uparrow}(3)$ to $\epsilon \cdot Id$, and therefore sends $\pm\mathscr{L}$ to $\epsilon \cdot Id$, with $\epsilon = \pm 1$ being the parity of the bosonic representation. A fermionic representation $\pi$ sends $-I$ to $-Id$, hence given that $\mathscr{L}^2 = -I$ in case (b),

$$\pi\left(\mathscr{L}^2\right) = \pi(\mathscr{L})\pi(\mathscr{L}) = \pi(-I) = -Id.$$

It follows that $\pi(\mathscr{L}) = \epsilon i \cdot Id$, where $\epsilon = \pm 1$ is the parity of the representation.

Thus, in the case of $m^2 > 0$, if the local symmetry group of space–time includes parity transformation, then the irreducible representations of $H \ltimes \mathbb{R}^{3,1}$ are not parameterized by mass and spin alone. Rather, they are parameterized by mass, spin and parity. Hence parity ('handedness') is an invariant property of a massive particle precisely when the local symmetry group of space–time includes parity transformations.

In the case of $m^2 > 0$, the method of induced representation, applied to the isochronous Poincaré group, yields vector bundles $E^{+}_{m,s,\epsilon}$ with the same base space and the same typical fibre as those produced for the restricted Poincaré group, but with different representations of an enlarged isotropy group upon the typical fibre.

Now consider the case of $m = 0$. The parity transformation $\mathscr{P} \in O^{\uparrow}(3,1)$, and therefore $\mathscr{L} \in \overline{SL}(2,\mathbb{C})$, do not leave fixed an arbitrary point in a light cone. Hence $\mathscr{L}$ does not belong to the isotropy group of a light cone. To label the different choices of isotropy group one must instead introduce an element $\mathscr{R} = \mathscr{U}\mathscr{L}$ with

$$\mathscr{U} = \begin{pmatrix} 0 & -1 \\ 1 & 0 \end{pmatrix}.$$

$\mathscr{R}$ *does* belong to the isotropy group of a light cone. Either $\mathscr{R}^2 = \mathscr{U}^2\mathscr{L}^2 = I$ or $\mathscr{R}^2 = \mathscr{U}^2\mathscr{L}^2 = -I$, depending upon the choice of isotropy group enlargement.

Whilst the irreducible representations of $\tilde{E}(2)$ are parameterized by $s \in \frac{1}{2}\mathbb{Z}$, the irreducible representations of an enlarged light cone isotropy group are parameterized by $t \in \frac{1}{2}\mathbb{Z}_+$. For $t \neq 0$, the $t$-representation can be decomposed into a direct sum of the $s = t$ and $s = -t$ representations of $\tilde{E}(2)$. The $t$-representation maps the group element $\mathscr{R} = \mathscr{U}\mathscr{L}$ to a linear transformation which sends the helicity $s$ representation of $\tilde{E}(2)$ into the helicity $-s$ representation. Hence, whilst the irreducible representations of $\tilde{E}(2)$ are 1-dimensional, the irreducible representations of an enlarged light cone isotropy group are 2-dimensional for $t \neq 0$.[31] In the case of $m = 0$, the method of induced representation, applied to the isochronous Poincaré group, yields vector bundles $E^+_{0,t}$ with the same base space, but different typical fibres, from those obtained for the restricted Poincaré group.

When $H = SL(2, \mathbb{C})$, i.e., when parity transformation is not a local space–time symmetry, the helicity $s$ representation and helicity $-s$ representation (assuming $s > 0$), for a zero mass particle, are deemed to represent right-handed and left-handed versions of the same particle. The particle associated with the helicity $s$ representation is said, for example, to have right-handed parity $\epsilon = 1$, whilst the particle associated with the helicity $-s$ representation is said to have left-handed parity $\epsilon = -1$. For example, if neutrinos are of zero mass, then all neutrinos are of negative helicity, and left-handed, while all anti-neutrinos are of positive helicity and right-handed. The choice of handedness is purely conventional; the important point is that the two representations correspond to opposite handedness. The interpretation that the $s$ representation and the $-s$ representation correspond to opposite parities is drawn from the fact that these representations are interchanged when the symmetry group is enlarged to include parity transformations.

Whilst the local space–time symmetry group of our universe is $SO_0(3,1) \ltimes \mathbb{R}^{3,1}$, recall that this restriction is determined by the violation of certain symmetries in weak force interactions. In contrast, other forces, such as the electromagnetic force, respect symmetries such as parity transformation. As a consequence, it is conventional to treat the interaction carrier of the electromagnetic force, the photon, as an irreducible representation of $O^\uparrow(3,1) \ltimes \mathbb{R}^{3,1}$. The photon is treated as a particle of mass $m = 0$ and spin $t = 1$. The bundle $E^+_{0,t} = E^+_{0,1}$ possesses sub-bundles of helicity $s = 1$ and $s = -1$ which are said to correspond to the right-handed and left-handed 'polarizations' of a photon. These sub-bundles correspond to the $E^+_{m,s} = E^+_{0,1}$ and $E^+_{m,s} = E^+_{0,-1}$ bundles used in the representations of the restricted Poincaré group.

When parity transformation is not a symmetry, the parity $\epsilon = \pm 1$ of a zero mass particle is an invariant property of the particle, determined by the irreducible unitary

[31] The case of $t = 0$ depends upon the choice of the enlarged isotropy group. For one choice there will be two 1-dimensional representations, whilst for the other choice, there will be four 1-dimensional representations.

representation of $SL(2, \mathbb{C}) \ltimes \mathbb{R}^{3,1}$ with which it is associated. Rotations and translations and Lorentz boosts cannot change the helicity of a zero mass particle. By contrast, when parity *is* a symmetry, parity is not an invariant property of a zero mass particle. When parity is a symmetry, the representation space of a zero-mass particle can be decomposed into a left-handed subspace and a right-handed subspace, hence such a particle can possess parity as a property, but the parity can change as the state of the particle changes, and it can change under a change of reference frame that involves the parity transformation $\mathscr{P}$. A change in the handedness of a Lorentz chart could flip the parity of a zero-mass particle. *Contra* Sternberg (1994, p. 155) it *does* make sense to speak of the handedness of a zero-mass particle when parity is a symmetry, but the point is that the handedness is not an *invariant* property under such a local space–time symmetry group.

Let us agree to define an intrinsic property of an object to be a property which the object possesses independently of its relationships to other objects. Let us also agree to define an extrinsic property of an object to be a property which the object possesses depending upon its relationships with other objects. These definitions of intrinsicness and extrinsicness may not be adequate to cover intrinsic and extrinsic properties in general (Weatherson, 2002, Section 2.1), but they are adequate for dealing with physical quantities and the values of physical quantities, the types of properties which philosophers refer to as 'determinables' and 'determinates' respectively (Swoyer, 2000). Being invariant under a change of reference frame is a necessary condition for a property to be possessed intrinsically; if the value of a quantity possessed by an object can change under a change of reference frame, then the value of that quantity must, instead, be an extrinsic property of the object. The value of such a quantity must be a relationship between the object and a reference frame, and under a change of reference frame, that relationship can change.

To reiterate, when parity-reversal is a space–time symmetry, the parity of a zero-mass particle can change under a change of reference frame. Hence, when parity-reversal is a space–time symmetry, parity is not possessed intrinsically by a zero-mass particle. This is distinct from the claim that parity itself, as an abstract property, is an extrinsic property.[32] Whether or not parity is possessed intrinsically depends upon both the symmetry group, and whether the particle in question is massive or not. Recall that when parity-reversal is a space–time symmetry, parity is an invariant property of massive particles. An invariant property cannot change under a change of reference frame, hence, if one assumes that invariance under a change of reference frame is also a sufficient condition for a property to be possessed intrinsically, then parity is possessed intrinsically by massive particles when parity-reversal is a space–time symmetry. Moreover, when parity-reversal does not belong to the local space–time symmetry group, the parity of a zero-mass particle is an invariant property, hence

[32] The distinction between a property *being* extrinsic/intrinsic, and being *possessed* extrinsically/intrinsically is made by Weatherson (2002, Section 1.2).

if one assumes again that an invariant property is possessed intrinsically, then a zero-mass particle possesses parity intrinsically under these circumstances.[33]

## 2.6. Free particles in other universes

Universes of a different dimension and/or geometrical signature, will possess a different local symmetry group, and will therefore possess different sets of possible elementary particles. Moreover, even universes of the same dimension and geometrical signature will not necessarily possess the same sets of possible particles. To reiterate, the dimension and geometrical signature merely determines the largest possible local symmetry group, and universes with different gauge fields, and different couplings between the gauge fields and matter fields, will possess different local symmetry groups, and, perforce, will possess different sets of possible particles. However, let us consider in this section the consequence of varying the dimension and geometrical signature.

Recall that for an arbitrary pseudo-Riemannian manifold $(\mathcal{M}, g)$, of dimension $n$ and signature $(p, q)$, the largest possible local symmetry group is the semi-direct product $O(p, q) \ltimes \mathbb{R}^{p,q}$. If there is no restriction to a subgroup of this, then one specifies the possible free elementary particles in such a universe by specifying the *projective*, unitary, irreducible representations of $O(p, q) \ltimes \mathbb{R}^{p,q}$. For a universe which has $p$ spatial dimensions and $q$ temporal dimensions, and which has, in addition, a local space orientation and a local time orientation, the local space–time symmetry group will be the semi-direct product $SO_0(p, q) \ltimes \mathbb{R}^{p,q}$. Given an arbitrary semi-direct product $H \ltimes N$, with $H$ a semi-simple group and $N$ a finite-dimensional real vector space, every projective, unitary, irreducible representation of $H \ltimes N$ lifts to a unitarizable irreducible representation of the universal cover $\tilde{H} \ltimes N$. For $p+q > 2$, $SO_0(p, q)$ is a connected simple group, hence every projective, unitary, irreducible representation of $SO_0(p, q) \ltimes \mathbb{R}^{p,q}$, for $p + q > 2$, corresponds to an ordinary, unitary, irreducible representation of the universal cover $\widetilde{SO}_0(p, q) \ltimes \mathbb{R}^{p,q}$.[34] Thus, the elementary particles in a universe which has $p$ spatial dimensions and $q$ temporal dimensions, and which has, in addition, a local space orientation and a local time orientation, will correspond to a subset of the irreducible, unitary representations of $\widetilde{SO}_0(p, q) \ltimes \mathbb{R}^{p,q}$. As before, the method of induced representation, applied to semi-direct products, can be used to classify these representations, and to explicitly generate the energy–momentum space representations. The first step is to find the orbits of the $\widetilde{SO}_0(p, q) \ltimes \mathbb{R}^{p,q}$-action on $\mathbb{R}^{p,q}$. Given that $\mathbb{R}^{p,q}$ is an Abelian

[33] The assumption that a property invariant under any change of reference frame must be intrinsic, or possessed intrinsically, has been challenged; for example, Pooley (2003) argues that the parity (or 'handedness') of a particle is an extrinsic property even in the event that it is invariant under the permitted space–time symmetries.

[34] Note that for $p, q > 2$, whilst $Spin(p, q)$ is a double cover of $SO_0(p, q)$, it is not the universal cover.

group, acting upon itself by conjugation, these orbits coincide with the orbits of the $\widetilde{SO}_0(p,q)$-action on $\mathbb{R}^{p,q}$. The orbits of the $\widetilde{SO}_0(p,q)$-action on $\mathbb{R}^{p,q}$ correspond to the orbits of $SO_0(p,q)$ on $\mathbb{R}^{p,q}$. As before, these orbits are hypersurfaces in $\mathbb{R}^{p,q}$, and, as before, they come in three types:

- For any $m \in \mathbb{R}_+$, there are the hypersurfaces $\mathscr{V}_m$, where $\mathscr{V}_m = \{p \in \mathbb{R}^{p,q}\colon \langle p,p\rangle = -m^2 < 0\}$.[35]
- There are the null hypersurfaces, $\mathscr{V}_0 = \{p \in \mathbb{R}^{p,q}\colon \langle p,p\rangle = m^2 = 0\}$.
- Then, for any $n \in \mathbb{R}_+$, there are the hypersurfaces $\{p \in \mathbb{R}^{p,q}\colon \langle p,p\rangle = n > 0\}$.

Unless $p = 1$ or $q = 1$, each such hypersurface consists of a single connected component. Hence, the special case of $SL(2,\mathbb{C}) \ltimes \mathbb{R}^{3,1}$, where the $m^2 > 0$ mass hyperboloids consist of two components (a forward mass hyperboloid and a backward mass hyperboloid), each of which is a separate orbit of the group action, is not typical. In addition, it is only in the case where $q = 1$ that the $\langle p,p\rangle < 0$ orbits are spacelike hypersurfaces. In general, both the $\langle p,p\rangle < 0$ orbits and the $\langle p,p\rangle > 0$ orbits will be timelike hypersurfaces. I.e., the tangent vector space at each point will be spanned by timelike and spacelike vectors. The spacelike submanifolds in $\mathbb{R}^{p,q}$ will be $p$-dimensional, and unless $q = 1$, they will not be hypersurfaces.

After identifying the orbits of the $\widetilde{SO}_0(p,q)$-action on $\mathbb{R}^{p,q}$, the next step is to identify the isotropy groups of the $\widetilde{SO}_0(p,q)$-action on $\mathbb{R}^{p,q}$. The isotropy group of the $\widetilde{SO}_0(p,q)$-action on the $\langle p,p\rangle < 0$ orbits is $\widetilde{SO}_0(p,q-1)$, and the isotropy group of the $\widetilde{SO}_0(p,q)$-action on the $\langle p,p\rangle > 0$ orbits is $\widetilde{SO}_0(p-1,q)$. This entails that the isotropy group of the $\widetilde{SO}_0(p,q) \ltimes \mathbb{R}^{p,q}$-action on the $\langle p,p\rangle < 0$ orbits is $\widetilde{SO}_0(p,q-1) \ltimes \mathbb{R}^{p,q}$, and the isotropy group of the $\widetilde{SO}_0(p,q) \ltimes \mathbb{R}^{p,q}$-action on the $\langle p,p\rangle > 0$ orbits is $\widetilde{SO}_0(p-1,q) \ltimes \mathbb{R}^{p,q}$. In the special case of $SL(2,\mathbb{C}) \ltimes \mathbb{R}^{3,1}$, the isotropy group of the action of $SL(2,\mathbb{C}) \cong Spin(3,1)$ on the $\langle p,p\rangle < 0$ orbits is $Spin(3) = SU(2)$. By virtue of being a compact group, $SU(2)$ has finite-dimensional irreducible unitary representations, which can be parameterized by a discrete parameter. However, in the general case, unless $p = 1$ or $q = 1$, the isotropy groups of the $\widetilde{SO}_0(p,q)$-action will be non-compact, and all the non-trivial irreducible unitary representations of such groups are infinite-dimensional. Hence, in addition to the first continuous parameter, one must deal with the additional continuous parameters (and the possible additional discrete parameters), required to classify the infinite-dimensional irreducible unitary representations of the isotropy groups, $\widetilde{SO}_0(p,q-1) \ltimes \mathbb{R}^{p,q}$ and $\widetilde{SO}_0(p-1,q) \ltimes \mathbb{R}^{p,q}$.[36]

In conclusion, the elementary particles which exist in universes with a different number of spatial and/or temporal dimensions to our own, are, in general, not parameterized by mass and spin. However, those cases in which $q = 1$ correspond to

[35] There is no significance here in the choice of '$p$' to denote both an element of energy–momentum space, and the number of spatial dimensions in the geometrical signature $(p,q)$.

[36] Private communications with Veeravalli Varadarajan and Shlomo Sternberg.

Lorentzian space–times with one temporal dimension and an arbitrarily high number $p$ of spatial dimensions, and in these cases one *can* define analogues of the elementary particles with which we are familiar. In such universes, one parameterizes the elementary particle representations by mass and the values of the discrete parameters necessary to classify the finite-dimensional irreducible unitary representations of the compact isotropy groups $\widetilde{SO}_0(p) = Spin(p)$. Note, however, that for $p > 3$ the representations are labelled by $n = [p/2]$ tuples[37] of discrete parameters called 'dominant weights'[38] $(\lambda_1, \ldots, \lambda_n)$, such that each $\lambda_i$ is either an integer or half-integer. In this notation, the 'vector' representation induced by $SO(p)$ is denoted as $(1, 0, \ldots, 0)$, the spinor representation is denoted as $(1/2, \ldots, 1/2)$ if $p$ is odd, and the spinor representations are denoted as $(1/2, \ldots, \pm 1/2)$ if $p$ is even. These weights are distinct from the 'spins' $(s_1, s_2)$ used to label the representations of $Spin(4) \cong SL(2, \mathbb{C})$. For example, the spinor representations of $Spin(4)$ are labelled as $(1/2, 1/2)$ and $(1/2, -1/2)$ using weights, and labelled as $(1/2, 0)$ and $(0, 1/2)$ using spins. One does, however, in this case have the relationship $s_1 = (\lambda_1 + \lambda_2)/2$ and $s_2 = (\lambda_1 - \lambda_2)/2$. It is only in the special case of $Spin(3) = SU(2)$ that one has a single discrete spin parameter $s \in \frac{1}{2}\mathbb{Z}_+$ to label the finite-dimensional irreducible unitary representations. In general, therefore, electrons, quarks, neutrinos, photons etc., as we define them with their characteristic mass and spin, cannot exist in universes with a different dimension or geometrical signature to our own.

This conclusion can be reinforced by considering the configuration space approach. Recall that when there are three spatial dimensions and one temporal dimension, one begins by defining a bundle which possesses a spin-$s$ representation of $SL(2, \mathbb{C}) \cong Spin(3, 1)$ upon its typical fibre. Whilst the complex, finite-dimensional, irreducible representations of $SL(2, \mathbb{C})$ are indexed by pairs of spins $(s_1, s_2)$, with $s = s_1 + s_2$, this does not generalise to the representations of $\widetilde{SO}_0(p, q)$, relevant to universes with an arbitrary number of space and time dimensions.

The representations of $\widetilde{SO}_0(p, q)$ correspond to the representations of the Lie algebra $\mathfrak{so}(p, q)$, and the complex representations of $\mathfrak{so}(p, q)$ correspond to the representations of its complexification $\mathfrak{so}(p, q) \otimes \mathbb{C}$. Moreover, $\mathfrak{so}(p, q) \otimes \mathbb{C} \cong \mathfrak{so}(n) \otimes \mathbb{C}$, for $p+q = n$. In the case of $\mathfrak{so}(3, 1)$, this entails that its complex, finite-dimensional, irreducible representations correspond to the complex, finite-dimensional, irreducible representations of $\mathfrak{so}(4)$. Now, it happens that $\mathfrak{so}(4) \cong \mathfrak{su}(2) \oplus \mathfrak{su}(2)$, hence the complex, finite-dimensional, irreducible representations of $\mathfrak{so}(3, 1)$ correspond to the finite-dimensional irreducible representations of $\mathfrak{so}(4) \otimes \mathbb{C} \cong (\mathfrak{su}(2) \otimes \mathbb{C}) \oplus (\mathfrak{su}(2) \otimes \mathbb{C})$ (Derdzinski, 1992, Section 3.4). The finite-dimensional irreducible representations of $\mathfrak{su}(2) \otimes \mathbb{C}$ are indexed by their spin, and the irreducible representations

[37] '$[\cdot]$' means the integral part.

[38] A weight vector in a Lie group representation is a simultaneous eigenvector of the operators representing a maximal commuting subalgebra (a 'Cartan' subalgebra) from the Lie algebra of the Lie group. The weights of such a vector are the simultaneous eigenvalues. The set of simultaneous eigenvalues taken over all the weight vectors in the representation space can be ordered, and a 'dominant' weight is closely related to the notion of a highest weight.

of a Lie algebra direct sum correspond to tensor products of the irreducible representations of the summands, hence the finite-dimensional irreducible representations of $(\mathfrak{su}(2) \otimes \mathbb{C}) \oplus (\mathfrak{su}(2) \otimes \mathbb{C})$ correspond to pairs of spins $(s_1, s_2)$. Sadly, $\mathfrak{so}(n)$ is not in general a direct sum of copies of $\mathfrak{su}(2)$, hence the complex, irreducible, finite-dimensional representations of $\mathfrak{so}(p, q)$ cannot, in general, be indexed by a tuple of spins. As a consequence, the complex, irreducible, finite-dimensional representations of $\widetilde{SO}_0(p, q)$ also cannot be indexed by a tuple of spins. Thus, the vector bundles possessing representations of $\widetilde{SO}_0(p, q)$ upon their typical fibres cannot be labelled as spin-$s$ free-particle bundles.

The dimension, signature, and orientation of a space–time determine the fundamental differential equations which govern the behaviour of free elementary particles. One cannot vary the dimension-signature-orientation of a space–time whilst holding fixed the fundamental laws of physics for free systems in that space–time. The laws of physics for free systems, such as the Klein–Gordon equation, the Dirac equation and the Weyl equation, are the configuration space expressions of the irreducible unitary representations of the local space–time symmetry group. For a fixed particle type in a fixed space–time, the differential equations governing that particle cannot be varied. In particular, the signature of the space–time metric is reflected in the signature of the partial differential equations governing the free elementary particles in that space–time. The Klein–Gordon equation, Dirac equation and Weyl equation in our universe are hyperbolic partial differential equations precisely because the signature of our space–time is Lorentzian. Universes with, for example, metric tensors of Riemannian signature, will possess elementary particles that satisfy elliptic partial differential equations in the configuration representation.

There is a history of argument which attempts to explain why we observe three spatial dimensions, by pointing out that if the laws of physics are held fixed, and the number of spatial dimensions are changed, then it is not possible for stable planetary orbits, or stable atoms and molecules to exist. Callender (2005) correctly emphasises that if the laws of physics are allowed to vary, then stable systems can exist in universes with a different number of spatial dimensions, and he asserts that "physical law is such a weak kind of necessity compared with conceptual or metaphysical necessity" (p. 115). However, Callender fails to recognize that the particle types in a universe are determined by the irreducible unitary representations of the local space–time symmetry group, and the fundamental laws of elementary particles are simply the configuration space expressions of those irreducible unitary representations. One is not making the glib assumption that the laws of physics in our universe must be the same as those in any other universe; rather, there is a conceptual link between the signature, dimension and orientation of a space–time and the fundamental laws of physics which govern particles within that space–time.

Let us briefly explain how the signature of space–time determines the differential equations on configuration space, via the inverse Fourier transform from the unitary representations on energy–momentum space. A function $\phi(p)$ with, for example, support on a forward mass hyperboloid in Minkowski energy–momentum space, will satisfy the equation $(\|p\|^2 - m^2)\phi(p) = 0$. Such a function is the Fourier transform of a function $\psi(x)$ on Minkowski configuration space which satisfies the equation $(\partial_0^2 - \partial_1^2 - \partial_2^2 - \partial_3^2 - m^2)\psi(x) = 0$. To see this, note first that for any function $\phi(p)$ on a mass hyperboloid in Minkowski energy–momentum space, there exists a function $\psi(x)$ on Minkowski configuration space which is the inverse Fourier transform of $\phi(p)$[39]:

$$\psi(x) = \int e^{i\langle p,x\rangle}\phi(p)\, d^4 p.$$

It follows that for $j = 1, 2, 3$, the Fourier transform of $\partial_j\psi$ is $ip_j\phi$ because

$$\begin{aligned}\partial_j\psi(x) &= \int \partial_j\big[e^{i\langle p,x\rangle}\big]\phi(p)\, d^4 p \\ &= \int e^{i\langle p,x\rangle} i p_j \phi(p)\, d^4 p.\end{aligned}$$

Similarly, for $j = 0$, the Fourier transform of $\partial_j\psi$ is $-ip_j\phi$. For $j = 0, 1, 2, 3$, the Fourier transform of $\partial_j^2\psi$ is $-p_j^2\phi$ because $(-ip_j)^2 = (ip_j)^2 = -p_j^2$.

Hence, the Fourier transform of $(\partial_0^2 - \partial_1^2 - \partial_2^2 - \partial_3^2 - m^2)\psi(x)$ is $(-p_0^2 + p_1^2 + p_2^2 + p_3^2 - m^2)\phi(p)$. Using the notation $(\partial_0^2 - \partial_1^2 - \partial_2^2 - \partial_3^2 - m^2)\psi(x) \equiv (\Box - m^2)\psi(x)$, one has:

$$\begin{aligned}(\Box - m^2)\psi(x) &= \int (\partial_0^2 - \partial_1^2 - \partial_2^2 - \partial_3^2 - m^2)\big[e^{i\langle p,x\rangle}\big]\phi(p)\, d^4 p \\ &= \int e^{i\langle p,x\rangle}\big((-ip_0)^2 - (ip_1)^2 - (ip_2)^2 - (ip_3)^2 - m^2\big)\phi(p)\, d^4 p \\ &= \int e^{i\langle p,x\rangle}\big(-p_0^2 + p_1^2 + p_2^2 + p_3^2 - m^2\big)\phi(p)\, d^4 p.\end{aligned}$$

Thus, the inverse Fourier transform of $(\|p\|^2 - m^2)\phi(p) = 0$ is the Klein–Gordon equation $(\partial_0^2 - \partial_1^2 - \partial_2^2 - \partial_3^2 - m^2)\psi(x) = 0$. Flip one of the metric signs from positive to negative, or vice versa, and the sign of the corresponding differential operator on configuration space must clearly also change.

Partial differential equations (PDEs) can be classified by the eigenvalues of their component matrices (Tegmark, 1998, p. 32). For example, consider the case of a second-order linear partial differential equation in an $n$-dimensional space–time:

$$\left[\sum_{i=1}^{n}\sum_{j=1}^{n} A_{ij}(x)\frac{\partial}{\partial x_i}\frac{\partial}{\partial x_j} + \sum_{i=1}^{n} b_i(x)\frac{\partial}{\partial x_i} + c(x)\right]u(x) = f(x).$$

Assuming that $A(x)$ is a symmetric real $n \times n$ matrix, it must have $n$ real eigenvalues, and one can classify a second-order linear partial differential equation in terms of the signs of the eigenvalues of $A(x)$. The PDE at $x$ is defined to be

[39] Note that, as in Section 2.2, the indefinite Minkowski space inner product $\langle p, x\rangle = -p_0 t + p_1 x_1 + p_2 x_2 + p_3 x_3$ is being used to define the Fourier transform here.

- elliptic if either all the eigenvalues of $A(x)$ are positive or all the eigenvalues of $A(x)$ are negative;
- hyperbolic if one eigenvalue is negative and all the others positive, or vice versa;
- ultrahyperbolic if at least two eigenvalues are positive and the others negative, or vice versa; and
- parabolic if at least one eigenvalue is zero, entailing that $A(x)$ is singular, $\det A(x) = 0$.

If the space–time signature is Riemannian, such as $(+ + \cdots +)$ or $(- - \cdots -)$, then the elementary particles in such a space–time will satisfy elliptic partial differential equations in the configuration representation. If the space–time signature is Lorentzian, such as $(+ + \cdots -)$ or $(- - \cdots +)$, then the elementary particles in such a space–time will satisfy hyperbolic partial differential equations in the configuration representation. If the space–time signature is like $(+ + \cdots - -)$, then the elementary particles in such a space–time will satisfy ultrahyperbolic partial differential equations in the configuration representation. Finally, if the space–time signature is degenerate, with at least one zero in the signature, then the elementary particles in such a space–time will satisfy parabolic partial differential equations in the configuration representation. The Schrodinger equation is a parabolic partial differential equation, which is consistent with the fact that space–time is represented by the degenerate metric tensor of Newton–Cartan space–time in non-relativistic quantum mechanics.

The problem of finding a solution $u(x)$ to a partial differential equation on a particular domain, with a particular set of boundary conditions or initial conditions, is said to be 'well-posed' if:

- At least one solution exists.
- At most one solution exists. I.e., the solution is unique.
- The solution depends continuously upon the boundary conditions and initial conditions.

If there are too many boundary or initial conditions, a solution will not exist, and the problem is said to be 'overdetermined.' If there are too few boundary or initial conditions, the solution will not be unique, and the problem is said to be 'underdetermined.' If the solution does not depend continuously upon the boundary and initial conditions, then the problem is said to be unstable. If the problem of solving a partial differential equation fails any one of the bullet-pointed conditions above, then it is said to be an 'ill-posed' problem.

Domains tend to be classified in this context as either open or closed, meaning non-compact or compact topology. One can treat initial conditions as special types of boundary conditions; initial conditions are the boundary conditions on the spacelike part of the boundary to a region of space–time. A region of space–time can be compact or non-compact, and the initial boundary can be compact or non-compact too. Boundary conditions tend to be classified as:

- Dirichlet — The solution $u$ is specified upon the boundary.
- Neumann — The derivative of the solution with respect to the normal vector field on the boundary is specified.
- Robin/Mixed/Cauchy — A conjunction or linear combination of the solution and its normal derivative is specified on the boundary.

The problem of solving the different types of partial differential equation is well-posed or ill-posed depending upon both the type of boundary conditions specified, and the type of domain. Whilst the situation is largely unresolved for ultrahyperbolic equations,[40] the following results are known for the other equation types (see, for example, Morse and Feshbach (1953, Section 6.1)):

- Dirichlet.
  - Open domain
    1. Hyperbolic: underdetermined
    2. Elliptic: underdetermined
    3. Parabolic: unique and stable in one direction
  - Closed domain
    1. Hyperbolic: overdetermined
    2. Elliptic: unique and stable
    3. Parabolic: overdetermined
- Neumann.
  - Open domain
    1. Hyperbolic: underdetermined
    2. Elliptic: underdetermined
    3. Parabolic: unique and stable in one direction
  - Closed domain
    1. Hyperbolic: overdetermined
    2. Elliptic: unique and stable with additional constraints
    3. Parabolic: overdetermined
- Robin/Mixed/Cauchy.
  - Open domain
    1. Hyperbolic: unique and stable
    2. Elliptic: ill-posed
    3. Parabolic: overdetermined

[40] Private communication with V.G. Romanov.

- Closed domain
  1. Hyperbolic: overdetermined
  2. Elliptic: overdetermined
  3. Parabolic: overdetermined

Hence, the well-posedness of the partial differential equations which govern free elementary particles in universes with a different number of spatial or temporal dimensions to our own, clearly depends upon the topology of the domains in question, and the type of the boundary conditions specified.

One qualification should be added to the results above: Cauchy data on a non-compact spacelike hypersurface provides a well-posed problem for a hyperbolic equation, but Asgeirsson's theorem implies that Cauchy data on a timelike hypersurface is ill-posed for such an equation. This theorem also has ramifications for ultrahyperbolic equations; because there are at least two spatial dimensions and at least two temporal dimensions for such equations, spacelike hypersurfaces do not exist, and Cauchy data on a timelike hypersurface does not provide a well-posed problem for an ultrahyperbolic equation. However, Cauchy data on a combination of null hypersurfaces and other types of hypersurface can provide a well-posed problem for a hyperbolic equation, so the same may prove to be the case for ultrahyperbolic equations.

Tegmark claims that in the universes where all $n$ dimensions are spatial, or all $n$ dimensions are temporal, the elliptic differential equations would not enable the physics in such universes to have predictive power. He acknowledges (1998, p. 33, paragraph d) that an elliptic equation on a closed domain leads to a well-posed problem, but seems to regard such a situation as a boundary value problem, distinct from an initial data problem. Tegmark assumes that an initial data problem must involve an open domain, and points out that elliptic PDEs are ill-posed on such domains: "specifying only $u$ on a non-closed surface gives an underdetermined problem, and specifying additional data, e.g., the normal derivative of $u$, generally makes the problem overdetermined" (1998, p. 33, footnote 11). However, there is no reason why initial data problems must have non-compact initial boundaries, and in a universe in which all dimensions are spatial, or in which all dimensions are temporal, it seems artificial to make a distinction between boundary value problems and initial data problems. Elliptic partial differential equations do not provide well-posed problems given Cauchy data on open or closed domains, and they do not provide well-posed problems on open domains for any type of boundary data, but they do provide well-posed problems on closed domains with either Dirichlet or Neumann boundary conditions. Thus, in these circumstances, it is still possible to make inferences about the interior of bounded regions from information on the boundary. Tegmark defines prediction as simply the use of local observations to make inferences about other parts of a pseudo-Riemannian manifold, whatever its dimension and signature, hence, under this definition, the differential equations which govern free elementary particles in universes where all $n$ dimensions are spatial, or all $n$ dimensions are temporal, would still have predictive power under certain circumstances.

There is also an implicit assumption in Tegmark's reasoning that the physics in a universe only has predictive power if the equations which govern the *elementary particles* have predictive power. The signature of the space–time geometry may determine the type of the equations which govern the elementary particles, but it doesn't determine the type of all the PDEs governing all the physical processes in a space–time. For example, in our own space–time, where the fundamental equations for free particles are hyperbolic, the diffusion equation, which governs certain higher-level statistical and thermodynamical processes, is parabolic.

## 2.7. Natural bundles

Mathematician Andrzej Derdzinski (1992) has presented a thorough development of the configuration space approach to the representation of free particles. Whilst Edward Nelson famously asserted that "first quantization is a mystery, but second quantization is a functor,"[41] Derdzinski treats *the first step* to first quantization, the selection of a mass-independent free-particle bundle over Minkowski space–time, as a functor as well. Working with the isochronous Lorentz group $O^{\uparrow}(3, 1)$ rather than $SO_0(3, 1)$, Derdzinski introduces free-particle bundles $\eta$, which are 'natural bundles' in the sense that for each point $x \in \mathcal{M}$ in Minkowski space–time, there is a representation $O^{\uparrow}(T_x\mathcal{M}) \to Aut(\eta_x)$ (Derdzinski, 1992, pp. 20–21). One therefore has a functor between the fibres of the tangent bundle over space–time, and the fibres of the free-particle bundle. Elementary particles, says Derdzinski, correspond to irreducible natural bundles, in the sense that the representation in each fibre is irreducible. The representations are permitted to be double-valued, hence they become irreducible representations of a double-cover $\overline{SL}(2, \mathbb{C})$ of $O^{\uparrow}(3, 1)$.

Derdzinski's use of irreducible representations might seem erroneous here because the Dirac spinor bundle $\sigma$, used in the first step to represent many elementary particles, and treated in detail in Section 2.8, possesses a *reducible*, direct sum representation of $SL(2, \mathbb{C})$ in each of its fibres. However, the corresponding representation of $\overline{SL}(2, \mathbb{C})$ is irreducible, and it is the latter group with which Derdzinski is working.

If first quantization consisted only of Derdzinski's functorial first step, then the symmetry group of first-quantized free particles would be the infinite-dimensional group of cross-sections of the automorphism bundle $Aut(\eta)$, the bundle of all automorphisms in each fibre of $\eta$. The first step to first quantization is, however, deceptive, and when the second step is taken into account, one obtains a Hilbert space $\mathscr{H}_{m,s}$,

[41] Nelson made this assertion in a lecture delivered at the mathematics department in Fine Hall, Princeton. Barry Simon and Mike Reed were in the audience, and used the quote in the heading for Chapter X.7 of *Methods of Modern Mathematical Physics Vol. II: Fourier Analysis, Self-Adjointness* (personal communication with Edward Nelson).

Table 2.2
Derdzinski's free-particle bundles

| Spin | Parity | Charged bundle | Neutral bundle |
|---|---|---|---|
| 0 | +1 | $\mathcal{M} \times \mathbb{C}$ | $\mathcal{M} \times \mathbb{R}$ |
| 0 | −1 | $\bigwedge^4 T^*\mathcal{M} \otimes \mathbb{C}$ | $\bigwedge^4 T^*\mathcal{M}$ |
| $\frac{1}{2}$ | +1 | $\sigma_L$ | – |
| $\frac{1}{2}$ | −1 | $\sigma_R$ | – |
| $\frac{1}{2}$ | – | $\sigma$ | – |
| $s \geqslant 1, s = k \in \mathbb{Z}$ | $(-1)^k$ | $S_0^k T^*\mathcal{M} \otimes \mathbb{C}$ | $S_0^k T^*\mathcal{M}$ |
| $s \geqslant 1, s = k \in \mathbb{Z}$ | $(-1)^{k+1}$ | $S_0^k T^*\mathcal{M} \otimes \bigwedge^4 T^*\mathcal{M} \otimes \mathbb{C}$ | $S_0^k T^*\mathcal{M} \otimes \bigwedge^4 T^*\mathcal{M}$ |
| $s = k + \frac{1}{2}, k \geqslant 1, k \in \mathbb{Z}$ | – | $\eta \subset S_0^k T^*\mathcal{M} \otimes \sigma$ | – |

and a unitary representation of the finite-dimensional group $SL(2, \mathbb{C}) \ltimes \mathbb{R}^{3,1}$ on $\mathscr{H}_{m,s}$. Such a Hilbert space is not invariant under the action of the infinite-dimensional group of cross-sections of $Aut(\eta)$, and the original functor is therefore not relevant here.

Derdzinski defines a mass-independent configuration space vector bundle for each possible combination of spin and parity. The space of sections of such a configuration space vector bundle represents many different particle species, each corresponding to the cross-sections for a specific mass. In addition, the space of cross-sections includes both the positive energy and negative energy cross-sections for each mass value. Hence, the space of cross-sections of a configuration space bundle $\eta$ includes the cross-sections for both the particles and anti-particles of each mass value. Derdzinski (1992, pp. 17–20), derives the vector bundles displayed in Table 2.2 to represent free particles in the first step of the configuration space approach.

The cross-sections of $S_0^k T^*\mathcal{M}$ consist of those cross-sections $\psi$ of the $k$-fold symmetric tensor product of $T^*\mathcal{M}$ which satisfy the equation $\Delta\psi = 0$. The Laplacian $\Delta$ is defined with respect to the pseudo-Riemannian metric $g$ upon the manifold $\mathcal{M}$. $S_0^k T_x^*\mathcal{M}$ is the space of pseudo-spherical harmonics in $T_x\mathcal{M}$.

In the case of a particle with $s = k + \frac{1}{2}$, for $k \geqslant 1$ and $k \in \mathbb{Z}$, the expression

$$\eta \subset S_0^k T^*\mathcal{M} \otimes \sigma$$

denotes the subset of $S_0^k T^*\mathcal{M} \otimes \sigma$ consisting of elements in the kernel of Clifford multiplication. In this context, Clifford multiplication $c$ is a map

$$c : \left(\bigotimes^k T^*\mathcal{M}\right) \otimes \sigma \to \left(\bigotimes^{k-1} T^*\mathcal{M}\right) \otimes \sigma.$$

## 2.8. Spinor bundles

In the configuration space approach, all quarks and leptons have state spaces constructed from cross-sections of either a Weyl spinor bundle or a Dirac spinor bundle.

Weyl spinor bundles and Dirac spinor bundles are complex vector bundles isomorphic to vector bundles associated to spin-frame bundles $S(\mathcal{M})$ over space–time.[42]

For any Lorentzian $n$-manifold $(\mathcal{M}, g)$, one can introduce an oriented Lorentz frame bundle $(L_0(\mathcal{M}), \Pi_L, \mathcal{M}, SO_0(n-1,1))$. The total space $L_0(\mathcal{M})$ of this bundle consists of all the orthonormal bases $\{e_\mu\colon \mu = 0, 1, \ldots, n-1\}$ of the tangent spaces at all the points of the manifold $\mathcal{M}$, such that each $e_0$ is a future-pointing, timelike vector, and such that each $\{e_i\colon i = 1, \ldots, n-1\}$ is a right-handed $(n-1)$-tuple of spacelike vectors. This principal fibre bundle has the restricted Lorentz group $SO_0(n-1,1)$ as its structure group.

A spin frame bundle $S(\mathcal{M})$ is a double cover of an oriented Lorentz frame bundle, satisfying certain conditions (Bleecker, 1981, pp. 81–82). A spin frame bundle $(S(\mathcal{M}), \Pi_S, \mathcal{M}, Spin(n-1,1))$ is a principal fibre bundle over $\mathcal{M}$, with structure group $Spin(n-1,1)$, the latter being the universal covering group of the restricted Lorentz group $SO_0(n-1,1)$. Denote the universal covering homomorphism as

$$\Lambda : Spin(n-1,1) \to SO_0(n-1,1),$$

so that an arbitrary $A \in Spin(n-1,1)$ is mapped to $\Lambda(A) \in SO_0(n-1,1)$.

The bundle of spin frames $S(\mathcal{M})$ is tied to the oriented Lorentz frame bundle $L_0(\mathcal{M})$ by a map $\lambda : S(\mathcal{M}) \to L_0(\mathcal{M})$ satisfying the conditions,

$$\Pi_L\big(\lambda(p)\big) = \Pi_S(p), \tag{i}$$

$$\lambda(pA) = \lambda(p)\Lambda(A) \tag{ii}$$

for any $p \in S(\mathcal{M})$, $A \in Spin(n-1,1)$ and $\Lambda(A) \in SO_0(n-1,1)$. In other words, the covering map $\lambda : S(\mathcal{M}) \to L_0(\mathcal{M})$ preserves the fibre structure of the bundles, and commutes with the right action of the structure groups upon the respective bundles.

In the case of Minkowski space–time, the oriented Lorentz frame bundle is $(L_0(\mathcal{M}), \Pi_L, \mathcal{M}, SO_0(3,1))$, and the spin-frame bundle has structure group $Spin(3,1) \equiv SL(2,\mathbb{C})$, the universal covering group of $SO_0(3,1)$. There are two inequivalent, irreducible representations of $SL(2,\mathbb{C})$ in $\mathbb{C}^2$ (Bleecker, 1981, p. 76). The Weyl spinor bundles, $\sigma_L$ and $\sigma_R$, can be obtained via these two 'mutually conjugate' representations of $SL(2,\mathbb{C})$:

$$\mathscr{D}^{1/2,0}A = A,$$

and

$$\mathscr{D}^{0,1/2}A = A^{*-1}.$$

[42] See Appendix C for a definition of an associated vector bundle.

These two inequivalent representations yield the 'right-handed' and 'left-handed' Weyl spinor bundles as associated vector bundles (Derdzinski, 1992, pp. 102–103):

$$\sigma_L = S(\mathcal{M}) \times_{\mathscr{D}^{1/2,0}} \mathbb{C}^2,$$
$$\sigma_R = S(\mathcal{M}) \times_{\mathscr{D}^{0,1/2}} \mathbb{C}^2.$$

The Dirac spinor bundle can be expressed as the direct sum of the two Weyl spinor bundles:

$$\sigma = \sigma_L \oplus \sigma_R.$$

The 'handedness' of a Weyl spinor bundle is often referred to as its 'chirality,' and the Dirac spinor bundle can be said to decompose into its two chiral halves.

The Dirac spinor bundle $\sigma$ can also be expressed as

$$\sigma = S(\mathcal{M}) \times_\tau \mathbb{C}^4,$$

the vector bundle associated to the spinor frame bundle by the reducible, direct sum representation $\tau$ on $\mathbb{C}^4$:

$$\tau = \mathscr{D}^{1/2,0} \oplus \mathscr{D}^{0,1/2},$$

where

$$\tau(A) = \begin{pmatrix} A & 0 \\ 0 & A^{*-1} \end{pmatrix}.$$

The spin-frame bundle $S(\mathcal{M})$ over Minkowski space–time admits a global cross-section. Hence, the spin-frame bundle is isomorphic to a trivial bundle, $S(\mathcal{M}) \cong \mathcal{M} \times SL(2, \mathbb{C})$. If the spin-frame bundle is trivial, then the spinor bundles must also be isomorphic to trivial bundles. With a global cross-section of the spin-frame bundle $S(\mathcal{M})$, one obtains a trivialization of the Weyl spinor bundles, $\sigma_L \cong \mathcal{M} \times \mathbb{C}^2 \cong \overline{\sigma_R}$ and a trivialization of the Dirac spinor bundle, $\sigma \cong \mathcal{M} \times \mathbb{C}^4$. Different global cross-sections of $S(\mathcal{M})$ give rise to different trivializations of $\sigma_L$, $\sigma_R$, and $\sigma$.

Because the Dirac spinor bundle over Minkowski space–time is trivial, one can treat the space of cross-sections $\Gamma(\sigma)$ as a space of $\mathbb{C}^4$-valued functions $\mathcal{F}(\mathcal{M}, \mathbb{C}^4)$. Similarly, one can treat the space of cross-sections of a Weyl spinor bundle, either $\Gamma(\sigma_L)$ or $\Gamma(\sigma_R)$, as a space of $\mathbb{C}^2$-valued functions $\mathcal{F}(\mathcal{M}, \mathbb{C}^2)$.

To understand spinors from a more general perspective, we digress momentarily to define Dirac gamma matrices and Clifford algebras. With each point $x = (x_0, x_1, x_2, x_3)$ of Minkowski configuration space $\mathbb{R}^{3,1}$, one can associate a self-adjoint $2 \times 2$ matrix (Sternberg, 1994, p. 230),

$$X = \begin{pmatrix} x_0 - x_3 & x_1 + ix_2 \\ x_1 - ix_2 & x_0 + x_3 \end{pmatrix},$$

its cofactor matrix[43]

$$X^a = \begin{pmatrix} x_0 + x_3 & -(x_1 + ix_2) \\ -(x_1 - ix_2) & x_0 - x_3 \end{pmatrix},$$

and a gamma matrix

$$\gamma(x) = \begin{pmatrix} 0 & X \\ X^a & 0 \end{pmatrix}.$$

In abstract terms, one has a map $\gamma : \mathbb{R}^{3,1} \to \mathfrak{gl}(\mathbb{C}^4)$ which satisfies the conditions:

$$\gamma(x)^2 = \langle x, x \rangle I,$$
$$\gamma(x)\gamma(y) + \gamma(y)\gamma(x) = 2\langle x, y \rangle I.$$

In the special case of a quadruple of orthonormal vectors $e_j$ in $\mathbb{R}^{3,1}$, the four matrices

$$\gamma_j = \gamma(e_j), \quad j = 0, 1, 2, 3,$$

are the Dirac gamma matrices. These Dirac gamma matrices satisfy the conditions

$$\gamma_i \gamma_j + \gamma_j \gamma_i = 0 \quad \text{for } i \neq j,$$
$$\gamma_0^2 = I \quad \text{and} \quad \gamma_1^2 = \gamma_2^2 = \gamma_3^2 = -I.$$

The Clifford algebra $C_{p,q}$ can be defined to be the associative algebra over the real numbers generated by $p + q$ elements, $p$ of which are square roots of $-1$, and $q$ of which are square roots of 1, all of which anticommute with each other. By forming all the possible products of these $p + q$ elements, and then taking all the real linear combinations of the products, one obtains the Clifford algebra $C_{p,q}$.

With any pseudo-Euclidean manifold $\mathbb{R}^{p,q}$ one has a corresponding Clifford algebra $C_{p,q}$, generated by $p + q$ Dirac gamma matrices $\{\gamma_j = \gamma(e_j)\}$. For example, the Clifford algebra $C_{3,1}$ of Minkowski space–time $\mathbb{R}^{3,1}$ is generated by four Dirac gamma matrices $\gamma_j = \gamma(e_j)$, $j = 0, 1, 2, 3$.

The complexification of $C_{p,q}$ can be treated as the set of complex linear combinations of the products of the $\gamma_j$, or it can simply be treated as $C_{p,q} \otimes \mathbb{C}$. The complexifications are isomorphic for all real Clifford algebras with the same number of generators $n = p + q$, hence one can index complexified Clifford algebras by the dimension $n$ alone of the pseudo-Euclidean manifold. Thus, we denote $C_n = C_{p,q} \otimes \mathbb{C}$ with $n = p + q$.

As promised, Clifford algebras provide us with a more general perspective upon spinors.[44] For a start, one can define the space of Dirac spinors in any dimension $n$ to be the smallest complex representation of the complex Clifford algebra $C_n$. In the case of $C_4$, the space of Dirac spinors is $\mathbb{C}^4$.

[43] For any matrix $X$, the cofactor matrix $X^a$ is such that $XX^a = (\det X)Id$.

[44] The facts presented here are largely derived from Baez (1996).

For any $n$, the real Clifford algebras $C_{n-1,1}$ and $C_{1,n-1}$ each have a smallest real representation. If the complexification of the smallest real representation of a real Clifford algebra is the space of Dirac spinors of the complexified Clifford algebra, then the smallest real representation is called a space of Majorana spinors.

In the case of Minkowski space–time, $C_{3,1}$ is $\mathbb{R}(4)$, the space of $4 \times 4$ real matrices, whilst $C_{1,3}$ is $\mathbb{H}(2)$, the space of $2 \times 2$ matrices with quaternionic coefficients. The smallest real representation of $C_{3,1}$ is $\mathbb{R}^4$, and given that $\mathbb{R}^4 \otimes \mathbb{C} \cong \mathbb{C}^4$, this is a space of Majorana spinors. The smallest real representation of $C_{1,3}$ is $\mathbb{R}^8$, which is not a spinor space of any type. The complexification $C_{3,1} \otimes \mathbb{C}$ is isomorphic to the complexification $C_{1,3} \otimes \mathbb{C}$, hence these complexifications share the same smallest complex representation, $\mathbb{C}^4$, but, as real Clifford algebras, $C_{3,1}$ and $C_{1,3}$ have different smallest real representations.

One can embed $Spin(n-1,1)$ in either $C_{n-1,1}$ or $C_{1,n-1}$, hence any representation of these Clifford algebras induces a representation of $Spin(n-1,1)$. It follows that one can also embed $Spin(n-1,1)$ in the complexification $C_n$. Obviously, a real representation of $C_{n-1,1}$ or $C_{1,n-1}$ induces a real representation of $Spin(n-1,1)$, and a complex representation of $C_n$ induces a complex representation of $Spin(n-1,1)$. A Weyl spinor space exists when the Dirac spinor representation of $C_n$ induces a reducible representation of $Spin(n-1,1)$. A Weyl spinor space is an irreducible direct summand of such a $Spin(n-1,1)$-representation. A Weyl–Majorana spinor space exists when a Majorana spinor representation of $C_{n-1,1}$ or $C_{1,n-1}$ induces a reducible representation of $Spin(n-1,1)$. A Weyl–Majorana spinor space is an irreducible direct summand of such a $Spin(n-1,1)$-representation.

Weyl spinor spaces exist in all even dimensions, but only in even dimensions. In the case of dimension 4, the Dirac spinor space $\mathbb{C}^4$ provides a reducible representation of $Spin(3,1)$, which splits into a pair of inequivalent irreducible representations, the Weyl spinor spaces.

In the special case of dimension 4, there is a Majorana spinor space $\mathbb{R}^4$, which is the 'realification' of a Weyl spinor space $\mathbb{C}^2$, and which complexifies into $\mathbb{C}^4$, the Dirac spinor space. In other words, there is a complex structure in $\mathbb{R}^4$ which renders it isomorphic with $\mathbb{C}^2$, and the tensor product $\mathbb{R}^4 \otimes \mathbb{C}$ is isomorphic with $\mathbb{C}^4$. These, however, are purely contingent relationships which do not hold in all dimensions. It would be wrong to define a Dirac spinor space as the complexification of a Majorana spinor space because, although Dirac spinor spaces exist in every dimension, there are dimensions in which Majorana spinors do not exist. Similarly, it would be wrong to define a Majorana spinor space as the realification of a Weyl spinor space, because there are dimensions in which Majorana spinors exist, but Weyl spinors do not.

Let us now present the Dirac equation and Weyl equation in terms of connections upon Dirac and Weyl spinor bundles. A connection upon the Dirac spinor bundle $\sigma$ which makes all of the ingredients of the spinor structure parallel, is referred to as a spinor connection. There is a natural bijective correspondence between spinor connections on $\sigma$, and so-called 'metric' connections upon the tangent bundle $T\mathcal{M}$.

A metric connection $\nabla$ is such that $\nabla g = 0$, with respect to the metric tensor field $g$ of the pseudo-Riemannian manifold $(\mathcal{M}, g)$. Amongst all the possible metric connections of a pseudo-Riemannian manifold $(\mathcal{M}, g)$, there is a unique connection which is torsion-free or symmetric (Torretti, 1983, p. 279). This connection is called the Levi-Civita connection. The bijective correspondence between spinor connections on $\sigma$ and metric connections on $T\mathcal{M}$, entails that $\nabla$, the Levi-Civita connection on $T\mathcal{M}$, corresponds to $\nabla^\sigma$ a unique spinor connection on $\sigma$.

A spinor connection on the Dirac spinor bundle $\sigma$, induces a spinor connection upon the Weyl sub-bundles, $\sigma_L$ and $\sigma_R$. Hence, the Levi-Civita spinor connection $\nabla^\sigma$ on $\sigma$, induces Levi-Civita spinor connections, $\nabla^{\sigma_L}$ and $\nabla^{\sigma_R}$, upon the Weyl sub-bundles.

The covariant derivative of the connection $\nabla^\sigma$ is a map:

$$\nabla^\sigma : \Gamma(\sigma) \to \Gamma(T^*\mathcal{M} \otimes \sigma).$$

Let $\psi = \psi^A E_A$ denote a Dirac spinor field, a cross-section of $\sigma$. $E_A$ is a cross-section of the general linear frame bundle of $\sigma$, providing a basis for each fibre, and $\psi^A$, $A = 1, 2, 3, 4$ denotes the components of $\psi$ with respect to these bases. The covariant derivative of $\psi \in \Gamma(\sigma)$ is

$$\nabla^\sigma \psi = \psi^A_{;\mu}\left(dx^\mu \otimes E_A\right) = dx^\mu \otimes \left(\psi^A_{;\mu} E_A\right),$$

with $\nabla^\sigma \psi \in \Gamma(T^*\mathcal{M} \otimes \sigma)$ and

$$\psi^A_{;\mu} = \frac{\partial \psi^A}{\partial x^\mu} + \psi^B \Gamma^A_{\mu B},$$

where $\Gamma^A_{\mu B}$ are the connection coefficients of $\nabla^\sigma$.[45]

With the combination of the covariant derivative and Clifford multiplication, one can define the Dirac operator. To understand Clifford multiplication, first note that with each point $p \in T^*_x\mathcal{M}$ in the fibre of the cotangent bundle over $x$, one has a corresponding element $\gamma(p)$ of the Clifford algebra $C(T^*_x\mathcal{M}) \cong C_{3,1}$, which can be treated as an endomorphism of $\sigma_x$, the fibre of the Dirac spinor bundle over $x$. Expressing $p$ as

$$p = p_\mu \, dx^\mu,$$

one has

$$\gamma(p) = p_\mu \gamma\left(dx^\mu\right) = p_\mu \gamma^\mu.$$

---

[45] See Torretti (1983, p. 273), for a definition of the coefficients of a connection.

The Dirac gamma matrices $\gamma^\mu$ can be written as

$$\gamma^0 = \begin{pmatrix} 0 & \sigma^0 \\ \sigma^0 & 0 \end{pmatrix}, \qquad \gamma^1 = \begin{pmatrix} 0 & \sigma^1 \\ -\sigma^1 & 0 \end{pmatrix},$$
$$\gamma^2 = \begin{pmatrix} 0 & \sigma^2 \\ -\sigma^2 & 0 \end{pmatrix}, \qquad \gamma^3 = \begin{pmatrix} 0 & \sigma^3 \\ -\sigma^3 & 0 \end{pmatrix},$$

when expressed in terms of the Pauli matrices,

$$\sigma^0 = \begin{pmatrix} 1 & 0 \\ 0 & 1 \end{pmatrix}, \qquad \sigma^1 = \begin{pmatrix} 0 & 1 \\ 1 & 0 \end{pmatrix},$$
$$\sigma^2 = \begin{pmatrix} 0 & -i \\ i & 0 \end{pmatrix}, \qquad \sigma^3 = \begin{pmatrix} 1 & 0 \\ 0 & -1 \end{pmatrix}.$$

Clifford multiplication provides a map (Derdzinski, 1992, p. 18),

$$c : T^*\mathcal{M} \otimes \sigma \to \sigma,$$

which is such that

$$c(p \otimes \psi) = \gamma(p)\psi.$$

By composing the covariant derivative of the Levi-Civita spinor connection $\nabla^\sigma$ with Clifford multiplication $c$, one obtains the Dirac operator on the Dirac spinor bundle (Derdzinski, 1992, p. 17):

$$\mathcal{D}\psi = c(\nabla^\sigma \psi).$$

The Dirac operator is a map

$$\mathcal{D} : \Gamma(\sigma) \to \Gamma(\sigma),$$

which is such that

$$\begin{aligned} \mathcal{D}\psi &= c(\nabla^\sigma \psi) \\ &= c(dx^\mu \otimes \psi^A_{;\mu} E_A) \\ &= \gamma(dx^\mu)(\psi^A_{;\mu} E_A) \\ &= \gamma^\mu(\psi^A_{;\mu} E_A). \end{aligned}$$

Expanding this expression for clarity,

$$\begin{aligned} c(\nabla^\sigma \psi) &= c(dx^0 \otimes \psi^A_{;0} E_A) + c(dx^1 \otimes \psi^A_{;1} E_A) \\ &\quad + c(dx^2 \otimes \psi^A_{;2} E_A) + c(dx^3 \otimes \psi^A_{;3} E_A) \\ &= \gamma^0(\psi^A_{;0} E_A) + \gamma^1(\psi^A_{;1} E_A) + \gamma^2(\psi^A_{;2} E_A) + \gamma^3(\psi^A_{;3} E_A). \end{aligned}$$

In the case of a flat spinor connection, $\Gamma^A_{\mu B} = 0$, hence

$$\psi^A_{;\mu} = \frac{\partial \psi^A}{\partial x^\mu},$$

and the Dirac operator reduces to

$$\mathcal{D}\psi = \gamma^0\left(\frac{\partial \psi^A}{\partial x^0}E_A\right) + \gamma^1\left(\frac{\partial \psi^A}{\partial x^1}E_A\right) + \gamma^2\left(\frac{\partial \psi^A}{\partial x^2}E_A\right) + \gamma^3\left(\frac{\partial \psi^A}{\partial x^3}E_A\right).$$

In other words, the covariant derivative of the 4-component field $\psi^A$ yields a quadruple of 4-component fields, each one obtained by taking $\frac{\partial \psi^A}{\partial x^\mu}$, $A = 1, 2, 3, 4$, the partial derivative of the original components with respect to a particular coordinate. One then applies each $4 \times 4$ gamma matrix $\gamma^\mu$ to the corresponding 4-component field $\frac{\partial \psi^A}{\partial x^\mu}$, $A = 1, 2, 3, 4$, to obtain another quadruple of 4-component fields. One then sums the quadruple of 4-component fields to obtain the final 4-component field.

In shorthand, for a flat spinor connection, one can write

$$\mathcal{D}\psi = \gamma^\mu \partial_\mu \psi.$$

Note that physicists often place a 'slash' through something to indicate that its components are being contracted with the Dirac gamma matrices. Thus, for example, $\not{\partial} = \gamma^\mu \partial_\mu$.

Free particles represented by cross-sections $\psi$ of the Dirac spinor bundle are required to satisfy the Dirac equation,

$$(\mathcal{D} + mc/\hbar)\psi = 0,$$

where $m$ is the mass of the particle in question.

Let us now turn to the Weyl equation. Given a cross-section $\psi_L$ of a Weyl spinor bundle $\sigma_L$, the covariant derivative of the Levi-Civita connection $\nabla^{\sigma_L}$ provides a map

$$\nabla^{\sigma_L} : \Gamma(\sigma_L) \to \Gamma(T^*\mathcal{M} \otimes \sigma_L).$$

With each point $p \in T^*_x\mathcal{M}$ in the fibre of the cotangent bundle over $x$, there is no Clifford algebra element on this occasion, but there is an endomorphism $P(p)$ of the fibre of the Weyl spinor bundle over $x$. In matrix terms, $P(p)$ is the following self-adjoint $2 \times 2$ matrix:

$$P = \begin{pmatrix} p_0 - p_3 & p_1 + ip_2 \\ p_1 - ip_2 & p_0 + p_3 \end{pmatrix}.$$

Expressing $p$ as $p = p_\mu dx^\mu$, $P(p)$ can also be treated as a linear combination of the Pauli matrices[46]:

$$P(p) = p_\mu P(dx^\mu) = p_\mu \sigma^\mu.$$

[46] The use of $\sigma^\mu$ to denote the Pauli matrices, and the use of $\sigma_L$ to denote a Weyl spinor bundle, is coincidental.

The endomorphism $P(p)$ enables one to define a map

$$c_L : T^*\mathcal{M} \otimes \sigma_L \to \sigma_L,$$

which is such that

$$c_L(p \otimes \psi_L) = P(p)\psi_L.$$

By composing the covariant derivative of the Levi-Civita spinor connection $\nabla^{\sigma_L}$ with the map $c_L$, one obtains the Dirac operator on the Weyl spinor bundle:

$$\mathcal{D}_L\psi_L = c_L\left(\nabla^{\sigma_L}\psi_L\right).$$

This Dirac operator is a map

$$\mathcal{D}_L : \Gamma(\sigma_L) \to \Gamma(\sigma_L).$$

In the case of a flat spinor connection, this Dirac operator reduces to

$$\mathcal{D}_L\psi_L = \sigma^0\left(\frac{\partial\psi_L^A}{\partial x^0}E_A\right) + \sigma^1\left(\frac{\partial\psi_L^A}{\partial x^1}E_A\right) + \sigma^2\left(\frac{\partial\psi_L^A}{\partial x^2}E_A\right) + \sigma^3\left(\frac{\partial\psi_L^A}{\partial x^3}E_A\right),$$

with $A = 1, 2$. In shorthand this can be written as

$$\mathcal{D}_L\psi_L = \sigma^\mu\partial_\mu\psi_L.$$

Free particles represented by cross-sections of a Weyl spinor bundle are required to satisfy the Weyl equation,

$$\mathcal{D}_L\psi_L = 0.$$

## 2.9. Electrons and neutrinos

In the configuration space approach, an electron is represented by the Hilbert space $\mathscr{H}^+_{m_e,1/2} \subset \mathscr{D}'(\mathcal{M}, \mathbb{C}^4)$ constructed from the space of mass $m_e$, positive-energy tempered solutions $\mathscr{S}^+_{m_e}(\mathcal{M}, \mathbb{C}^4)$ of the Dirac equation (or the cross-sectional analogue). Similarly, a neutrino is represented by a Hilbert space $\mathscr{H}^+_{0,1/2} \subset \mathscr{D}'(\mathcal{M}, \mathbb{C}^2)$ constructed from the space of mass 0, positive-energy tempered solutions $\mathscr{S}^+_0(\mathcal{M}, \mathbb{C}^2)$ of the Weyl equation (or the cross-sectional analogue). Both of these spaces provide a unitary, irreducible, infinite-dimensional representation of $SL(2, \mathbb{C}) \ltimes \mathbb{R}^{3,1}$.

To build a simple bridge between the representation of electrons and neutrinos in the configuration space approach, and their representation in the Wigner approach, start by supposing that we have a function space $\mathcal{F}(\mathcal{M}, V)$.

For each value of mass $m$, one has a subspace of tempered functions $\mathscr{S}_m(\mathcal{M}, V)$, the elements of which, under Fourier transform, become identifiable with elements of $\mathscr{S}(\mathscr{V}_m, V)$, the space of tempered $V$-valued functions with support on a mass hyperboloid/light cone $\mathscr{V}_m$ in Minkowski energy–momentum space. For each value of mass

$m$, there is a further subspace $\mathscr{S}_m^+(\mathcal{M}, V)$ of positive-energy functions, the elements of which become, under Fourier transform, elements of $\mathscr{S}(\mathscr{V}_m^+, V)$, the space of tempered $V$-valued functions on the forward hyperboloid/light cone. However, whilst the elements of $\mathscr{S}_m^+(\mathcal{M}, V)$ can be identified with elements of $\mathscr{S}(\mathscr{V}_m^+, V)$, it is not correct to identify $\mathscr{S}_m^+(\mathcal{M}, V)$ with the whole of $\mathscr{S}(\mathscr{V}_m^+, V)$; the Fourier transform of $\mathscr{S}_m^+(\mathcal{M}, V)$ is only a subspace of $\mathscr{S}(\mathscr{V}_m^+, V)$.

Take the case of the electron as an example. In the Wigner approach, the vector bundle for an electron $E_{m_e,1/2}^+$ has typical fibre $\mathbb{C}^2$. The space of square-integrable cross-sections $\Gamma_{L^2}(E_{m_e,1/2}^+)$ can be identified with the space $L^2(\mathscr{V}_{m_e}^+, \mathbb{C}^2)$ of square-integrable $\mathbb{C}^2$-valued functions on the forward mass hyperboloid $\mathscr{V}_{m_e}^+$. The relationship between the Minkowski configuration space approach and the Wigner approach is that the elements of $\mathscr{S}_{m_e}^+(\mathcal{M}, \mathbb{C}^4)$, tempered positive energy solutions of the Dirac equation for mass $m_e$, are identified with the elements of $\mathscr{S}(\mathscr{V}_{m_e}^+, \mathbb{C}^2)$. Note the transition from $\mathbb{C}^4$-valued functions to $\mathbb{C}^2$-valued functions. Elements of $\mathscr{S}_{m_e}^+(\mathcal{M}, \mathbb{C}^4)$ are firstly identified with elements of $\mathscr{S}(\mathscr{V}_{m_e}^+, \mathbb{C}^4)$ by Fourier transform, and although, obviously, $\mathscr{S}(\mathscr{V}_{m_e}^+, \mathbb{C}^2) \neq \mathscr{S}(\mathscr{V}_{m_e}^+, \mathbb{C}^4)$, elements of $\mathscr{S}(\mathscr{V}_m^+, \mathbb{C}^4)$ which satisfy the Dirac equation in Minkowski energy–momentum space,

$$\gamma(p) f(p) = m f(p),$$

have only two independent degrees of freedom at each point. Hence, elements of $\mathscr{S}(\mathscr{V}_{m_e}^+, \mathbb{C}^4)$ which satisfy the Dirac equation for mass $m_e$ can be identified with elements of $\mathscr{S}(\mathscr{V}_{m_e}^+, \mathbb{C}^2)$.[47]

Whilst an electron, and for that matter, any type of quark, can be represented by a $\mathbb{C}^4$-valued function on Minkowski configuration space, a neutrino on Minkowski configuration space can be represented by a $\mathbb{C}^2$-valued function. Whilst an electron or quark field is constrained to satisfy the Dirac equation, a neutrino field is constrained to satisfy the Weyl equation. Elements of $\mathscr{S}(\mathcal{M}, \mathbb{C}^2)$ which are constrained to satisfy the Weyl equation (i) have Fourier transforms which are concentrated on the light-cone $\mathscr{V}_0$ in Minkowski energy–momentum space, and (ii) have only one degree of freedom in their value at each point $p \in \mathscr{V}_0$. The Fourier transforms of the elements of $\mathscr{S}_0^+(\mathcal{M}, \mathbb{C}^2)$, the $\mathbb{C}^2$-valued functions which satisfy the Weyl equation with positive energy, all belong to $\mathscr{S}(\mathscr{V}_0^+, \mathbb{C}^2)$. However, the Fourier transform of $\mathscr{S}_0^+(\mathcal{M}, \mathbb{C}^2)$ is a subspace of $\mathscr{S}(\mathscr{V}_0^+, \mathbb{C}^2)$, not the whole space. The $f \in \mathscr{S}(\mathscr{V}_0^+, \mathbb{C}^2)$ which satisfy the Weyl equation are constrained to have only one independent degree of freedom at each point, hence the Fourier transform of $\mathscr{S}_0^+(\mathcal{M}, \mathbb{C}^2)$ corresponds to $\mathscr{S}(\mathscr{V}_0^+, \mathbb{C}^1)$.[48]

---

[47] This derivation is obtained in detail by Sternberg (1994, pp. 229–231).

[48] Again, this is derived in detail by Sternberg (1994, pp. 231–234).

## 2.10. Particles and anti-particles

Recall from Section 2.1 that, in the Wigner approach, if a free particle of mass $m$ and spin $s$ is represented by the Hilbert space of cross-sections $\Gamma_{L^2}(E^+_{m,s})$ of the vector bundle $E^+_{m,s}$ over the forward mass hyperboloid/light cone $\mathscr{V}^+_m$, then the free anti-particle is represented by the Hilbert space of cross-sections $\Gamma_{L^2}(E^-_{m,s})$ of the vector bundle $E^-_{m,s}$ over the backward mass hyperboloid/light cone $\mathscr{V}^-_m$. These Hilbert spaces are related by an anti-unitary transformation, and this mapping between the particle and anti-particle spaces can also be implemented in the configuration representation.

Suppose that the state space of a particle in the configuration representation can be provisionally identified with $\mathscr{S}^+_m(\mathcal{M}, V)$, the space of tempered, positive-energy, mass-$m$, $V$-valued functions on $\mathcal{M}$. (As before, one can apply what follows to a space $\mathscr{S}^+_m(\eta)$ of tempered, positive-energy, mass-$m$ cross-sections of a vector bundle $\eta$ with typical fibre $V$.) Each $\psi^+(x) \in \mathscr{S}^+_m(\mathcal{M}, V)$ is the inverse Fourier transform of a $V'$-valued function $a(p)$ which is concentrated upon the forward mass hyperboloid $\mathscr{V}^+_m$. I.e.,

$$\psi^+(x) = \int e^{i\langle p,x\rangle} a(p)\, d^4p.$$

Now, the infinite-dimensional complex vector space $\mathscr{S}^+_m(\mathcal{M}, V)$ possesses a conjugate vector space $\overline{\mathscr{S}^+_m(\mathcal{M}, V)}$, consisting of all the tempered, positive-energy, mass-$m$, $\bar{V}$-valued functions on $\mathcal{M}$, where $\bar{V}$ is the conjugate vector space to the finite-dimensional complex vector space $V$. There is a bijective anti-linear mapping between a complex vector space and its conjugate vector space.

Given the function $\psi^+(x) \in \mathscr{S}^+_m(\mathcal{M}, V)$, the conjugate function $\bar{\psi}^+(x)$ will belong to the conjugate vector space $\overline{\mathscr{S}^+_m(\mathcal{M}, V)}$. Given the expression of $\psi^+(x)$ as an inverse Fourier transform, the conjugate function $\bar{\psi}^+(x)$ can be expressed as

$$\bar{\psi}^+(x) = \int e^{-i\langle p,x\rangle} \bar{a}(p)\, d^4p.$$

This integral should be invariant under a change of variable from $p = (p_0, \mathbf{p})$ to $-p = (-p_0, -\mathbf{p})$. Hence,

$$\begin{aligned}\bar{\psi}^+(x) &= \int e^{-i\langle -p,x\rangle} \bar{a}(-p)\, d^4p \\ &= \int e^{i\langle p,x\rangle} \bar{a}(-p)\, d^4p.\end{aligned}$$

Given that $a(p)$ is concentrated upon the forward mass hyperboloid $\mathscr{V}^+_m$, $\bar{a}(-p) \equiv b(p)$ will be concentrated upon the backward mass hyperboloid $\mathscr{V}^-_m$. Thus, the conjugate function $\bar{\psi}^+(x)$ is the inverse Fourier transform of a function $b(p)$ which is concentrated upon the backward mass hyperboloid. In other words, $\bar{\psi}^+(x)$ is an element of $\mathscr{S}_m^-(\mathcal{M}, V)$, the space of tempered, negative-energy, mass-$m$, $V$-valued functions on $\mathcal{M}$, the provisional state space of the anti-particle. Hence, in this simple picture, the anti-particle space $\mathscr{S}_m^-(\mathcal{M}, V)$ is the conjugate vector space to the particle space $\mathscr{S}_m^+(\mathcal{M}, V)$ in the configuration representation.[49]

Similarly, in the case of a negative energy function

$$\psi^-(x) = \int e^{i\langle p,x\rangle} c(p)\, d^4 p,$$

the conjugate function $\bar{\psi}^-(x)$ can be expressed as

$$\int e^{i\langle p,x\rangle} \bar{c}(-p)\, d^4 p,$$

with the function $d(p) \equiv \bar{c}(-p)$ being concentrated on the forward mass hyperboloid.

A general version of this simple picture can be found in Derdzinski (1992, p. 7), who asserts that if a free-particle is represented by a complex vector bundle $\eta$, then the anti-particle is represented by the conjugate bundle $\bar{\eta}$. Recall that Derdzinski defines a configuration space vector bundle $\eta$ for each possible combination of spin and parity, and the space of sections of such a bundle represents many different particle species, each species corresponding to the cross-sections for a specific mass. In addition, the space of mass-$m$ cross-sections includes both the positive energy and negative energy cross-sections. Hence, the space of cross-sections of a configuration space bundle $\eta$ already includes the cross-sections for both the particles and anti-particles of each mass value. The conjugate bundle $\bar{\eta}$ does, however, exchange the particle and anti-particle cross-sections for each value of mass.

The simple picture outlined above requires some modification in the case of spinor fields. In the case of Dirac spinor fields there is an operator $C$ which maps mass-$m$ positive energy solutions $\psi^+(x)$ of the Dirac equation on Minkowski configuration space, to mass-$m$ negative energy solutions $\psi^-(x) = (C\psi^+)(x)$, but its expression is not as simple as merely taking the complex conjugate.

The operator $C$ is often called the charge conjugation operator, but really it is the particle–anti-particle conjugation operator for a spinor field, executing both charge conjugation and parity reversal. Letting $\mathcal{C}$ denote the operator which maps $\psi(x)$ to its complex conjugate $\bar{\psi}(x)$, the operator $C$ can be defined as (Dautray and Lions, 1988, Chapter IX, §1, 6h):

$$C = \begin{pmatrix} 0 & i\sigma^2 \\ -i\sigma^2 & 0 \end{pmatrix} \mathcal{C} = i\gamma^2 \cdot \mathcal{C}.$$

[49] Note that in the case where $V = \mathbb{C}^n$, the conjugate function is simply the complex conjugate function.

The anti-particle state is therefore $i\gamma^2\bar{\psi}$. Where $\psi^+(x)$ denotes an electron spinor field, one can write the corresponding positron spinor field as follows (Dautray and Lions, 1988):

$$C\psi^+(x) = \int e^{i\langle p,x\rangle} i\gamma^2 \bar{a}(-p)\, d^4 p.$$

There are, however, different ways of breaking down the conjugation operator $C$. For example, elsewhere one finds $C = i\gamma^2\gamma^0$, and the anti-particle state written as $C(\psi^*\gamma^0)^T$, where $(\cdot)^*$ denotes the adjoint, and $(\cdot)^T$ denotes the transpose. Bearing in mind that the adjoint is the transpose of the complex conjugate, one can expand this expression as follows to obtain the same anti-particle state,

$$\begin{aligned} C(\psi^*\gamma^0)^T &= C(\gamma^0)^T\bar{\psi} \\ &= i\gamma^2\gamma^0(\gamma^0)^T\bar{\psi} \\ &= i\gamma^2\bar{\psi}, \end{aligned}$$

given that $(\gamma^0)^T = \gamma^0$ and $(\gamma^0)^2 = I$.

Given that the Dirac spinor bundle is a direct sum $\sigma = \sigma_L \oplus \sigma_R$ of left-handed and right-handed Weyl spinor bundles, the action of the particle–anti-particle conjugation operator upon the cross-sections of the Dirac spinor bundle induces conjugation operators on the cross-sections of the Weyl spinor bundles. Right-handed Weyl spinor particle states $\psi_R$ are mapped to left-handed anti-particle states $i\sigma^2\bar{\psi}_R$, and left-handed states $\psi_L$ are mapped to right-handed anti-particle states $-i\sigma^2\bar{\psi}_L$. The anti-particle of a fermion with charge and parity possesses both opposite charge *and* opposite parity. The anti-particle of a left-handed electron $e_L$ is a right-handed positron $e^+_R$, and the anti-particle of a left-handed neutrino $\nu_L$ is a right-handed anti-neutrino $\bar{\nu}_R$. Note carefully that the use of the bar in $\bar{\nu}$ denotes the anti-particle, whilst the use of the bar in a particle state such as $\bar{\psi}$, denotes the complex conjugate; the two things are not the same.

## 2.11. Fock space and second quantization

Whilst this text provides an account of matter fields, gauge fields, and their associated particles in the first-quantized theory, it is worth casting a glance at the second-quantized theory to see if the first-quantized notions of field and particle are inherited, revised, or replaced with something else.

We need to begin, however, by understanding why the first-quantized theory is considered to be empirically inadequate, and why it was supplanted by the second-quantized theory in the first place. The first reason is the existence of negative-energy solutions to the relevant differential equations. Negative energy states are not physically observed, and applied to scattering processes the first-quantized theory predicts a non-zero probability that an incoming positive energy state will transition to a negative energy outgoing state. This is often called the Klein paradox.

Secondly, although the matter fields $\psi(x)$ in the first-quantized theory are the upshot of quantizing classical particles, and therefore provisionally interpretable as wave-functions, coding probabilistic information about a particle, there is no guarantee that these solutions $\psi(x)$ yield a positive-definite probability density, and there is no well-defined notion of localisability. Some texts incorrectly equate localisability with the existence of a 'current,' hence we shall digress to explain both the notion of a current, and the debate over localisability.

Let $\mathcal{M}$ denote Minkowski space–time, and let $Z = Z^3(\mathcal{M})$ denote the space of closed three-forms on $\mathcal{M}$ which vanish with sufficient rapidity towards spacelike infinity. Given the one-particle Hilbert space $\mathscr{H}$ for some species of particle, equipped with a unitary representation of the restricted Poincaré group $G$, and given a finite-dimensional representation of $G$ upon a vector space $V$, one can define a $V$-valued current on $\mathscr{H}$ to be a sesquilinear mapping $j : \mathscr{H} \times \mathscr{H} \to V \otimes Z$ which is equivariant under the action of $G$ (Sternberg, 1994, p. 301). For any $(\phi, \psi) \in \mathscr{H} \times \mathscr{H}$, the integral of a $V$-valued current $\int_\Sigma j(\phi, \psi)$ over any spacelike hypersurface $\Sigma$ in Minkowski space–time converges, and because $j(\phi, \psi)$ is closed, the integral is independent of the choice of spacelike hypersurface (Sternberg, 1994). The integral is therefore a $V$-valued sesquilinear form on $\mathscr{H}$. In the case where $V$ is the trivial one-dimensional representation of $G$, $j$ is simply called a current, and if $\mathscr{H}$ is an irreducible representation, the integral $\int_\Sigma j(\phi, \psi)$ must be a multiple of the inner product $\langle \phi, \psi \rangle$ on $\mathscr{H}$.

Now, it happens that currents exist on Minkowski space–time for the matter fields in the first-quantized relativistic quantum theory, but there is no current for the Minkowski space–time representation of photon 'wave-functions.' For the mass-zero one-particle representations of the Poincaré group, the representations for $s = 0$ and $s = 1/2$ possess currents, but the representations with spin $s \geqslant 1$ do not (Sternberg, 1994). Sternberg asserts that "for $s = 1$ this expresses the well-known fact that the photon cannot be localized" (Sternberg, 1994). A photon can, however, be localised in energy–momentum space, a fact which Sternberg equates to the existence of a $T^*_x\mathcal{M}$-valued current for the photon, where $T^*_x\mathcal{M}$ is energy–momentum space.

It is, however, an error to think that the existence of a current is a sufficient condition for localisability. In Sections 4.6 and 4.8 we shall have reason to introduce interacting currents. In this section, however, we shall confine our attention to the free Klein–Gordon current and the free Dirac current. In the case of a mass $m$, spin $s = 0$ system, there exists a Klein–Gordon current on Minkowski space–time. Given a spacelike hyperplane $\Sigma$ in Minkowski space–time, specified by the points with $t = 0$ in a global Lorentz chart, there is a unique Lorentz-invariant and time-independent inner product in the space of solutions to the Klein–Gordon equation,

$$\langle \psi_1, \psi_2 \rangle = i \int_\Sigma \left[ \overline{\psi_1(x)} \partial^\mu \psi_2(x) - \overline{\partial^\mu \psi_1(x)} \psi_2(x) \right] d\mathbf{x},$$

and this is compatible with the fact that a solution to the Klein–Gordon equation over Minkowski space–time has the following Lorentz-invariant current

$$j^\mu(x) = (i/2m)\big[\overline{\psi(x)}\partial^\mu\psi(x) - \overline{\partial^\mu\psi(x)}\psi(x)\big].$$

The timelike component $j^0(x)$ is provisionally interpretable as the probability of observing the spin zero, mass $m$ particle at the point $x$ in Minkowski space–time. However, even for positive-energy solutions of the Klein–Gordon equation, and indeed, for *any* positive-energy solution, there are points at which $j^0(x) < 0$ and points at which $j^0(x) > 0$ (Prugovecki, 1995, p. 98). The Klein–Gordon current is therefore not interpretable as a probability current.

Given a solution $\psi(x)$ of the Dirac equation, one can also introduce a Lorentz-invariant Dirac current $j^\mu(x) = \bar{\psi}(x)\gamma^0\gamma^\mu\psi(x)$, and whilst the timelike component $j^0 = \bar{\psi}(x)\gamma^0\gamma^0\psi(x) = \bar{\psi}(x)\psi(x)$ is positive-definite for all solutions of the Dirac equation, all positive-energy solutions of the Dirac equation are non-zero at all points of Minkowski space–time (Prugovecki, 1995, p. 99). If a particle such as an electron is represented as a positive-energy solution of the Dirac equation, if such particles are localised within bounded regions of space–time when they are detected, and if one assumes that the localisation of such a particle within a bounded region of space–time requires the solution to vanish outside that region, then there is an inconsistency here.

Either one has positive-energy solutions which are non-zero in every open subset of space–time, or one has solutions confined to an open bounded subset which contain both positive-energy and negative-energy components (Streater, 1988, p. 138). Unless one could argue that the detection of a particle doesn't correspond to its localisation in space–time, empirical adequacy requires the first-quantized theory to provide positive energy solutions which are confined to bounded subsets of space–time.

Hegerfeldt demonstrated that if a quantum state is localised either within a bounded subset of Minkowski space–time, or a bounded subset of a spacelike hyperplane in Minkowski space–time, and if there is a lower bound on the energy of the system, then the requirement that the state satisfy, say, the Dirac equation, entails that the state will superluminally propagate outside the causal future of the bounded region. This result also applies to the case of 'approximate localisation,' where the state doesn't vanish outside the bounded region, but rather has 'exponentially bounded tails' outside the region (Prugovecki, 1992, p. 7).

Consider the localisability of a particle of mass $m$ and spin $s = 0$ in a bounded subset of a spacelike hyperplane. One can use the inner product defined above in a Hilbert space $\mathscr{H}_{m,0}$ of solutions to the Klein–Gordon equation on the whole of Minkowski space–time, to define an isomorphic Hilbert space $(\mathscr{H}_{m,0})_{t_0}$ of solutions restricted to the time $t_0$ spacelike hyperplane $\Sigma$ (Schwartz, 1968, p. 129). Given an element $\psi_A$ of this space which is localised inside a bounded open subset $A \subset \Sigma$, and given another element $\psi_B$, which is localised inside a disjoint bounded open subset $B \subset \Sigma$, the inner product of these two states is non-zero. I.e., the two states are not orthogonal. This indicates that a state which appears to be momentarily localised inside a bounded open subset, actually has a non-zero probability of being found in any other bounded open subset of the same hyperplane. Although $\psi_A$ and $\psi_B$ are each of compact support on the spacelike hyperplane, their time derivatives are not of compact support, and these time derivatives feed into the inner product as specified above, with the result that the two states are not orthogonal. Moreover, given a positive-energy state which is localised in a bounded region of a spacelike hyperplane in Minkowski space–time, the complement of that bounded region in the hyperplane is still a subset of measure zero in the whole of Minkowski space–time, and the state will be non-zero outside the causal future of the bounded region. Hence, positive energy states which appear to be momentarily spatially localised, also seem to propagate superluminally throughout space. In addition, if one has a positive energy state which is spatially localised in the constant time spacelike hyperplane of one inertial Lorentz frame, and one uses a Lorentz boost to transition to another Lorentz frame, then the state will be completely dispersed across the constant time $t_0$ spacelike hyperplane of the other Lorentz frame (Teller, 1995, pp. 85–91; Fleming and Butterfield, 1999).

If the first-quantized matter fields $\psi(x)$ are not to be interpreted as the wave-functions of particles, then there is no need for them to generate a positive-definite probability density. However, if the first-quantized matter fields $\psi(x)$ are to be alternatively interpreted as fields, then one must accept that they represent a species of field which is either of positive-energy and non-localisable, or which is localisable, but contains positive-energy and negative-energy components. Given that photons are just as detectable as electrons and mesons, despite the fact that one-particle photon states are not even candidates for localisability in space–time, one could suggest that the detection of a particle does not correspond to its localisation in space–time. One could suggest, instead, that the detection of a particle merely corresponds to the localisation in energy–momentum space of a system which remains non-local in space–time. However, it then remains to be explained why a particle detection event occurs over a bounded period of time within a bounded region of space. Perhaps, even if the particle is a non-local field in space–time, and even if the particle detector is also a non-local field in space–time, the interaction between them only rises above a crucial threshold in a bounded region of space–time, and hence it is the particle detection event which is localisable within a bounded region of space–time.

If free fields are not localisable in space–time, but observational phenomena clearly are localised in space–time, then it is reasonable to suggest that it is the interactions between fields which are localisable, not the fields themselves. Consider, for example, the particle tracks in a Wilson cloud chamber. Such a chamber contains air molecules and supersaturated water vapour (see Ne'eman and Kirsh, 1996, p. 24). The passage of a charged, ionizing particle through such a chamber leaves ionized air molecules in its wake, and water droplets form around the line of charged ions. The line of water droplets can then be photographed. Prima facie, one infers the motion of the particle from the line of water droplets, rather like one infers the motion of a high-altitude aircraft from the vapour trail it leaves in the sky. However, in the case of a particle, each condensation droplet is the result of a measurement-like interaction (Teller, 1995, pp. 138–139) between the ionizing particle, the ionized air molecule, and the water vapour molecules which form the droplet. Individually, both the ionizing particle, the air molecule, and the water vapour molecules are non-local fields in Minkowski space–time, but the interaction between them is localised in space–time, and localises the energy–momentum of the ionizing particle. The detection event occurs where it does because the strength of interaction between the ionizing particle field, the air molecule field, and the droplet field, rises above a crucial threshold in that bounded region of space–time. The perceptible droplet is a manifestation of the localised interaction event in space–time.

Let us turn now to the second-quantized theory. Wigner's classification of the unitary irreducible representations of the Poincaré group supplies an infinite-dimensional Hilbert space for a particle with any possible combination of mass $m$ and spin/helicity $s$. These Hilbert spaces are the so-called single-particle Hilbert spaces from which the Fock spaces of the second-quantized theory can be built. Given the single-particle Hilbert space $\mathscr{H}_{m,s}$ for a bosonic system, the Fock space is[50]

$$\mathscr{F}_{m,s} = \mathscr{F}(\mathscr{H}_{m,s}) = \bigoplus_{n=0}^{\infty} \mathscr{H}_{m,s}^{\odot n},$$

where $\mathscr{H}_{m,s}^{\odot n}$ is the $n$-fold symmetric tensor product of $\mathscr{H}_{m,s}$. The symmetric tensor product is the image of the tensor product under the following projection mapping:

$$S_n^+(f_1 \otimes \cdots \otimes f_n) = (n!)^{-1/2} \sum_{\sigma} f_{\sigma(1)} \otimes \cdots \otimes f_{\sigma(n)}.$$

The sum here is over all the permutations $\sigma$ of the indices $(1, 2, \ldots, n)$. The symmetric tensor product is $\mathscr{H}_{m,s}^{\odot n} = S_n^+(\mathscr{H}_{m,s}^{\otimes n})$.

Given the single-particle Hilbert space $\mathscr{H}_{m,s}$ for a fermionic system, the Fock space is

$$\mathscr{F}_{m,s} = \mathscr{F}(\mathscr{H}_{m,s}) = \bigoplus_{n=0}^{\infty} \mathscr{H}_{m,s}^{\wedge n},$$

where $\mathscr{H}_{m,s}^{\wedge n}$ is the $n$-fold anti-symmetric tensor product of $\mathscr{H}_{m,s}$. The anti-symmetric tensor product is the image of the tensor product under the following projection mapping:

$$S_n^-(f_1 \otimes \cdots \otimes f_n) = (n!)^{-1/2} \sum_{\sigma} \chi(\sigma) f_{\sigma(1)} \otimes \cdots \otimes f_{\sigma(n)},$$

[50] There is an alternative approach, in which the Fock space is obtained as the space of complex-valued polynomial functionals defined upon the complex infinite-dimensional single-particle space. In this approach, the degree $n$ polynomials correspond to the $n$-particle subspace of Fock space (Derdzinski, 2002, Section 3.6).

where $\chi(\sigma)$ is the sign of the permutation. Each permutation of a set of elements can be expressed as a sequence of transpositions of pairs of elements, and, in this sense, $\chi(\sigma) = 1$ for permutations obtained from an equal number of transpositions, and $\chi(\sigma) = -1$ for permutations obtained from an odd number of transpositions. The anti-symmetric tensor product is $\mathscr{H}_{m,s}^{\wedge n} = S_n^-(\mathscr{H}_{m,s}^{\otimes n})$.

The irreducible, strongly continuous, unitary representation $U(a, A)$ of $SL(2, \mathbb{C}) \ltimes \mathbb{R}^{3,1}$ on a single-particle Hilbert space extends to a *reducible*, strongly continuous, unitary representation on either the corresponding bosonic Fock space,

$$\bigoplus_{n=0}^{\infty} U(a, A)^{\odot n},$$

or on the corresponding fermionic Fock space,

$$\bigoplus_{n=0}^{\infty} U(a, A)^{\wedge n}.$$

One can define creation and annihilation operators upon a Fock space. Suppose one has a bosonic Fock space. For each $(n-1)$-particle Hilbert space $\mathscr{H}_{m,s}^{\odot n-1}$, the creation of a particle with a state $f \in \mathscr{H}_{m,s}$ corresponds to the operator $a_n^*(f) : \mathscr{H}_{m,s}^{\odot n-1} \to \mathscr{H}_{m,s}^{\odot n}$ defined by

$$a_n^*(f)(f_1 \odot \cdots \odot f_{n-1}) = S_n^+(f \otimes f_1 \otimes \cdots \otimes f_{n-1}),$$

where $S_n^+$ is the projection operator onto the symmetric $n$-particle subspace, so that

$$S_n^+(f \otimes f_1 \otimes \cdots \otimes f_{n-1}) = f \odot f_1 \odot \cdots \odot f_{n-1}.$$

For each $n$-particle Hilbert space $\mathscr{H}_{m,s}^{\odot n}$ the annihilation of a particle with a state $f$ corresponds to the adjoint operator $a_n(f) : \mathscr{H}_{m,s}^{\odot n} \to \mathscr{H}_{m,s}^{\odot n-1}$, defined by

$$a_n(f)(f_1 \odot \cdots \odot f_n) = \sum_{j=1}^{n} \langle f, f_j \rangle f_1 \odot \cdots \odot f_{j-1} \odot f_{j+1} \odot \cdots \odot f_n.$$

Creation and annihilation operators are then defined upon the entire Fock space:

$$a^*(f) = 0 \oplus \bigoplus_{n=1}^{\infty} \sqrt{n} a_n^*(f),$$

$$a(f) = \bigoplus_{n=1}^{\infty} \sqrt{n} a_n(f).$$

These creation and annihilation operators are operator-valued distributions in the sense that they assign operators to functions $f$. So-called 'field operators' are constructed from these creation and annihilation operators, as we will see below.

The annihilation and creation operators on the bosonic Fock space satisfy canonical commutation relations (CCRs):

$$a(f)a(g) - a(g)a(f) = 0,$$
$$a^*(f)a^*(g) - a^*(g)a^*(f) = 0,$$
$$a(f)a^*(g) - a^*(g)a(f) = \langle f, g\rangle I.$$

In the case of a fermionic Fock space, for each $(n-1)$-particle Hilbert space $\mathcal{H}_{m,s}^{\wedge n-1}$, the creation of a particle with a state $f \in \mathcal{H}_{m,s}$ corresponds to the operator $a_n^*(f) : \mathcal{H}_{m,s}^{\wedge n-1} \to \mathcal{H}_{m,s}^{\wedge n}$ defined by

$$a_n^*(f)(f_1 \wedge \cdots \wedge f_{n-1}) = S_n^-(f \otimes f_1 \otimes \cdots \otimes f_{n-1}),$$

where $S_n^-$ is the projection operator onto the anti-symmetric $n$-particle subspace, so that

$$S_n^-(f \otimes f_1 \otimes \cdots \otimes f_{n-1}) = f \wedge f_1 \wedge \cdots \wedge f_{n-1}.$$

As before, the adjoint $a_n(f)$ is an annihilation operator, which in the case of a fermionic Fock space provides a mapping $a_n(f) : \mathcal{H}_{m,s}^{\wedge n} \to \mathcal{H}_{m,s}^{\wedge n-1}$, defined by

$$a_n(f)(f_1 \wedge \cdots \wedge f_n) = \sum_{j=1}^{n} (-1)^{j+1} \langle f, f_j\rangle f_1 \wedge \cdots \wedge f_{j-1} \wedge f_{j+1} \wedge \cdots \wedge f_n.$$

The creation operators and annihilation operators on fermionic Fock space satisfy the canonical anti-commutation relations (CARs):

$$a(f)a(g) + a(g)a(f) = 0,$$
$$a^*(f)a^*(g) + a^*(g)a^*(f) = 0,$$
$$a(f)a^*(g) + a^*(g)a(f) = \langle f, g\rangle I.$$

The creation and annihilation operators on a bosonic or fermionic Fock space generate other commutation relations. For example, one can define the self-adjoint Segal field operators,

$$\Phi_S(f) = 2^{-1/2}\big(a^*(f) + a(f)\big).$$

These operators are defined on the subspace of finite-particle vectors $F_0$, where a finite particle vector is an element of Fock space $\psi = \bigoplus_{n=0}^{\infty} \psi_n$ in which all but a finite number of the $\psi_n$ are zero. In the case of a bosonic field,[51] the Segal field operators satisfy the following commutation relations for finite-particle vectors $\psi$ (Reed and Simon, 1975, p. 210):

$$\Phi_S(f)\Phi_S(g)\psi - \Phi_S(g)\Phi_S(f)\psi = i \operatorname{Im}\langle f, g\rangle \psi.$$

[51] See Saunders (1991, p. 96), for the case of a fermionic field.

$f$ and $g$ here both come from the single-particle Hilbert space. If one defines (bounded) unitary operators $W(f) = e^{i\Phi_S(f)}$ from the Segal field operators, then these unitary operators satisfy a Weyl form of the commutation relations:

$$W(f+g) = e^{-i\,\mathrm{Im}\langle f,g\rangle/2}W(f)W(g).$$

Conventional field operators, however, are distinct from the Segal field operators. Given the creation and annihilation operators, one can try to define field operators at each point $x$ of Minkowski space–time by expressions such as

$$\Phi(x) = \sum_{\alpha=1}^{\infty}\big[f_\alpha(x)a(f_\alpha) + \bar{f_\alpha}(x)a^*(\bar{f_\alpha})\big],$$

where the $\{f_1, f_2, \ldots\}$ provide an orthonormal basis of the single-particle Hilbert space. However, the second part of this series diverges, and, in general, quantum field operators cannot be well-defined at individual points of Minkowski space–time. Instead, one must treat the field operators as operator-valued distributions $\Phi(f)$ by 'smearing' them over functions $f$ from a test-function space as follows,

$$\begin{aligned}\Phi(f) &= \int_{\mathcal{M}} f(x)\Phi(x)\,d^4x \\ &= \text{w-}\lim_{n\to\infty}\int_{\mathcal{M}} f(x)\sum_{\alpha=1}^{n}\big[f_\alpha(x)a(f_\alpha) + \bar{f_\alpha}(x)a^*(\bar{f_\alpha})\big]\,d^4x,\end{aligned}$$

where both of the terms exist as weak limits[52] on a dense set in Fock space as $n\to\infty$ (Prugovecki, 1995, p. 155). The test function space is typically taken to be the Schwartz space $\mathcal{S}(\mathbb{R}^4)$. Such functions $f(x)$ are infinitely differentiable, and both they, and all their partial derivatives, tend towards zero faster than a polynomial function $x^{-n}$, for any $n\in\mathbb{N}$, as $|x|\to\infty$. The test functions, denoted here as $f$, should not be confused with the elements of the single-particle Hilbert space, which are also denoted as $f$ elsewhere. In particular, for any bounded open subset of Minkowski space–time, there is a space of Schwartz functions which have support in this bounded region, and one can smear field operators over these Schwartz functions.

In both fermionic and bosonic Fock spaces the zero-particle subspace is $\mathscr{H}^0 = \mathbb{C}^1$, the so-called vacuum sector. This subspace contains a distinguished non-zero vector $1\in\mathbb{C}^1$, called the vacuum vector. The vacuum vector is denoted as $\mathbf{0}$ or $|\mathbf{0}\rangle$ in some of the quantum field theory literature, despite the fact that it is not the zero vector in Fock space. To distinguish it from the zero vector, it is more usefully denoted as $\Omega$.

[52] "The weak limit of a sequence of operators $A_1, A_2, \ldots$ in a Hilbert space is determined by the weak limits of the corresponding sequences $A_1 f, A_2 f, \ldots$ for all vectors $f$ in their domains. In turn, a vector $h$ is the weak limit of a sequence of vectors $h_1, h_2, \ldots$ in a Hilbert space if $\langle g|h\rangle$ is the limit of $\langle g|h_1\rangle, \langle g|h_2\rangle, \ldots$ for *any* vector $g$ in that space" (Prugovecki, 1995, p. 155, footnote 2).

The Fock space representations of the canonical commutation relations are defined by the requirement that there is a unique vector $\Omega$ which is such that $a(f)\Omega = 0$ for all $f$. The Fock space vacuum vector is cyclic with respect to the algebra generated by the representation of the commutation relations. In addition to the Fock spaces for inertial observers in Minkowski space–time, one can introduce distinct Fock spaces for accelerated observers in Minkowski space–time, and distinct Fock spaces for quantum field theory in curved space–time. The Fock spaces for inertial observers in Minkowski space–time can be termed 'Minkowski vacuum' Fock spaces, and these are defined by the further requirement that the Fock space possesses a unitary representation of the Poincaré group, under whose action the vacuum vector is invariant.

On each bosonic and fermionic Fock space there is an operator called the number operator,

$$N = 0 \oplus 1I \oplus 2I \oplus 3I \oplus \cdots.$$

The eigenstates of this operator are often considered to represent the states of the second-quantized theory in which there are a definite number of particles. The utility of Fock space is that it enables one to represent situations where there is (i) a variable number of particles, or (ii) an indefinite number of particles. Yet, because the one-particle subspace of the second-quantized Fock space theory is composed of the vector bundle cross-sections from the first-quantized theory, any notion of a particle in the second-quantized Fock space theory must be constructed from the notion of a particle in the first-quantized theory. The states which Fock space quantum field theory calculates the transition probabilities between, are constructed from the one-particle states of the first-quantized theory, and this is a primary link which quantum field theory has with the empirical world. Such considerations can be used to temper any suggestion that the second-quantized theory *completely* replaces the particle concept, if any, to be found within the first-quantized theory.

In the second-quantized theory, the negative energy states of the first-quantized theory are re-interpreted as positive-energy states of anti-particles (Teller, 1995, p. 77), and possible transitions to negative-energy states are reinterpreted as the probability of a scattering process producing anti-particles (Wightman, 1973, p. 445). This is best seen in the heuristic expressions for the field operators $\hat{\Psi}(x)$.[53] Suppose that one has a first-quantized field possessing complex-valued components. We need to consider both the particle field $\psi(x)$, and the anti-particle field $\bar{\psi}(x)$. The first-quantized particle field $\psi(x)$, expressed as an inverse Fourier transform, contains positive energy

[53] The hat here is conventionally used in quantum theory to indicate an operator, at least in those circumstances where there might be some ambiguity as to whether an operator is being denoted.

and negative energy components:

$$\begin{aligned}\psi(x) &= \psi^+(x) + \psi^-(x) \\ &= \int e^{i\langle p,x\rangle} a(p)\, d^4p + \int e^{i\langle p,x\rangle} c(p)\, d^4p \\ &= \int e^{i(\mathbf{p}\cdot\mathbf{x} - p_0 t)} a(p)\, d^4p + \int e^{i(\mathbf{p}\cdot\mathbf{x} - p_0 t)} c(p)\, d^4p.\end{aligned}$$

$a(p)$ has support upon the forward mass hyperboloid $\mathscr{V}_m^+$, and $c(p)$ has support upon the backward mass hyperboloid $\mathscr{V}_m^-$. Hence, as specified in Section 2.2, the positive-energy component can be expressed as

$$\begin{aligned}\psi^+(x) &= \frac{1}{(2\pi)^{3/2}} \int_{\mathscr{V}_m^+} e^{i(\mathbf{p}\cdot\mathbf{x} - p_0 t)} a(\mathbf{p})\, d^3\mathbf{p}/2\omega(\mathbf{p}) \\ &= \frac{1}{(2\pi)^{3/2}} \int_{\mathscr{V}_m^+} e^{i(\mathbf{p}\cdot\mathbf{x} - \omega(\mathbf{p})t)} a(\mathbf{p})\, d^3\mathbf{p}/2\omega(\mathbf{p}),\end{aligned}$$

and the negative-energy component can be expressed as

$$\begin{aligned}\psi^-(x) &= \frac{1}{(2\pi)^{3/2}} \int_{\mathscr{V}_m^-} e^{i(\mathbf{p}\cdot\mathbf{x} - p_0 t)} c(\mathbf{p})\, d^3\mathbf{p}/2\omega(\mathbf{p}) \\ &= \frac{1}{(2\pi)^{3/2}} \int_{\mathscr{V}_m^-} e^{i(\mathbf{p}\cdot\mathbf{x} + \omega(\mathbf{p})t)} c(\mathbf{p})\, d^3\mathbf{p}/2\omega(\mathbf{p}),\end{aligned}$$

where $\omega(\mathbf{p}) = +(m^2 + \|\mathbf{p}\|^2)^{1/2}$. In order to quantize the negative energy component, one expresses the integral over the backward mass hyperboloid as an integral over the forward mass hyperboloid. To do this, one uses the fact that the expression for the inverse Fourier transform is invariant under replacing $p = (p_0, \mathbf{p})$ with $-p = (-p_0, -\mathbf{p})$:

$$\begin{aligned}\psi^-(x) &= \frac{1}{(2\pi)^{3/2}} \int_{\mathscr{V}_m^+} e^{i(-\mathbf{p}\cdot\mathbf{x} + p_0 t)} c(-\mathbf{p})\, d^3\mathbf{p}/2\omega(\mathbf{p}) \\ &= \frac{1}{(2\pi)^{3/2}} \int_{\mathscr{V}_m^+} e^{-i(\mathbf{p}\cdot\mathbf{x} - p_0 t)} c(-\mathbf{p})\, d^3\mathbf{p}/2\omega(\mathbf{p}) \\ &= \frac{1}{(2\pi)^{3/2}} \int_{\mathscr{V}_m^+} e^{-i(\mathbf{p}\cdot\mathbf{x} - \omega(\mathbf{p})t)} c(-\mathbf{p})\, d^3\mathbf{p}/2\omega(\mathbf{p}).\end{aligned}$$

Next, one substitutes annihilation operators $\hat{a}(\mathbf{p})$ and $\hat{c}(-\mathbf{p})$ in place of the functions $a(\mathbf{p})$ and $c(-\mathbf{p})$ to obtain the field operators:

$$\begin{aligned}\hat{\psi}^+(x) &= \frac{1}{(2\pi)^{3/2}} \int_{\mathscr{V}_m^+} e^{i(\mathbf{p}\cdot\mathbf{x} - \omega(\mathbf{p})t)} \hat{a}(\mathbf{p})\, d^3\mathbf{p}/2\omega(\mathbf{p}), \\ \hat{\psi}^-(x) &= \frac{1}{(2\pi)^{3/2}} \int_{\mathscr{V}_m^+} e^{-i(\mathbf{p}\cdot\mathbf{x} - \omega(\mathbf{p})t)} \hat{c}(-\mathbf{p})\, d^3\mathbf{p}/2\omega(\mathbf{p}).\end{aligned}$$

$\hat{a}(\mathbf{p})$ is the annihilation operator for a positive-energy particle with energy–momentum $(\omega(\mathbf{p}), \mathbf{p})$, and $\hat{c}(-\mathbf{p})$ is the annihilation operator for a positive-energy particle with energy–momentum $(\omega(\mathbf{p}), -\mathbf{p})$.

Recall that in Section 2.10 it was shown that $d(p) \equiv \bar{c}(-p)$, hence $\bar{d}(p) \equiv c(-p)$. As a general rule of quantization, the conjugate of a function is replaced by the adjoint of the operator substituted for the function, hence

$$\hat{c}(-\mathbf{p}) = \hat{d}^*(\mathbf{p}).$$

$\hat{d}^*(\mathbf{p})$ is the creation operator for a positive-energy anti-particle of energy–momentum $(\omega(\mathbf{p}), \mathbf{p})$. Hence, the annihilation of a particle with momentum $-\mathbf{p}$ is considered to be equivalent in this context to the creation of an anti-particle with momentum $\mathbf{p}$. Replacing $\hat{c}(-\mathbf{p})$ with $\hat{d}^*(\mathbf{p})$, one obtains:

$$\hat{\Psi}^-(x) = \frac{1}{(2\pi)^{3/2}} \int_{\mathscr{V}_m^+} e^{-i(\mathbf{p}\cdot\mathbf{x}-\omega(\mathbf{p})t)} \hat{d}^*(\mathbf{p})\, d^3\mathbf{p}/2\omega(\mathbf{p}).$$

It is in this sense that the negative energy states of first-quantized particle fields are forged anew as the positive-energy states of anti-particles. The expression for the field operator is then $\hat{\Psi}(x) = \hat{\Psi}^+(x) + \hat{\Psi}^-(x)$.

By following a similar procedure for the anti-particle field $\bar{\psi}(x)$, one obtains an anti-particle field operator $\hat{\bar{\Psi}}(x)$ which contains creation operators for particles, and annihilation operators for anti-particles.

Recall that we previously associated creation and annihilation operators with single-particle states $f$. One can think of heuristic expressions such as $\hat{a}(\mathbf{p})$ as shorthand for annihilation operator-valued distributions $a(f_p)$, where $f_p$ is a single-particle state of energy–momentum $p = (\omega(\mathbf{p}), \mathbf{p})$. The notion of such a state is, however, slightly problematical, and requires a brief digression.

The genuine eigenfunctions of the energy–momentum operators $\hat{p}_\mu = -i\partial_\mu$, such as the plane waves on Minkowski space–time $f(x) = e^{i\langle p,x\rangle}$, are not square-integrable, and consequently do not belong to the one-particle Hilbert space $\mathscr{H}$. However, where $S \subset \mathscr{H}$ is a dense subset of $\mathscr{H}$ in the norm topology of $\mathscr{H}$, such as the domain of the energy–momentum operators, the eigenfunctions belong to the anti-dual[54] $S^*$, and the triple of normed vector spaces $S \subset \mathscr{H} \subset S^*$ is called a 'rigged' or 'equipped' Hilbert space, depending on some additional specifics of the construction.

A quantum system possesses a definite value for a quantity if and only if the state of the system is an eigenstate of the operator which represents that quantity. Hence, whilst there are functions in $\mathscr{H}$ whose energy–momentum expectation values equal $p$, there are none which possess $p$ as a property. If the energy–momentum of a system in a particular state has an expectation value of $p$, it simply means that if the energy–momentum of such a system in such a state were to be measured on many

[54] The anti-dual is the set of continuous anti-linear functionals.

occasions, then the measured values would have a statistical distribution with a mean value of $p$. Many physicists argue that because it is impossible to measure a continuous quantity with arbitrary precision, it is acceptable for quantum theory to only use states in which the value of such a quantity is localised within a bounded subset of its range of possible values. However, this only makes sense if one endorses the operationalist notion that theoretical entities should be defined in terms of observational and measurement procedures. Under a realist approach, a theory should be capable of representing continuous quantities to possess precise values, as specific real numbers. The inability of measurement to exactly detect those precise values should then follow from the representation of measurement in the theory as a special type of interaction between systems, in which the measurement system cannot be infinitely sensitive to the value of a continuous quantity possessed by the measured system.

So, as an operationalist, one can think of $f_p$ as an element of $\mathscr{H}$ whose energy–momentum expectation value is $p$, but as a realist, one should enlarge the one-particle state spaces to include, amongst other things, the energy–momentum eigenfunctions in the rigged Hilbert space, and one should associate creation and annihilation operators with the elements of this enlarged single-particle space. Given the enlarged single-particle space, one can construct an enlarged Fock space, which these creation and annihilation operators should act upon.

Although the states in Fock space are the states of a free system, by defining a scattering operator $S$ on Fock space one can calculate the transition probabilities $\langle \psi_{\text{out}}, S\psi_{\text{in}} \rangle$ between the asymptotically free incoming states $\psi_{\text{in}}$ and the asymptotically free outgoing states $\psi_{\text{out}}$ in a collision or decay process, a special type of interaction in which the interaction is transient and spatially localised. Such calculations provide a primary link between quantum field theory and the empirical world. The scattering operator is defined in terms of an interaction Hamiltonian density operator $\hat{H}_I(x)$. In the classical canonical formulation of a theory, the Hamiltonian is the function on the phase space of a system which represents the energy of the system. In the classical canonical formulation of a field theory, the Hamiltonian density is a field on space–time which represents the energy density of the field. The total Hamiltonian of an interacting system can be broken up into the 'free' Hamiltonian, the part of the total Hamiltonian which is present in the absence of the interaction, and an additional term called the interaction Hamiltonian. The interaction Hamiltonian density of a field theory represents the field interaction energy density. The operator (or operator-valued distribution) which represents the Hamiltonian in the quantized theory can correspondingly be broken up into the free Hamiltonian operator $\hat{H}_0$ and the interaction Hamiltonian operator $\hat{H}_I$. For example, a self-interacting field of mass $m$ and spin 0 can be represented by the solutions of an equation with the form

$$\left(\Box + m^2\right)\phi(x) = -\mathscr{P}'\left(\phi(x)\right),$$

where $\mathscr{P}$ is a polynomial expression of some degree which specifies the self-interaction potential, and $\mathscr{P}'$ denotes the derivative of the potential. The corresponding classical Hamiltonian density can be written as

$$\begin{aligned} H(x) &= H_0(x) + H_I(x) \\ &= H_0(x) + \mathscr{P}\big(\phi(x)\big) \\ &= \frac{1}{2}\left[\left(\frac{\partial \phi(x)}{\partial t}\right)^2 + \big(\nabla \phi(x)\big)^2 + m^2 \phi(x)^2\right] + \mathscr{P}\big(\phi(x)\big). \end{aligned}$$

In canonical quantum field theory (see Appendix F), the use of scattering theory typically relies upon the so-called 'interaction picture,' in which the field operators evolve under the free Hamiltonian operator $\hat{H}_0$, and the states evolve under the interaction Hamiltonian operator $\hat{H}_I$.

The scattering operator $S$ can be expressed in a Dyson perturbation expansion as

$$\begin{aligned} S &= 1 + \sum_{n=1}^{\infty} \frac{(-i)^n}{n!} \int_{\mathbb{R}^4} d^4 x_1 \cdots \int_{\mathbb{R}^4} d^4 x_n \, T\big[\hat{H}_I(x_1) \cdots \hat{H}_I(x_n)\big] \\ &= \sum_{n=0}^{\infty} S_n. \end{aligned}$$

$\hat{H}_I(x)$ is the interaction Hamiltonian density operator, and $T[\hat{H}_I(x_1) \cdots \hat{H}_I(x_n)]$ is a time-ordered permutation[55] of the interaction Hamiltonian density operators in the sense that

$$T\big[\hat{H}_I(x_1) \cdots \hat{H}_I(x_n)\big] = \hat{H}_I(x_{i_1}) \cdots \hat{H}_I(x_{i_n}),$$

with $t(x_{i_1}) \geqslant t(x_{i_2}) \geqslant \cdots \geqslant t(x_{i_n})$. The interaction Hamiltonian density operator is expressed in terms of the field operators relevant to the interaction in question. In the case of quantum electrodynamics (QED), for example, one has field operators $\hat{\Psi}(x)$ for the second-quantized Dirac field, field operators $\hat{\bar{\Psi}}(x)$ for the second-quantized conjugate Dirac field, and field operators for the second-quantized electromagnetic field $\hat{A}_\mu(x)$ (Teller, 1995, pp. 124–125).

The creation operators, annihilation operators and field operators of a Dirac field are defined upon the electron–positron Fock space $\mathscr{F}_{e,e^+}$. This can be obtained either by taking the Fock space of the direct sum of the electron and positron single-particle spaces,

$$\mathscr{F}_{e,e^+} = \mathscr{F}(\mathscr{H}^+_{m_e,1/2} \oplus \mathscr{H}^-_{m_{e^+},1/2}),$$

or by taking the direct sum of the Fock spaces for the electron and positron,

$$\mathscr{F}_{e,e^+} = \mathscr{F}(\mathscr{H}^+_{m_e,1/2}) \oplus \mathscr{F}(\mathscr{H}^-_{m_{e^+},1/2}).$$

[55] When the Hamiltonian includes fermion field operators, one must introduce the sign of the permutation into the expression for the time-ordering.

The creation operators, annihilation operators and field operators for an electromagnetic field are defined upon a photonic Fock space $\mathscr{F}_\gamma$. For quantum electrodynamics, one takes the tensor product $\mathscr{F}_\gamma \otimes \mathscr{F}_{e,e^+}$ of these two Fock spaces. The operators of the individual fields are extended to the entire tensor product space in a trivial manner. For example, the Dirac field operator $\hat{\Psi}(x)$ defined upon $\mathscr{F}_{e,e^+}$ is extended to the operator $I \otimes \hat{\Psi}(x)$ defined upon $\mathscr{F}_\gamma \otimes \mathscr{F}_{e,e^+}$. This fact can be conveniently suppressed from most of the notation, but is important to bear in mind given that the interaction Hamiltonian density operator is a product of field operators from $\mathscr{F}_{e,e^+}$ and field operators from $\mathscr{F}_\gamma$. The interaction Hamiltonian density operator is only defined upon the tensor product Fock space $\mathscr{F}_\gamma \otimes \mathscr{F}_{e,e^+}$.

The interaction Hamiltonian density operator is obtained by substituting field operators into the expression for the classical interaction Hamiltonian density, and subjecting them to 'normal ordering' (see Teller, 1995, pp. 127–129). The classical interaction Hamiltonian density, in turn, may be derived from the interaction term of the classical Lagrangian density. As we will see in Section 4.1, a matter field interacting with a gauge field can be represented by a pair $(\psi, \nabla)$ which satisfy the interacting matter field equation and the coupled Yang–Mills equation. $\psi$ is a cross-section of a fibre bundle, and $\nabla$ is a connection on a fibre bundle. These coupled equations can be derived from a Lagrangian, as explained in Section 4.6. The Lagrangian for a matter field interacting with a gauge field is a map from each pair $(\psi, \nabla)$ to a real-valued function on space–time $\mathcal{L}(\psi, \nabla) : \mathcal{M} \to \mathbb{R}$. The Lagrangian for a Dirac field $\psi$ interacting with an electromagnetic gauge field potential $A_\mu$ is:

$$\begin{aligned}\mathcal{L}(\psi, A_\mu) &= -\frac{1}{4}F_{\mu\nu}F^{\mu\nu} + i\bar{\psi}\gamma^0\gamma^\mu\nabla_\mu\psi - m_e\bar{\psi}\gamma^0\psi \\ &= -\frac{1}{4}F_{\mu\nu}F^{\mu\nu} + i\bar{\psi}\gamma^0\gamma^\mu\partial_\mu\psi - m_e\bar{\psi}\gamma^0\psi + q\bar{\psi}\gamma^0\gamma^\mu\psi A_\mu.\end{aligned}$$

This Lagrangian is usually called the QED Lagrangian by physicists, despite the fact that it is really a part of the classical theory. $q$ here is the charge of the electron, which provides the 'coupling constant' that measures the strength of the interaction; $m_e$ is the mass of an electron; $F_{\mu\nu}$ is the gauge field strength (see Section 3.5); $\nabla_\mu = \partial_\mu - iqA_\mu$ is the gauge covariant derivative (as explained in Section 4.5); and $\gamma^\mu$ are the Dirac gamma matrices (see Section 2.8). One can split this Lagrangian into three parts: $\mathcal{L}(\psi) = i\bar{\psi}\gamma^0\gamma^\mu\nabla_\mu\psi - m_e\bar{\psi}\gamma^0\psi$ is obtained from the Lagrangian density of a free Dirac field by substituting the gauge covariant derivative $\nabla_\mu$ in place of the ordinary partial derivative $\partial_\mu$; $\mathcal{L}(A_\mu) = -\frac{1}{4}F_{\mu\nu}F^{\mu\nu}$ is the Lagrangian density of a free electromagnetic field; and $q\bar{\psi}\gamma^0\gamma^\mu\psi A_\mu$ is the interaction term by virtue of the fact that it contains products of different fields.

One can obtain a classical Hamiltonian density $H(x)$ from a classical Lagrangian density by employing certain rules, such as the following for the case of a matter field Lagrangian (Derdzinski, 1992, p. 169):

$$H(\psi) = \dot{\psi}^a\frac{\partial\mathcal{L}}{\partial\dot{\psi}^a} - \mathcal{L}(\psi).$$

In the case of a Dirac field interacting with an electromagnetic field, the interaction term in the Lagrangian density provides the interaction Hamiltonian density. Thus, in the case of quantum electrodynamics, and prior to normal ordering, one obtains the following interaction Hamiltonian density operator:

$$\hat{H}_I(x) = q\hat{\bar{\Psi}}(x)\gamma^0\gamma^\mu\hat{\Psi}(x)\hat{A}_\mu(x).$$

Even with the interaction Hamiltonian density operator for quantum electrodynamics inserted into the expression for the scattering operator, it is still impossible to *rigorously* calculate the transition amplitudes between the asymptotically free incoming states $\psi_{\text{in}}$ and the asymptotically free outgoing states $\psi_{\text{out}}$ of electrons, positrons and photons, because the following perturbation series is divergent:

$$\langle\psi_{\text{out}}, S\psi_{\text{in}}\rangle = \langle\psi_{\text{out}}, I\psi_{\text{in}}\rangle + \langle\psi_{\text{out}}, S_1\psi_{\text{in}}\rangle + \langle\psi_{\text{out}}, S_2\psi_{\text{in}}\rangle + \cdots.$$

Physicists respond to this difficulty by assuming, rather than proving, that the series is a divergent asymptotic series. The significance of an asymptotic series is that the first few terms can approximate the 'true value' one is trying to calculate, even if the series as a whole diverges. Hence, assuming the series is asymptotic, physicists calculate only, say, the first and second-order terms:

$$\langle\psi_{\text{out}}, S_1\psi_{\text{in}}\rangle = -i\int_{\mathbb{R}^4} d^4x_1 \big\langle\psi_{\text{out}}, \hat{H}_I(x_1)\psi_{\text{in}}\big\rangle,$$

$$\langle\psi_{\text{out}}, S_2\psi_{\text{in}}\rangle = -\frac{1}{2}\int_{\mathbb{R}^4} d^4x_1 \int_{\mathbb{R}^4} d^4x_2 \big\langle\psi_{\text{out}}, T\big[\hat{H}_I(x_1)\hat{H}_I(x_2)\big]\psi_{\text{in}}\big\rangle.$$

The Feynman diagrams beloved of textbooks on quantum field theory, offer a graphical mnemonic for the algorithmic procedure involved in calculating each term in such a perturbation series. However, each term in the perturbation series is itself a divergent integral, hence even the calculation of the first and second-order terms requires the use of so-called 'renormalization' to obtain finite results. Renormalization basically amounts to introducing factors into the integrands which enable the integrals to approach a finite value as the limits of the integrals are taken to infinity (see Teller, 1988).

Thus, as Berestetskii *et al.* comment, "The lack of complete logical consistency in this theory [QED] is shown by the occurrence of divergent expressions when the mathematical formalism is directly applied, although there are quite well-defined ways of eliminating those divergencies. Nevertheless, such methods remain to a considerable extent, semiempirical rules, and our confidence in the correctness of the results is ultimately based only on their excellent agreement with experiment, not on the internal consistency or logical ordering of the fundamental principles of the theory" (1982, p. 4).

Note that the Fock space vacuum is the vacuum state of a free field, and, with respect to the Fock space number operator, a state of zero particles. The free field vacuum of Fock space is a distinct concept from what is often called the 'dressed vacuum' of interacting fields. The dressed vacuum purportedly contains an infinite number of 'virtual' particles; particles which, if they existed, would violate the relativistic relationship $\langle p, p \rangle = -m^2$ between mass $m$ and energy–momentum $p$. The experimentally detectable effects ascribed to the dressed vacuum are referred to as 'vacuum fluctuations.' The notion of virtual particles is invoked to explain and justify otherwise ad-hoc renormalization procedures, which are used to obtain finite results from the perturbation series in the second-quantized scattering theory. However, the free-field vacuum is the only vacuum which is theoretically well-defined in quantum field theory. As Prugovecki states, "the actual computations performed in perturbation theory actually begin with expressions for asymptotic states, ... formulated in Fock space, and then progress through a chain of computations dictated by Feynman rules, which have no direct bearing to a mathematically rigorous realization of a non-Fock representation of the canonical commutation relations... Hence, in conventional QFT [quantum field theory] the existence of such a representation, and of a corresponding unique and global 'dressed vacuum,' is merely a conjecture rather than a mathematical fact" (1995, pp. 198–199). Rugh and Zinkernagel concur, arguing that the popular picture of the production and annihilation of virtual particles in the 'interacting' vacuum, "is actually misleading as no production or annihilation takes place in the vacuum. The point is rather that, in the ground state of the full interacting field system, the number of quanta (particles) for any of the fields is not well-defined" (2002, Note 27).

Haag's theorem demonstrates that a free-field Fock space cannot directly represent an interacting field system[56]; it basically proves that a Fock space cannot possess the vacuum vector of a free-field and the vacuum vector of an interacting field. A vacuum vector is required to be invariant under space–time translations, and a Fock space possesses, up to phase, a unique translation-invariant vector. The vacuum state $\Omega_0$ of a free field[57] must be the ground state of the free-field Hamiltonian $\hat{H}_0$, in the sense that $\hat{H}_0 \Omega_0 = 0$, and no vector in the ray spanned by $\Omega_0$, $\{c\Omega_0 \colon |c| = 1\}$, can also be the ground state of the full Hamiltonian for an interacting system $\hat{H}$. If $\hat{H}_0 \Omega_0 = 0$, as required, then the requirement that $\hat{H} \Omega_0 = 0$ cannot also be satisfied. For example, in the case of a self-interacting scalar field, with an interaction Hamiltonian density $H_I(x) = \mathscr{P}(\phi(x)) = \phi^4(x)$, when field operators are substituted into this expression to obtain the interaction Hamiltonian density operator, it contains, at the very least, one term with four creation operators, which is not cancelled out by any other term. As a consequence, $\hat{H}_I \Omega_0 \neq 0$, and, in fact, $\Omega_0$ is not even an eigenstate of $\hat{H}_I$, hence one cannot render $\Omega_0$ as the interaction vacuum by adding to $\hat{H}_I$ a term containing a finite constant (Fraser, 2005).

In more precise terms, Haag's theorem demonstrates that the full Hamiltonian for an interacting system cannot be well-defined on a Hilbert space representation of the

[56] See Earman and Fraser (2006) for an excellent discussion.

[57] Here we use $\Omega_0$ rather than $\Omega$ to denote the *free-field* vacuum state, just in case there might also be the vacuum vector of an interacting field in the same Hilbert space.

CCRs unitarily equivalent to that on which the free Hamiltonian is defined. Thus, from Haag's theorem it is often argued that the representation of an interacting field system requires a unitarily inequivalent representation of the CCRs from any one of the free field Fock space representations. It has been proposed that the dynamics of the interaction would then determine the choice of representation for the CCR-algebra. The idea is that the dynamics are treated, in the Heisenberg picture, as a one-parameter group of automorphisms of the CCR-algebra, and in the chosen representation of the CCRs, these automorphisms are implemented as a one-parameter group of unitary operators $U(t)$, whose self-adjoint generator would be the Hamiltonian of the interacting field system.

Let us return to the question of particles and fields in second-quantized quantum field theory. One can define the particle states of matter fields and gauge fields to be the states of specific energy and momentum, and second-quantized quantum field theory can indeed be used to calculate the transition probabilities $\langle \psi_{\text{out}}, S\psi_{\text{in}} \rangle$ between matter field states and gauge field states of a specific energy and momentum, as well as a specific particle number. For example, consider Compton scattering, in which a single electron collides with a single photon (Teller, 1995, pp. 132–133). An incoming state in which the electron has 4-momentum $(\omega(\mathbf{p}_i), \mathbf{p}_i)$, with $\omega(\mathbf{p}_i) = +(m_e^2 + \|\mathbf{p}_i\|^2)^{1/2}$, and the photon has 4-momentum $(\omega(\mathbf{k}_i), \mathbf{k}_i)$, with $\omega(\mathbf{k}_i) = +\|\mathbf{k}_i\|$, is represented as $\psi_{\text{in}} = \hat{Z}_\mu^*(\mathbf{k}_i)\hat{c}^*(\mathbf{p}_i)\,\Omega$, where $\Omega$ is the vacuum vector, $\hat{c}^*(\mathbf{p}_i)$ is the creation operator for an electron of 4-momentum $(\omega(\mathbf{p}_i), \mathbf{p}_i)$, and $\hat{Z}_\mu^*(\mathbf{k}_i)$ is the creation operator for a photon of 4-momentum $(\omega(\mathbf{k}_i), \mathbf{k}_i)$.[58] An outgoing state in which the electron has 4-momentum $(\omega(\mathbf{p}_o), \mathbf{p}_o)$ and the photon has 4-momentum $(\omega(\mathbf{k}_o), \mathbf{k}_o)$ is represented as $\psi_{\text{out}} = \hat{Z}_\mu^*(\mathbf{k}_o)\hat{c}^*(\mathbf{p}_o)\Omega$, where $\hat{c}^*(\mathbf{p}_o)$ is the creation operator for an electron of 4-momentum $(\omega(\mathbf{p}_o), \mathbf{p}_o)$, and $\hat{Z}_\mu^*(\mathbf{k}_o)$ is the creation operator for a photon of 4-momentum $(\omega(\mathbf{k}_o), \mathbf{k}_o)$.

Note that in terms of the tensor product Fock space $\mathscr{F}_\gamma \otimes \mathscr{F}_{e,e^+}$, these incoming and outgoing states should strictly be represented as

$$\begin{aligned}\psi_{\text{in}} &= \hat{Z}_\mu^*(\mathbf{k}_i) \otimes \hat{c}^*(\mathbf{p}_i)(\Omega_\gamma \otimes \Omega_{e,e^+}) \\ &= \big(\hat{Z}_\mu^*(\mathbf{k}_i)\Omega_\gamma\big) \otimes \big(\hat{c}^*(\mathbf{p}_i)\,\Omega_{e,e^+}\big),\end{aligned}$$

and

$$\begin{aligned}\psi_{\text{out}} &= \hat{Z}_\mu^*(\mathbf{k}_o) \otimes \hat{c}^*(\mathbf{p}_o)(\Omega_\gamma \otimes \Omega_{e,e^+}) \\ &= \big(\hat{Z}_\mu^*(\mathbf{k}_o)\Omega_\gamma\big) \otimes \big(\hat{c}^*(\mathbf{p}_o)\Omega_{e,e^+}\big),\end{aligned}$$

where $\Omega_\gamma$ is obviously the vacuum vector of the photonic Fock space, and $\Omega_{e,e^+}$ is the vacuum vector of the electron–positron Fock space.

However, recall that the states of specific energy and momentum do not strictly belong to Fock space. Rather, these states belong to the enlarged Fock space constructed

[58] As a notational convenience, in the case of a photon we shall use '$\mathbf{k}$' rather than '$\mathbf{p}$' to denote the 4-momentum.

from the 'rigged' or 'equipped' one-particle space, as explained earlier. Neither the incoming electron state of 4-momentum $(\omega(\mathbf{p}_i), \mathbf{p}_i)$, nor the outgoing electron state of 4-momentum $(\omega(\mathbf{p}_o), \mathbf{p}_o)$ belong to the single-particle subspace of the proper Fock space for an electron. Similarly, neither the incoming photon state of 4-momentum $(\omega(\mathbf{k}_i), \mathbf{k}_i)$, nor the outgoing photon state of 4-momentum $(\omega(\mathbf{k}_o), \mathbf{k}_o)$ belong to the single-particle subspace of the proper Fock space for a photon.

Even if one can deal with this technical difficulty, the second-quantized theory still relegates particles to particular states of a quantum field. Whilst the first-quantized theory aspired to represent particles as objects in their own right, the second-quantized theory represents fields first and foremost, and particles are merely considered to be 'excitation modes' of quantum fields. As Wallace (2005) comments, "it is normal in modern quantum field theory to regard the field as the primary concept and the particles as secondary, derivative entities." If one regards quantum fields as prior, then the Fock space vacuum state, the state of zero particles, is the vacuum state of a free quantum field, the state of minimum energy. The states in Fock space with a non-zero number of particles are then excitations of the free quantum field, states of the free quantum field with energy levels greater than the minimum.

The localisability problems do not evaporate on moving to the second-quantized theory, but localisability is not required of a field theory in the way it is required of a particle theory. Whilst a first-quantized field is a cross-section $\psi(x)$ of a fibre bundle over space–time, satisfying a differential equation such as the Klein–Gordon equation or Dirac equation, a quantum field is often said 'to be' a field operator, or operator valued distribution, $\hat{\Psi}(x)$, such as the one defined above, satisfying a differential equation such as the Klein–Gordon or Dirac equation. Aitchison, for example, asserts that "a free (non-interacting) quantum field $[\hat{\Psi}(x)]$ is conventionally represented as a sum of an infinite number of modes" (1991, p. 171), where the expression $\hat{\Psi}(x) = 1/2\pi^{3/2} \int e^{i(\mathbf{p}\cdot x-\omega(\mathbf{p})t)}\hat{a}(\mathbf{p}) + e^{-i(\mathbf{p}\cdot x-\omega(\mathbf{p})t)}\hat{d}^*(\mathbf{p})\, d^3\mathbf{p}/2\omega(\mathbf{p})$ is understood as the decomposition of the quantum field into an infinite number of modes. Aitchison asserts that "states with particles present correspond to the states of the field in which some modes are excited; the vacuum state $|0\rangle$ is the state in which all modes are unexcited, and the field is in its ground state" (1991). In this context, then, a mode simply corresponds to a particular value of 4-momentum $p$, and the states with particles present are simply states such as $\hat{c}^*(\mathbf{p})\Omega$. Such states belong to (the anti-dual of dense subspaces of) the single-particle Hilbert spaces of the first-quantized theory, hence the particle concept in the second-quantized theory, if any, is built from the particle concept, if any, which resides in the first-quantized theory.

The excitation modes of a quantum field are often referred to as 'quanta,' to distinguish them from any pre-quantum notion of particles. As Halvorson writes, "philosophical reflection on quantum field theory has tended to focus on how it revises our conception of what a particle is. For instance, though there is a self-adjoint operator in the theory representing the total number of particles of a field, the standard 'Fock space' formalism does not individuate particles from one another. Thus, Teller (1995, Chapter 2) suggests that we speak of *quanta* that can be 'aggregated,' instead of (enumerable) *particles* — which implies that they can be distinguished and labelled. Moreover, because the theory does contain a total number of quanta observable (which, therefore, has eigenstates corresponding to different values of this number), a field state can be a nontrivial superposition of number eigenstates that fails to predict any particular number of quanta with certainty. Teller (1995, pp. 105–106) counsels that we think of these superpositions as not actually containing any quanta, but only propensities to *display* various numbers of quanta when the field interacts with a 'particle detector' " (Halvorson, 2001, p. 109).

This conclusion is reinforced when one considers the existence of 'Rindler particles' (or 'Rindler quanta'). Whilst every inertial (i.e., non-acceleratory) reference frame in Minkowski space–time shares a common vacuum state, a common number operator, and a common decomposition of the solution space to a relativistic differential equation into positive-energy and negative-energy states, accelerated observers have their own vacuum state, their own number operator, and their own notion of positive-energy and negative-energy. In the inertial Minkowski vacuum state, an accelerated particle detector will detect a flux of particles (the so-called 'Unruh effect'). These problems recur for quantum field theory in curved space–time, where each local freely-falling reference frame, defined by a geodesic of the curved space–time geometry, possesses its own notion of a vacuum, a number operator, and a decomposition into positive-energy and negative-energy states.

The second-quantized theory therefore presents problems for the traditional concept of a particle because:

- When there are multiple particles present, they cannot be individuated.
- For any class of reference frame, there are states of the quantum field which are not eigenstates of the number operator for that class of reference frame. In such states one cannot say that a definite number of particles are present.
- Any state of the quantum field fails to be an eigenstate of the number operator for some class of reference frame.

Suppose that one defined particles not to be objects in their own right, but to be properties of an underlying quantum field, and suppose, in particular, that one defined the particle states of a quantum field to be those states which are eigenstates of the number operator. Suppose, in addition, that one defined superpositions of the number operator eigenstates to be non-particle states of the underlying quantum field, states in which there is merely a propensity to detect a certain definite number of particles if a measurement-like interaction takes place. Although all the states of the first-quantized theory are eigenstates of the number operator in the second-quantized theory, belonging as they do to the single-particle subspace, one cannot use this fact to argue that every state of the first-quantized theory is a particle state. Eigenstates of the number operator are only eigenstates of the number operator associated with a particular inertial class of reference frame. States which are single-particle states for inertial reference frames in Minkowski space–time, are superpositions of eigenstates of the number operator for an accelerated class of reference frame. Hence, so-called single-particle states for an inertial reference frame, are non-particle states for an accelerated reference frame. The second-quantized theory treats a particle not merely as a property of an underlying quantum field, but as a relationship between a quantum field and a class of reference frame. Whether or not there is a single particle present is dependent upon the observer's state of motion, and is therefore a relationship between the reference frame of the observer and the quantum field.

As Cao states, "in the field ontology framework, the number of particles is only a parameter characterizing the state of a free field" (1999, p. 10) and one can add that, because the state of a free-field is reference frame dependent, the number of particles is also reference frame dependent. Particles must be properties of quantum fields, rather than objects in their own right because, whilst properties can be reference frame dependent, and recast as relationships, the existence or non-existence of objects cannot.

Let us now consider how algebraic quantum field theory (AQFT) enters the discussion of particle concepts. It is well known that the canonical commutation relations admit an uncountable infinity of unitarily inequivalent representations for systems with an infinite number of degrees of freedom, of which field systems are one such example. Not only does each mass $m$ and spin $s$ free field Fock space provide a unitarily inequivalent representation of the Poincaré group, but it also provides a unitarily inequivalent representation of the CCRs. Moreover, even if the mass $m$ and spin $s$ are fixed, the Fock space with the inertial vacuum is unitarily inequivalent to a Fock space with an accelerated vacuum. In algebraic quantum field theory, one therefore abstracts from the Hilbert space formulation to an abstract $C^*$-algebra $\mathscr{A}$, which possesses a multiplicity of unitarily inequivalent Hilbert space representations. In particular, advocates of AQFT propose a 'net' of local $C^*$-algebras over the bounded open subsets $\mathscr{O} \subset \mathcal{M}$ of space–time, such that each local algebra $\mathscr{A}(\mathscr{O})$ is a subalgebra of the abstract $C^*$-algebra $\mathscr{A}$. A net of algebras over the bounded open subsets is such that if $\mathscr{O}_1 \subset \mathscr{O}_2$, then $\mathscr{A}(\mathscr{O}_1)$ is a subalgebra of $\mathscr{A}(\mathscr{O}_2)$. These local algebras are considered to be algebras of observables, in the sense that the self-adjoint elements are supposed to represent experimentally measurable quantities. The observables are considered to be functionals of the underlying fields.

If one starts with the Fock space formalism, then there is an algebra $\mathscr{F}(\mathscr{O})$ associated with each open subset $\mathscr{O}$, called the field algebra. This is constructed from field operators 'smeared' over test functions that have their support in $\mathscr{O}$.[59] This is distinct from the notion of an algebra of local observables. The smeared field operators tend

[59] The Reeh–Schlieder theorem demonstrates that the vacuum vector is cyclic for polynomials in the field operators smeared over test functions with support in an arbitrarily small subset of space–time. See Halvorson (2001) for a discussion of the interpretational implications of this theorem.

to be unbounded, so one must define the local field algebra $\mathscr{F}(\mathscr{O})$ to be, say, the algebra of all bounded functions of the local field operators. One might then define the algebra of local observables $\mathscr{A}(\mathscr{O})$ to be the subalgebra of gauge-invariant elements in $\mathscr{F}(\mathscr{O})$ (Bogolubov *et al.*, 1990, p. 327).

The closure of the union of all the local algebras $\mathscr{A}(\mathscr{O})$ equals the 'quasilocal' algebra $\mathscr{A}$. It is then often claimed that there is a different notion of particle associated with each inequivalent representation of the abstract algebra. However, the impact of this is weakened by AQFT's lack of physical applicability. As David Wallace points out, "the major problem with AQFT is that very few concrete theories have been found which satisfy the AQFT axioms. To be precise, the only known theories in four dimensions which do satisfy the axioms are interaction-free: no examples are known of AQFT-compatible interacting field theories, and in particular the standard model cannot at present be made AQFT-compatible" (2001, p. 1). Moreover, as Prugovecki emphasises, "no interpretation of the operators within these $C^*$-algebras as representing *measurable* quantities was ever provided" (Prugovecki, 1995, p. 156). Bogolubov *et al.* draw attention to the fact that "the energy and the momentum (or their spectral projections) are not local observables. The same is true of all conserved quantities such as the total angular momentum and the electric charge of the system" (1975, pp. 595–596). These global observables do not even belong to the quasilocal algebra $\mathscr{A}$, although they do belong to the von Neumann algebra associated to $\mathscr{A}$.

To return to the main thread of this section, we can now establish some interpretational conclusions. Whilst the second-quantized quantum theory is empirically adequate, it sits uncomfortably with a realist interpretation. One could argue that the second-quantized theory provides little more than the transition probabilities between free states in scattering processes, the transition probabilities to free states in decay processes, and, in general, the relationships between free states $\psi$ and observables of the free field $\hat{O}$, expressed as expectation values $\langle \psi, \hat{O}\psi \rangle$. The relationships between states and observables in quantum theory code probabilistic information about empirical phenomena. This level of structure in the second-quantized theory captures the structure possessed by the empirical data, not the structure of the physical world beyond the empirical data, hence this accords with structural empiricism, but not with structural realism. The only level of structure in the second-quantized theory which potentially captures the structure of the physical world, is the level of structure inherited from the first-quantized theory, the structure of the fibre bundle cross-sections over space–time, from which the one-particle subspaces of the second-quantized theory can be constructed.

As already remarked, quantum field theory calculates the transition probabilities $\langle \psi_{\text{out}}, S\psi_{\text{in}} \rangle$ between incoming and outgoing free states in scattering processes, and these incoming and outgoing states are matter field and gauge field states of a specific particle number, and a specific energy and momentum. As Huang states, "from an experimental point of view, particles are detectable packets of energy and momentum"

(1992, p. 1). However, these states must be constructed from the one-particle states of the first-quantized theory which, as we have seen, do not admit a wave-function interpretation, and cannot be localised in space–time without introducing negative-energy components. Streater asserts that quantum field theorists avoid the non-localisability of the one-particle theory "by claiming that physicists do not actually measure the projection operator onto the subspace of one-particle states in a region; instead, they measure an average... of a field strength over the region. If this is large, the physicist infers that there is something in the region... Usually, the large value of the field... in a region can be used to trigger a counter or leave a track on a photograph" (Streater, 1988, pp. 141–142). This accords with the suggestion made earlier, that the detection of a particle does not localise the particle in space–time; it is merely the manifestation of an interaction between the detector and the non-local particle field, and it is the interaction which is localised in space–time.

Incorporating the points made in this chapter, the basic interpretational doctrine in this text can be stated as follows:

*Each type of matter field or gauge field in first-quantized relativistic quantum theory, is the type of thing which is represented by a cross-section of such-and-such a bundle over space–time, satisfying such-and-such conditions. In the guise of one-particle states, and with the negative energy states re-interpreted as the positive-energy states of anti-particles, these are, more or less, the types of thing which are created or annihilated in second-quantized relativistic quantum theory, but these things are there re-cast as properties of a quantum field, rather than things in their own right. Relative to a specific class of reference frame, these properties can be deemed to be particle states of the quantum field. These particle states are not localisable in space–time, but one can still equate particles with such states because particle detection events do not require the associated particle states to be localisable. Thus, relative to a specific class of reference frame, each type of particle is a set of non-localisable cross-sections of such-and-such a bundle over space–time, satisfying such-and-such conditions. Whilst a free quantum field creates and annihilates the cross-sections of free-particle bundles, an interacting quantum field creates and annihilates the cross-sections of interacting-particle bundles, although in this case there is no linear space structure, and no Fock space of states. Lacking a mathematically well-defined theory of interacting quantum fields, we may not be able to define, from a structural viewpoint, what an interacting quantum field is, but we can assume that the interacting quantum fields create and annihilate interacting particles, which constitute excitations of the interacting quantum fields. Thus, we can assume that the interacting particles created and annihilated by interacting quantum fields are, more or less, the cross-sections of the interacting-particle bundles considered in the first-quantized treatment of interacting fields.*

Contrary to this, Fraser argues that interacting quantum fields do not possess quanta (i.e., particle states): "It is not possible to produce a mathematical representation for an interacting system that is relevantly similar to the Fock representation for a free system" (2005, p. 22). However, what Fraser's arguments actually demonstrate is that there is no *Hilbert space* representation for an interacting system which satisfies the necessary requirements to admit particle states.

Fraser considers three possible means of obtaining a Hilbert space representation for an interacting quantum field:

1. Under the canonical approach to field quantization, define interacting quantum field operators $\phi_I$ and $\pi_I$ (see Appendix F for a definition of the canonically conjugate field operators $\phi$ and $\pi$), and an interaction Hamiltonian $H_I$, in the Fock space of a free field.
2. Take the Fourier decomposition of a classical field satisfying an interacting field equation, and substitute creation and annihilation operators in place of the Fourier coefficients, to mimic the quantization procedure for a free field.
3. Assume that the interacting quantum field operator-valued distribution $\phi_I(f, t)$ satisfies a version of the interacting field equation for the classical field, and define smeared annihilation operators[60] as

$$a(f, t) = \frac{1}{2}\big[\phi_I(f, t) + i\pi_I(f, t)\big].$$

Then obtain the irreducible representation of the canonical commutation relations which is unique up to unitary equivalence, by the condition that there exists a vector $\Omega$ such that $a(f, t)\Omega = 0$ for all $f$.

Fraser argues that Haag's theorem rules out the first approach; that the creation and annihilation operators obtained in the second approach are not Lorentz covariant; and that, whilst a number operator can be defined in the third approach, it fails to possess a no-particle state which is also a vacuum state, in the sense of being the lowest energy eigenstate of the Hamiltonian.

Fraser's arguments, however, involve *linear* field operators, *linear* Hamiltonian operators, vacuum *vectors* and *linear* representations of the canonical commutation relations. These arguments do not exclude the possibility of representing interacting quantum fields in terms of non-linear mathematical structures which admit interacting particle states. Just as the second-quantization of first-quantized free fields is a functor from a category of linear vector spaces into another category of linear vector spaces, one would expect the second-quantization of first-quantized interacting fields to be a functor from a category of non-linear spaces, into another category of non-linear spaces. Whilst free-particle states are eigen-vectors of both the free linear Hamiltonian operator and the number operator on the Fock space for a free field, interacting particle states could be represented, perhaps, as eigen-elements of a non-linear Hamiltonian operator and a non-linear number operator on the non-linear space of an interacting field.

[60] Note that $a(f, t)$ is a smeared version of $a(x, t)$, which, in turn, is the inverse Fourier transform of an annihilation operator $\hat{a}(p)$.

To conclude, then, in this section we have strengthened the rationale for interpreting the first-quantized standard model in terms of structural realism. It is not inconsistent with empirical observation to be realist about the structures in the first-quantized theory because the first-quantized structures provide only part of the empirical structure of the second-quantized theory, and only some aspects of the first-quantized structures are linked with the empirical world. There are only partial correspondence rules between the structures in the first-quantized theory and the empirical world. The structures in the first-quantized theory alone cannot be used to organise, predict or explain empirical phenomena, but that does not preclude a realist interpretation of these structures. Structural realism only requires structures to exist, not for them to be measured or even measurable.[61]

The inability of (Fock space) second-quantized relativistic quantum theory to provide a mathematically rigorous representation of interacting fields should be taken as a pathological condition which, from an realist perspective, entails the need for a new theory. The empirical success of second-quantized quantum theory does not entail that it represents the real structure of the world. An alternative theory is required which is both empirically and ontologically adequate. Such a theory may be built from the first-quantized structures in a different way, or may be built differently from the ground-up.

[61] Note that the Galilei-invariant theory of quantum mechanics ('non-relativistic' quantum mechanics) also admits a fibre bundle formulation. One can apply a realist interpretation to the fibre bundle cross-sections of this theory also, but in this case the cross-sections *can* be interpreted as wave-functions, coding probabilistic information about empirical phenomena. Hence, in the Galilei-invariant theory, this level of structure codes both empirical and realist information.

Chapter 3

# Gauge Fields

## 3.1. Principal fibre bundles and $G$-structures

In the standard model, each gauge force field corresponds to a compact connected Lie Group $G$, called the gauge group. A gauge field with gauge group $G$ can either be represented by a connection on a principal fibre bundle $P$ with structure group $G$, or by a connection on a vector bundle $\delta$ equipped with a so-called '$G$-structure.'

Given a complex vector bundle $\delta$ of fibre dimension $n$, any matrix sub-group $G \subset GL(n, \mathbb{C})$ acts freely, from the right, upon the set of bases in each fibre. Treating a basis as a row vector, $(e_1, \dots, e_n)$, it is mapped by $g \in G$ to another basis $(e'_1, \dots, e'_n)$ by matrix multiplication:

$$(e'_1, \dots, e'_n) = (e_1, \dots, e_n)g = (g_{j1}e_j, \dots, g_{jn}e_j).$$

Needless to say, one sums over repeated indices in this expression.

In general, the $G$-action upon the set of bases in each fibre will possess multiple orbits. The selection of one particular orbit of this $G$-action, in each fibre of $\delta$, is called a $G$-structure in $\delta$ (Derdzinski, 1992, pp. 81–82). Because $G$ acts freely, it acts simply transitively within each orbit. A vector bundle $\delta$ equipped with a $G$-structure is sometimes referred to as a '$G$-bundle.'

Geometrical objects in each fibre of a vector bundle, such as inner products and volume forms, can be used to select a $G$-structure. For example, suppose $\delta$ is a complex vector bundle of fibre dimension $n$: the unitary group $U(n)$ acts freely upon the set of bases in each fibre, and there are multiple orbits of the $U(n)$-action in each fibre, but if each fibre is equipped with a positive-definite Hermitian inner product, then the inner product singles out the orbit consisting of orthonormal bases. By stipulating that an inner product selects the orbit of orthonormal bases, one defines a bijection between inner products and $U(n)$-structures. For any orbit of the $U(n)$-action, there is an inner product with respect to which that orbit consists of orthonormal bases. Given any basis $(e_1, \dots, e_n)$, one can define an inner product which renders that basis an orthonormal basis by stipulating that the matrix of inner products between the vectors in the basis has the form $\text{diag}\{1, 1, \dots, 1\}$. Given any pair of vectors $v, w$, they can be expressed as $v = c_1e_1 + \cdots + c_ne_n$ and $w = a_1e_1 + \cdots + a_ne_n$ in this basis, and their inner product is now defined to be

$$\langle v, w\rangle = \langle c_1e_1 + \cdots + c_ne_n, a_1e_1 + \cdots + a_ne_n\rangle = c_1a_1 + \cdots + c_na_n.$$

Once an inner product has been defined which renders $(e_1, \ldots, e_n)$ orthonormal, all the other bases which can be obtained from $(e_1, \ldots, e_n)$ under the action of $U(n)$ must themselves be orthonormal.[1]

Following Derdzinski, we shall refer to a vector bundle $\delta$ equipped with a $G$-structure as an *interaction bundle*. The interaction bundles employed in the standard model are complex vector bundles.

The selection of a $G$-structure in a vector bundle $\delta$ determines the selection of a principal $G$-subbundle of the general linear frame bundle of $\delta$. Given the selection of a matrix sub-group $G \subset GL(n, \mathbb{C})$, there are multiple principal $G$-subbundles of the general linear frame bundle. These correspond to the different orbits of the $G$-action upon the set of bases in each fibre of $\delta$. For example, a vector bundle $\delta$ equipped with a $U(n)$-structure is equivalent to the selection of a principal $U(n)$-subbundle $Q$ of the general linear frame bundle of $\delta$.

Whilst the selection of a $G$-structure in a vector bundle $\delta$ determines a specific principal $G$-bundle $P$, the converse is also true. Given a principal $G$-bundle $P$, with $G \subset GL(n, \mathbb{C})$, the associated vector bundle $P \times_G \mathbb{C}^n$, defined by the standard representation of $G$ on $\mathbb{C}^n$, is a vector bundle of fibre dimension $n$, which is equipped with a distinguished $G$-structure, namely the one provided by the principal $G$-bundle to which it is associated.

For a fixed $G$, all the principal $G$-subbundles of a general linear frame bundle are mutually isomorphic; they share the same collection of transition functions as the general linear frame bundle, and they share the same typical fibre as each other, namely $G$. However, these mutually isomorphic principal $G$-bundles do not determine mutually isomorphic $G$-structure vector bundles. For example, although all the principal $U(n)$-subbundles of the general linear frame bundle of a complex vector bundle $\delta$, are mutually isomorphic, the Hermitian vector bundles they correspond to are not isomorphic as Hermitian vector bundles. Whilst they all share the same vector bundle structure, namely $\delta$, the inner product spaces of their respective fibres are not unitarily isomorphic. By design, the different principal $G$-subbundles of a general linear frame bundle are obtained by assigning different inner products and/or volume forms to the fibres of the complex vector bundle $\delta$. The mutually isomorphic principal $G$-bundles correspond to a single principal fibre bundle under the classification scheme of the next section, yet they correspond to multiple interaction bundles. However, for a fixed vector bundle and a fixed group $G$, the specific $G$-structure chosen is not of physical significance; what matters is that the vector bundle has *some* $G$-structure.

---

[1] This bijection between inner products and $U(n)$-structures is, however, merely conventional (private communication with Andrzej Derdzinski). Given the specification of an inner product, the convention is that a basis belongs to the $U(n)$-structure if the matrix of inner products between its constituent vectors has the form $\text{diag}\{1, 1, \ldots, 1\}$. Given the specification of an inner product, one could alternatively fix an arbitrary positive-definite Hermitian matrix, and stipulate that a basis belongs to the $U(n)$-structure if the matrix of inner products between its constituent vectors equals the chosen positive-definite Hermitian matrix. This would provide an alternative bijection between inner products and $U(n)$-structures.

## 3.2. Classification of principal $G$-bundles

Whilst each gauge field corresponds to a particular compact and connected Lie group $G$, the choice of a particular $G$ does not, in general, determine a unique principal fibre bundle $P$ with structure group $G$, even up to isomorphism. In terms of interaction bundles, this means that even the vector bundle structure is not determined up to isomorphism. Hence, the choice of a gauge group does not uniquely determine the mathematical object upon which the representation of a gauge field is dependent. Gauge fields on non-isomorphic principal bundles or interaction bundles are capable of being empirically distinct, hence this isn't a case of 'surplus structure.' Rather, these are distinct structures, capable of describing physically distinct objects.

In the case of a 4-dimensional manifold $\mathcal{M}$, it is possible, for any Lie group $G$, to classify all the principal $G$-bundles over $\mathcal{M}$ (DeWitt, Hart and Isham, 1979, pp. 199–201). Although the purview of this text extends to space–times of arbitrary dimension, let us consider the classification over 4-manifolds for illustrative purposes.

Suppose that $G$ is a simply connected Lie group.[2] In this case, the principal $G$-bundles over a four-dimensional manifold $\mathcal{M}$ are classified by the elements of the fourth cohomology group[3] over the integers $H^4(\mathcal{M}; \mathbb{Z})$. In the event that $\mathcal{M}$ is compact and orientable, $H^4(\mathcal{M}; \mathbb{Z}) = \mathbb{Z}$, hence the principal $G$-bundles, for a simply connected Lie group $G$ over a compact and orientable 4-manifold, are in one-to-one correspondence with the integers. In the event that $\mathcal{M}$ is either non-compact or non-orientable, $H^4(\mathcal{M}; \mathbb{Z}) = \{Id\}$. This entails that for a simply connected Lie group $G$, all the principal $G$-bundles over a non-compact or non-orientable 4-manifold are trivial bundles, isomorphic to $\mathcal{M} \times G$.

In the special case where the simply connected Lie group $G$ is a special unitary group $SU(n)$, the element of $H^4(\mathcal{M}; \mathbb{Z})$ which corresponds to a particular principal $SU(n)$-bundle is the second Chern class of that bundle.[4] For different principal $SU(n)$-bundles, the second Chern class of the bundle corresponds to different cohomology equivalence classes of the base manifold $\mathcal{M}$. The case of a special unitary group is of relevance to the standard model, where $SU(2)$ is involved with the electroweak force, and $SU(3)$ is the gauge group of the strong force.

---

[2] See Appendix A for a definition of simply connected.

[3] One can think of the cohomology groups of a manifold as constructions from the set of forms $\Lambda(\mathcal{M})$. A form $\omega$ such that $d\omega = 0$ (a closed form) can be deemed a cocycle, and a form $\omega$ such that $\omega = d\theta$ (an exact form), can be deemed a coboundary. Letting $Z^p$ denote the space of degree $p$ cocycles, and $B^p$ the space of degree $p$ coboundaries, the quotient $H^p(\mathcal{M}) = Z^p/B^p$ is the $p$th-cohomology group. Two closed forms which belong to the same equivalence class are said to belong to the same cohomology class. One can generalise by varying the coefficients to, say, integers, in which case one obtains the $p$th-cohomology group over the integers $H^p(\mathcal{M}; \mathbb{Z})$ (Choquet-Bruhat *et al.*, 1982, pp. 222–223).

[4] Given a principal bundle over a manifold $\mathcal{M}$, equipped with a connection, the $k$th Chern class is the cohomology class of a closed $2k$-form on $\mathcal{M}$, which is constructed from the curvature of the connection, but independent of the choice of connection. Whilst it is a cohomology class of the base manifold $\mathcal{M}$, it is a topological invariant of the bundle (Choquet-Bruhat *et al.*, 1982, pp. 390–393).

Turning to non-simply connected Lie groups, take the case where $G$ is a unitary group $U(n)$. In the event that $G = U(1)$, the set of inequivalent principal $U(1)$-bundles over any 4-manifold $\mathcal{M}$ is in one-to-one correspondence with the elements of the second cohomology group over the integers $H^2(\mathcal{M};\mathbb{Z})$. The element of $H^2(\mathcal{M};\mathbb{Z})$ which corresponds to a particular principal $U(1)$-bundle is the first Chern class of that bundle. This case is relevant to the electromagnetic force, which has gauge group $U(1)$.

In the case of $U(n)$, with $n > 1$, the set of inequivalent principal $U(n)$-bundles over any 4-manifold $\mathcal{M}$ is in one-to-one correspondence with the elements of $H^2(\mathcal{M};\mathbb{Z}) \oplus H^4(\mathcal{M};\mathbb{Z})$. The case of relevance to the standard model is $G = U(2)$, the gauge group of the electroweak force.

To reiterate, these results demonstrate that the choice of principal fibre bundle or interaction bundle is not determined by the gauge group. In the case of the electromagnetic force, there are many principal $U(1)$-bundles, $\{P_i \colon i \in H^2(\mathcal{M};\mathbb{Z})\}$, and for each different bundle $P_i$, the standard representation of $U(1)$ on $\mathbb{C}^1$ defines a different interaction bundle $\lambda_i = P_i \times_{U(1)} \mathbb{C}^1$. Similarly, in the case of the electroweak force, there are many principal $U(2)$-bundles, $\{Q_i \colon i \in H^2(\mathcal{M};\mathbb{Z}) \oplus H^4(\mathcal{M};\mathbb{Z})\}$, and for each different bundle $Q_i$, the standard representation of $U(2)$ on $\mathbb{C}^2$ defines a different interaction bundle $\iota_i = Q_i \times_{U(2)} \mathbb{C}^2$.

In the case of the strong force, with simply connected gauge group $SU(3)$, there are, in general, many principal $SU(3)$-bundles $\{S_i \colon i \in H^4(\mathcal{M};\mathbb{Z})\}$, and for each different bundle $S_i$, the standard representation of $SU(3)$ on $\mathbb{C}^3$ defines a different interaction bundle $\rho_i = S_i \times_{SU(3)} \mathbb{C}^3$. However, in the case of a non-compact or non-orientable 4-dimensional manifold, $H^4(\mathcal{M};\mathbb{Z}) = \{Id\}$, and the only principal $SU(3)$-bundle is therefore $S = \mathcal{M} \times SU(3)$. In this case there is, therefore, up to isomorphism, only one vector bundle over space–time which can be equipped with an $SU(3)$-structure, namely $\rho = S \times_{SU(3)} \mathbb{C}^3$. One can select many different $SU(3)$-structures in $\rho$ by assigning different inner products and volume forms to the fibres of $\rho$, and each such $SU(3)$-structure corresponds to a different principal $SU(3)$-subbundle of the general linear frame bundle of $\rho$, but each such principal $SU(3)$-bundle is isomorphic to $\mathcal{M} \times SU(3)$. Whilst it is not true to say that all $SU(3)$-interaction bundles are mutually isomorphic as oriented Hermitian vector bundles, they are all mutually isomorphic as vector bundles.

We will see in Section 3.5 that under a choice of gauge, the gauge field strength $F_{\mu\nu}$ of a gauge field corresponds to a two-form on the base space $\mathcal{M}$. As such, the gauge field strength determines an element of a *real* cohomology group such as $H^2(\mathcal{M};\mathbb{R})$, and the antisymmetric tensor product $F \wedge F$ determines an element of a real cohomology group such as $H^4(\mathcal{M};\mathbb{R})$. However, the determination of the real cohomology class of the gauge field strength does not uniquely determine elements in the corresponding integral cohomology groups, and therefore does not uniquely determine a choice of principal fibre bundle or interaction bundle. If, instead, one begins by choosing a principal fibre bundle or interaction bundle, then the corresponding element in the relevant integral cohomology group has a unique image in the corresponding real cohomology group, and this uniquely determines the cohomology class of the gauge field strength.

Because Minkowski space–time is contractible (meaning it can be continuously deformed to an individual point), all its cohomology groups are trivial. This entails that in the special case of the standard model over Minkowski space–time the choice of a gauge group $G$ determines a unique principal $G$-bundle, and all the interaction bundles are trivial.

## 3.3. Gauge connections

The purpose of this and the following sections is to analyse the concepts of a gauge connection, a choice of gauge, and a gauge transformation, in terms of both the principal fibre bundle $P$ picture, and the interaction bundle $\delta$ picture. Let's begin with the principal fibre bundle picture, and the notion of a gauge connection.

A gauge force field potential can be represented by a connection $\nabla$ on a principal fibre bundle $(P, \pi, \mathcal{M}, G)$, where $G$ is an $m$-dimensional Lie Group, and $\mathcal{M}$ is $n$-dimensional space–time. A connection on a bundle can be defined in terms of the assignment of a 'horizontal' subspace to the tangent vector space at each point of the bundle (Torretti, 1983, pp. 269–271). At each point $p \in P$ in the total space of a principal fibre bundle, the vectors in the tangent vector space $T_pP$ which are tangent to the fibre of $P$ over $\pi(p)$, constitute an $m$-dimensional subspace called the space of vertical vectors $V_p$. There are an infinite number of possible $n$-dimensional subspaces of $T_pP$ which, together with $V_p$, span the tangent vector space. Such an $n$-dimensional subspace is called a horizontal subspace, $H_p$, and is such that $T_pP = V_p \oplus H_p$. A connection on the principal fibre bundle $P$ smoothly assigns a horizontal subspace to each point, $p \mapsto H_p$, in a manner which respects the right action $R_g$ of each $g \in G$ on $P$. Thus, a connection satisfies the condition[5]:

$$H_{pg} = R_{g*}(H_p).$$

Each connection $\nabla$ corresponds to a Lie-algebra valued one-form $\omega$ on $P$. For each point $p \in P$, there is a natural isomorphism of $V_p$ onto the Lie algebra $\mathfrak{g}$ of $G$, and the selection of a horizontal subspace $H_p$ enables one to extend this mapping to the entire tangent vector space $T_pP$ using the stipulation that $\ker \omega_p = H_p$. At each point $p \in P$, a connection one-form is a mapping $\omega_p : T_pP \to \mathfrak{g}$. Such a Lie-algebra valued one-form satisfies the condition

$$\omega_{pg}(R_{g*p}v) = Ad(g^{-1})\omega_p(v),$$

where $Ad$ denotes the adjoint representation, the representation of a group upon its own Lie algebra.

---

[5] The asterisk subscript here indicates a differential map, a map between tangent vector spaces, and in this case the differential map of $R_g$. See Torretti (1983, p. 261).

Although each connection $\nabla$ corresponds to a Lie-algebra valued one-form $\omega \in \Lambda^1(P, \mathfrak{g})$, not every Lie-algebra valued one-form on $P$ corresponds to a connection on $P$. An element of $\Lambda^1(P, \mathfrak{g})$ must satisfy the condition above so that it respects the right action $R_g$ for each $g \in G$, and must agree on $V_p$ with the natural isomorphism to $\mathfrak{g}$.

A connection on a principal fibre bundle $P$ enables one to define parallel transport between the fibres of $P$, to define parallel cross-sections of $P$, and to define a covariant derivative upon the cross-sections of $P$.

Given a curve in the base space $\gamma : I \to \mathcal{M}$, for an interval $I \subset \mathbb{R}$, which passes through a pair of points $x_1, x_2 \in \mathcal{M}$, then for each $p \in \pi^{-1}(x_1)$ there is a unique *horizontal* curve in the total space $\hat{\gamma} : I \to P$, which is such that $\pi(\hat{\gamma}) = \gamma$, and which passes through $p$. This is called the lift of $\gamma$ through $p$. The lift of $\gamma$ through $p$ intersects the fibre $\pi^{-1}(x_2)$ at a specific point $q$. The point $q$ is said to be parallel to $p$ along the curve $\gamma$. This defines an isomorphism between $\pi^{-1}(x_1)$ and $\pi^{-1}(x_2)$, which is dependent upon the curve $\gamma$. This path-dependent isomorphism,

$$\tau^{\gamma}_{x_1 x_2} : \pi^{-1}(x_1) \to \pi^{-1}(x_2),$$

is referred to as parallel transport along the curve $\gamma$. The existence of the horizontal subspaces, i.e., the existence of a connection in the total space, determines the existence of parallel transport along curves.

Parallel transport enables one to define the covariant derivative $\nabla_X \sigma$ of a cross-section $\sigma$ of $P$ with respect to a vector field $X$ on the base space $\mathcal{M}$. At a point $x$ of the base space, one is interested in how the cross-section $\sigma$ is changing with respect to the vector field $X$. Hence, one takes the maximal integral curve[6] $\gamma$ of $X$ through the point $x$, and one uses parallel transport to define the rate of change of the cross-section $\sigma$ along the lift of the integral curve:

$$\nabla_X \sigma\big|_x = \lim_{t \to 0} \frac{1}{t}\Big(\big(\tau^{\gamma}_{\gamma(t)\gamma(0)} \sigma_{\gamma(t)}\big) - \sigma_{\gamma(0)}\Big).$$

Whilst the covariant derivative $\nabla_X \sigma$ of a cross-section of a fibre bundle $P$ is also a cross-section of the same bundle, there is a related operator, sometimes called the covariant differential $\nabla \sigma$, but also often called the covariant derivative, which maps a cross-section of $P$ to a cross-section of $T^*\mathcal{M} \otimes P$.

A cross-section $\sigma : \mathcal{M} \to P$ can be defined to be parallel if the differential map $\sigma_*$ maps the tangent space at each point $x$ onto the horizontal subspace at $\sigma(x) \in P$. Alternatively, it can be defined to be parallel if $\nabla \sigma = 0$.

The horizontal subspaces of a connection on the principal fibre bundle $P$ induce horizontal subspaces upon the associated vector bundle $\delta \cong P \times_G \mathbb{C}^n$. Given any $v \in \mathbb{C}^n$, there is a surjective mapping $f_v$ of $P$ onto the associated vector bundle $P \times_G \mathbb{C}^n$, defined by $p \mapsto [(p, v)]$. The horizontal subspaces of $\delta$ are the images of

[6] See Torretti (1983, p. 260).

the horizontal subspaces of $P$ under the differential map of $f_v$:

$$H_{[(p,v)]} = f_{v*}H_p.$$

This definition is independent of the choice of $(p, v)$ in $[(p, v)]$.

The existence of the horizontal subspaces enables one to define parallel transport between the fibres of $\delta$, to define parallel cross-sections of $\delta$, and to define a covariant derivative upon the cross-sections of $\delta$. In the case where $P$ is a bundle of bases in the fibres of the associated bundle $\delta$, one parallel transports a vector from $\delta_{x_1}$ to $\delta_{x_2}$ by expressing it as a linear combination in some basis at $x_1$, parallel transporting that basis to a basis at $x_2$, and then expressing the vector at $x_2$ as a linear combination in the parallel transported basis with the same coefficients as the linear combination in the basis at $x_1$.

## 3.4. Choice of gauge and gauge transformations

The purpose of this section is to define and clarify the notions of a choice of gauge, and a gauge transformation. *En route*, we will introduce the distinction between active and passive gauge transformations, and explain the sense in which a gauge transformation can be thought of either as an automorphism linking physically equivalent theoretical elements, or as a change of 'internal reference frame.'

A cross-section $\sigma$ of the principal $G$-bundle $P$ of some gauge field is called a *choice of gauge*. There is a correspondence between cross-sections and trivializations of a principal fibre bundle: A local cross-section $\sigma : U \to P$ picks out an element $\sigma(x)$ from the fibre $P_x$ over each point $x \in U \subset \mathcal{M}$, and thereby establishes an isomorphism between the points in each fibre and the elements of the structure group $G$. Each $p \in P_x$ is mapped to the unique $g \in G$ which is such that $p = \sigma(x)g$. This defines an isomorphism between $\pi^{-1}(U)$ and $U \times G$.

A section of a principal fibre bundle determines both a trivialization of the principal fibre bundle, and a trivialization of any associated vector bundle. If a global cross-section exists for the principal bundle, then the principal bundle and its associated vector bundles must be trivial. Although the principal $G$-bundles and associated bundles over Minkowski space–time are indeed trivial, there are no canonical trivializations; for each global cross-section of $P$ there is a different global trivialization of $P$. Each different cross-section $\sigma$ can pick out a different element $\sigma(x)$ from the fibre $P_x$ over a point $x$, and establish a different isomorphism between the points in the fibre and the elements of the structure group $G$.

Picking out an element in each fibre of a principal fibre bundle $P$ is equivalent to picking out a basis for each fibre of the associated vector bundles. Doing so for a vector bundle $\delta \cong P \times_G V$ establishes an isomorphism between each fibre $\delta_x$ and the typical fibre $V$. A cross-section $\sigma : U \to P$ enables each fibre to be expressed as $\delta_x = \{[\sigma(x), v] : v \in V\}$, and there is the natural isomorphism $[\sigma(x), v] \mapsto v \in V$

between $\delta_x$ and $V$. As a consequence, the representation $r$ of $G$ on $V$ induces a representation upon each fibre. The representation $r_x$ of $G$ on $\delta_x$ is given by

$$r_x(g) : [\sigma(x), v] \mapsto [\sigma(x), gv],$$

and this induces a representation of $G$ upon the space of sections $\Gamma(\delta)$. Although the representation of $G$ on $\Gamma(\delta)$ changes with a different choice of gauge, it merely changes to an equivalent representation.

A *gauge transformation* of a principal fibre bundle $(P, \pi, \mathcal{M}, G)$ can be defined to be a diffeomorphism $f : P \to P$ which satisfies the conditions:

$$f(pg) = f(p)g, \quad p \in P, \ g \in G, \tag{i}$$

$$\pi\big(f(p)\big) = \pi(p). \tag{ii}$$

Condition (i) means that a gauge transformation is an automorphism of the total space $P$. The stipulation that $f(pg) = f(p)g$ means that a gauge transformation $f$ is '$G$-equivariant.' Condition (ii) means that the fibres of $P$ remain fixed in a gauge transformation. The elements of each fibre are re-arranged, but the fibres themselves are not exchanged. A gauge transformation re-arranges the elements in each fibre of the principal $G$-bundle in a way which preserves the $G$-relationships between the different elements of the fibre.

In general, a bundle automorphism involves an automorphism of both the total space and the base space. Given that the base space is diffeomorphic with the set of fibre spaces, an automorphism of the total space induces an automorphism of the base space. An automorphism of a bundle which leaves the points of the base space fixed is referred to as a vertical automorphism. Given that a gauge transformation does not permute the set of fibre spaces, this is the same as saying that the induced map upon the base space is the identity map. Hence, a gauge transformation is a vertical automorphism of a principal fibre bundle.

A gauge transformation can also be treated as a map $\tau : P \to G$ such that

$$\tau(pg) = g^{-1}\tau(p)g.$$

Each such map $\tau$ from $P$ into $G$ corresponds to an automorphism $f_\tau : P \to P$ defined by

$$f_\tau(p) = p\tau(p).$$

This automorphism is $G$-equivariant, for

$$\begin{aligned} f_\tau(pg) &= (pg)\tau(pg) \\ &= (pg)g^{-1}\tau(p)g \\ &= p\tau(p)g \\ &= f_\tau(p)g. \end{aligned}$$

Hence, each map $\tau : P \to G$ corresponds to a gauge transformation in the first sense defined above.

In general, merely selecting an element of $G$ at each space–time point $x$ is not sufficient to specify a gauge transformation $f : P \to P$. In general one also requires a choice of gauge to obtain an explicit expression for a gauge transformation. In the special case of an Abelian structure group $G$, the right action $R_g$ of $G$ on $P$ does provide a set of gauge transformations. However, in the standard model, an Abelian structure group only occurs in the case of the electromagnetic structure group $U(1)$. In the case of a non-Abelian structure group, the right action $R_g$ of an arbitrary $g \in G$ does not necessarily commute with the right action of all the other elements in $G$, hence the right action $R_g$ of an arbitrary $g \in G$ is not $G$-equivariant, and therefore does not define a gauge transformation. Only if $g$ belongs to the centre of the group $G$ would it define a gauge transformation in this manner.

To reiterate, the introduction of a choice of gauge $\sigma : U \to P$ establishes an isomorphism between each fibre over $U$ and the structure group $G$. This enables one to express $\pi^{-1}(U)$ as a product $U \times G$. In this event, the natural left action $L_g$ of an arbitrary $g \in G$ *does* define a $G$-equivariant transformation of each fibre:

$$L_g : (x, g') \mapsto (x, g \circ g').$$

In the absence of a trivialization, with a non-Abelian structure group, there is no natural left action of $G$ on $P$ which is $G$-equivariant. Hence, in general, selecting an element of $G$ at each space–time point $x$ is insufficient to specify a gauge transformation of a principal $G$-bundle $P$.

Physicists make a distinction between global gauge transformations and local gauge transformations. Expressed as automorphisms of principal $G$-bundles, the distinction only makes sense with respect to a particular choice of gauge. With respect to the corresponding trivialization, a *global* gauge transformation uses the left action $L_g$ of a single element $g \in G$ to define an automorphism:

$$L_g : (x, g') \mapsto (x, g \circ g'),$$

whereas a *local* gauge transformation $g(x)$ selects an element from $G$ at each space–time point $x$, and uses the product structure to define the automorphism:

$$L_{g(x)} : (x, g') \mapsto \big(x, g(x) \circ g'\big).$$

As Bleecker points out (1981, pp. 46–47), the distinction between local and global gauge transformations only has meaning for product bundles, or within a local trivialization, hence a gauge transformation could be global in one trivialization, but local in another.

A gauge transformation can be treated as a cross-section of the associated group bundle $P \times_G G$, where it is understood that the action of $G$ upon itself is the conjugate action. A cross-section of $P \times_G G$ selects the equivalence class $[(\sigma(x), g(x))]$ of a pair $(\sigma(x), g(x))$ at each point $x$. A choice of gauge $\sigma(x)$ determines an isomorphism between each fibre $P_x$ and $x \times G$, and a gauge transformation is then defined by the left action of $g(x)$ upon each fibre $x \times G$. When the choice of gauge is $\sigma(x)$, each

$q \in P_x$ is such that $q = \sigma(x)g'$ for a unique $g' \in G$, and $q$ is mapped by the gauge transformation to

$$f(q) = \sigma(x)\big(g(x) \circ g'\big).$$

By selecting an equivalence class $[(\sigma(x), g(x))]$ over each point $x$, a cross-section of $P \times_G G$ specifies the gauge transformation for any choice of gauge. For an alternative choice of gauge $\sigma(x)h(x)$, with $h(x) \in G$, one uses the equivalence relationship:

$$\big(\sigma(x), g(x)\big) \sim \big(\sigma(x)h(x), h(x)^{-1}g(x)h(x)\big).$$

Hence, if the choice of gauge $\sigma(x)h(x)$ determines the isomorphism between each fibre $P_x$ and $x \times G$, then the gauge transformation is defined by the left action of $h(x)^{-1}g(x)h(x)$ upon each fibre $x \times G$. For this choice of gauge, each $q \in P_x$ is such that $q = \sigma(x)h(x) \circ g''$ for a unique $g'' = h(x)^{-1}g' \in G$, and $q$ is mapped by the gauge transformation to

$$\begin{aligned} f(q) &= \sigma(x)h(x) \circ h(x)^{-1}g(x)h(x) \circ g'' = \sigma(x)\big(g(x)h(x) \circ g''\big) \\ &= \sigma(x)\big(g(x)h(x) \circ h(x)^{-1}g'\big) \\ &= \sigma(x)\big(g(x) \circ g'\big). \end{aligned}$$

In other words, it's the same gauge transformation, expressed in a different gauge.

A gauge transformation of a principal $G$-bundle $P$ induces a vertical automorphism of any associated vector bundle $\delta \cong P \times_G V$. A cross-section of $P \times_G G$ defines an automorphism of each fibre of $\delta$ by means of supplying the equivalence class $[(\sigma(x), g(x))]$ of a pair $(\sigma(x), g(x))$ at each point $x$. $\sigma(x) \in P_x$ is a basis in the fibre $\delta_x$, $g(x)$ is an element of the structure group $G$, and a basis is required in each fibre $\delta_x$ to establish an isomorphism between $G \subset GL(n, \mathbb{C}^n)$ and $Aut(\delta_x)$. Hence, a section of $P \times_G G$ pairs elements of $G$ with bases in $\delta_x$ to specify a $G$-automorphism of each fibre. A cross-section of $P \times_G G$ can be used to define either a $G$-equivariant automorphism of $P$, or a $G$-automorphism of $\delta \cong P \times_G V$.

Given a Lie-algebra valued connection 1-form $\omega$ on the total space of a principal fibre bundle $P$, a choice of gauge $\sigma$ enables one to pull-down $\omega$ to a Lie-algebra valued 1-form $A = \sigma^*\omega$ on the domain $U \subset \mathcal{M}$ of the cross-section. I.e.,

$$A_x(v) = (\sigma^*\omega)_x(v) = \omega_{\sigma(x)}(\sigma_* v), \quad v \in T_x U.$$

Given a coordinate chart $(x_1, \ldots, x_n)$ on $U$, each cotangent vector space $T_x^*U$ is spanned by the differentials $(dx^1, \ldots, dx^n)$. Hence, the pull-down can be expressed as

$$A = A_\mu(x)\, dx^\mu, \quad A_\mu(x) \in \mathfrak{g}.$$

A gauge transformation can be understood in terms of the effect it has upon such a pull-down. In this sense, one can treat a gauge transformation as a change of the connection 1-form, $\omega \mapsto \omega'$, or as a change of gauge, $\sigma \mapsto \sigma'$, but not both. Consider it first as a change in the choice of gauge: Given a gauge transformation $f : P \to P$, a choice of gauge $\sigma$ can be changed to $\sigma' = f \circ \sigma$. Suppose that we treat a gauge transformation as a cross-section $[(\sigma(x), g(x))]$ of $P \times_G G$. As before, when the choice of gauge is $\sigma(x)$, each $q \in P_x$ is such that $q = \sigma(x)g'$ for a unique $g' \in G$, and $q$ is mapped by the gauge transformation to

$$f(q) = \sigma(x)\big(g(x) \circ g'\big).$$

Hence, in the special case of $q = \sigma(x) \circ Id$ we have

$$\begin{aligned} f(q) &= \sigma(x)\big(g(x) \circ Id\big) \\ &= \sigma(x)g(x). \end{aligned}$$

Therefore, the left action of $g(x)$ has the same effect on the choice of gauge as the right action of $g(x)$:

$$\sigma'(x) = \sigma(x)g(x).$$

Curiously, then, a choice of gauge is required to express a gauge transformation, but a gauge transformation can be treated as changing the choice of gauge.

Given that a gauge transformation of $P$ can be treated as changing the choice of gauge, and given that a choice of gauge $\sigma(x)$ picks out a basis in each fibre of $\delta \cong P \times_G V$, a gauge transformation corresponds to a change in the choice of basis in each fibre of an associated bundle $\delta$. Moreover, the coordinates on each fibre of $P$ are derived from the coordinates on the typical fibre $G$, and the gauge-determined isomorphism between each fibre and $G$. Hence, a change of gauge changes the coordinates on each fibre of $P$. Both as a change of basis in the fibres of $\delta$, and as a change of fibre coordinates on $P$, a change of gauge can be legitimately described as a 'change of internal reference frame.'

A gauge transformation can also be treated as changing the connection 1-form on $P$. A gauge transformation $f$ changes $\omega$ to the pull-back $f^*\omega$. I.e.,

$$f^*\omega_p(v) = \omega_{f(p)}\big(f_*(v)\big), \quad v \in T_pP.$$

This has the same effect as a change of gauge upon the pull-down of the connection one-form onto the base space. Consider first the case where the choice of gauge $\sigma$ is kept fixed, but the connection 1-form changes to $\omega' = f^*\omega$. The pull-down to a connection 1-form on $U \subset \mathcal{M}$ is

$$\begin{aligned} \big(\sigma^*(f^*\omega)\big)_x(v) &= (f^*\omega)_{\sigma(x)}\big(\sigma_*(v)\big) \\ &= \omega_{f(\sigma(x))}\big(f_*\big(\sigma_*(v)\big)\big) \\ &= \omega_{(f\circ\sigma)(x)}\big((f \circ \sigma)_*(v)\big), \quad v \in T_xU. \end{aligned}$$

Consider now the case where the connection 1-form $\omega$ is kept fixed, but the choice of gauge changes to $\sigma' = f \circ \sigma$. The pull-down to a connection 1-form on $U \subset \mathcal{M}$

is

$$(\sigma'^*\omega)_x(v) = \big((f \circ \sigma)^*\omega\big)_x(v)$$
$$= \omega_{(f\circ\sigma)(x)}\big((f \circ \sigma)_*(v)\big).$$

Hence one obtains the same connection 1-form on $U \subset \mathcal{M}$ in both cases.

These two treatments of a gauge transformation are sometimes referred to as the 'active' and 'passive' points of view. The active point of view is the one in which the connection $\omega$ on $P$ is changed by a gauge transformation to another connection $\omega' = f^*\omega$, but the choice of gauge $\sigma$ is kept fixed. The passive point of view is the one in which the connection $\omega$ is kept fixed, and a gauge transformation changes the choice of gauge from $\sigma$ to $\sigma' = f \circ \sigma$. Changing the choice of gauge changes the coordinates of each fibre of $P$, and in this sense a passive gauge transformation is merely a change of coordinates, whilst an active gauge transformation is an automorphism of $P$ which re-arranges the points in each fibre.

The existence of gauge transformations is often interpreted as a consequence of surplus structure in the theory. This is clearly understood in the active viewpoint, where, if $\mathcal{C}$ denotes the space of connections on a principal fibre bundle $P$, and $\mathcal{G}$ denotes the group of gauge transformations, then the quotient space $\mathcal{C}/\mathcal{G}$, the space of gauge equivalence classes, is interpreted as the space of physically distinct objects. If $P$ is a principal bundle over a 3-dimensional hypersurface $\Sigma$, then $\mathcal{C}/\mathcal{G}$ is referred to as the physical configuration space, whilst if $P$ is a principal bundle over a 4-dimensional space–time $\mathcal{M}$, then $\mathcal{C}/\mathcal{G}$ is referred to as the space of physically distinct histories. In this sense, a gauge transformation is an automorphism between different members of a physical equivalence class.

Not only does the pull-down $A = \sigma^*\omega$ of a connection one-form change under the 'internal' symmetry of a gauge transformation, but it also changes under an 'external' space–time symmetry. As a Lie-algebra valued one-form on the base space, a cross-section of the bundle $T^*\mathcal{M} \otimes (\mathcal{M} \times \mathfrak{g})$ over Minkowski configuration space, the pull-down $A = A_\mu(x)\,dx^\mu$ transforms under an infinite-dimensional representation of (the universal cover of) the restricted Poincaré group $SL(2,\mathbb{C}) \ltimes \mathbb{R}^{3,1}$ as follows (Prugovecki, 1992, p. 309, Eq. (1.4)):

$$A_\mu(x) \mapsto A'_\mu(x) = \Lambda^\nu_\mu A_\nu\big(\Lambda^{-1}(x-a)\big), \quad (\Lambda, a) \in SO_0(3,1) \ltimes \mathbb{R}^{3,1}.$$

This is the same type of expression as that obtained for the transformation of a matter field on Minkowski configuration space under the Poincaré group. Once again it is understood that $\Lambda$ is the image of an element of $SL(2,\mathbb{C})$ under the covering homomorphism $\Lambda : SL(2,\mathbb{C}) \to SO_0(3,1)$. In the case of a spinor field, one required a finite-dimensional representation of $SL(2,\mathbb{C})$ upon the typical fibre of the spinor bundle to define an infinite-dimensional representation upon the space of cross-sections of the spinor bundle. In the case of a gauge field pull-down, the standard representation of $SO_0(3,1)$ on $\mathbb{R}^4$, the typical fibre of $T^*\mathcal{M}$, determines an infinite-dimensional representation of $SL(2,\mathbb{C}) \ltimes \mathbb{R}^{3,1}$ on the cross-sections of $T^*\mathcal{M} \otimes (\mathcal{M} \times \mathfrak{g})$.

If we choose a basis $\{E_b\}$ for the Lie algebra $\mathfrak{g}$, and write the Lie-algebra valued one-form as $A = A^b_\mu \, dx^\mu \otimes E_b$, then a transformation under the restricted Poincaré group is simply written as

$$A^b_\mu(x) \mapsto A'^b_\mu(x) = \Lambda^\nu_\mu A^b_\nu\big(\Lambda^{-1}(x - a)\big), \quad (\Lambda, a) \in SO_0(3, 1) \ltimes \mathbb{R}^{3,1}.$$

Note carefully that the group of gauge transformations $\mathcal{G}$ is an infinite-dimensional group, whilst the universal cover of the Poincaré group, $SL(2, \mathbb{C}) \ltimes \mathbb{R}^{3,1}$, is a finite-dimensional group. In this text, our use of language will be to treat 'gauge transformations' and 'internal symmetries' as synonymous phrases. Given a fibre-bundle representation of a gauge field, an internal symmetry is exclusively a fibre-level symmetry, whilst an external symmetry is at least derived from a symmetry of the base space, if not confined there.

A Kaluza–Klein interpretation of internal symmetries asserts that gauge transformations are the external symmetries of compactified, homogeneous, extra spatial dimensions. Active gauge transformations are active external symmetries of these extra dimensions, and passive gauge transformations are passive external symmetries of these extra spatial dimensions. A passive external symmetry is the change of coordinates that corresponds to a change of observational standpoint, hence, under a Kaluza–Klein interpretation, a passive gauge transformation is, in some sense, a change of observational standpoint. The compactification of the hypothetical extra dimensions entails that such changes of standpoint only exist on an acutely small scale, hence macroscopic systems such as human beings cannot undertake these particular changes of standpoint! We raise this interpretation not necessarily to endorse it, but to offer an empirical interpretation of a gauge transformation in terms of a change of standpoint. Under a Kaluza–Klein interpretation, a gauge transformation is not *merely* an automorphism between different members of a physical equivalence class.

Given that a change of gauge corresponds to a change of coordinates in the fibres of $P$, and a change of basis in the fibres of an associated vector bundle $\delta$, we have used the phrase 'a change of internal reference frame' to describe a passive gauge transformation. Note, however, that the selection of a basis in each fibre of $\delta$ is not comparable to the issue of selecting a basis in quantum mechanics, where the measurement of each different quantity by means of some detection apparata, corresponds to a different choice of basis in the quantum state space. The fibres of vector bundles such as $\eta$ and $\delta$ are not quantum state spaces; rather, the quantum state spaces are constructed from cross-sections of such bundles. A passive gauge transformation does not, therefore, correspond empirically to a change in measurement apparata.

## 3.5. Gauge field curvature

Given a gauge potential connection one-form $\omega$ on $P$, the field strength curvature two-form $\Omega$ is defined to be the exterior *covariant* derivative of the connection,

$$\Omega^\omega = D^\omega \omega = (d\omega)^H,$$

where

$$(d\omega)^H(X, Y) = d\omega\left(X^H, Y^H\right),$$

for any pair $X, Y$ of vector fields on $P$. The '$H$' superscript denotes the horizontal component of a vector, and $d\omega$ is the exterior derivative of the one-form $\omega$. The exterior covariant derivative is such that

$$\Omega^\omega = d\omega + \frac{1}{2}[\omega, \omega].$$

For any pair $X, Y$ of vector fields, the connection one-form $\omega$ and curvature two-form $\Omega^\omega$ satisfy the Cartan 'equation of structure,'

$$\Omega^\omega(X, Y) = d\omega(X, Y) + \frac{1}{2}\left[\omega(X), \omega(Y)\right],$$

and the Bianchi identity

$$D^\omega \Omega^\omega = d\Omega^\omega + \left[\omega, \Omega^\omega\right] = 0,$$

which is obviously equivalent to

$$d\Omega^\omega = \left[\Omega^\omega, \omega\right],$$

given that the bracket [ , ] is an anti-symmetric product operation.

The Yang–Mills equation for a free gauge field represented by a curvature two-form is

$$\operatorname{div} \Omega^\omega = 0.$$

This generalises the vacuum (source-free) Maxwell equation.[7] The curvature two-form $\Omega^\omega$ is a $\mathfrak{g}$-valued two-form on $P$, which corresponds to a cross-section of

$$\bigwedge^2 T^*\mathcal{M} \otimes P \times_G \mathfrak{g}.$$

The gauge connection on $P$ induces a connection upon the associated bundle $P \times_G \mathfrak{g}$, and the tensor bundle $\bigwedge^2 T^*\mathcal{M}$ inherits the Levi-Civita space–time connection. The covariant derivative $\nabla$ of a cross-section of $\bigwedge^2 T^*\mathcal{M} \otimes P \times_G \mathfrak{g}$ is therefore defined with respect to both the gauge covariant derivative and the Levi-Civita covariant derivative. The divergence operator is defined to be the contraction of the covariant differential, $\operatorname{div} = C \cdot \nabla$, hence the divergence of the curvature two-form in the Yang–Mills equation, $\operatorname{div} \Omega^\omega = 0$, involves both the gauge covariant derivative and the Levi-Civita covariant derivative.

Given a cross-section $\sigma$ of $P$, one can pull the curvature two-form $\Omega^\omega$ on $P$ down to a two-form on the base-space, $F = \sigma^* \Omega^\omega$. The selection of a particular gauge

[7] If one likes to understand equations in terms of a variational principle, then the Yang–Mills equation is the Euler–Lagrange equation obtained from the Yang–Mills Lagrangian for a free field. See Section 4.6 for an account of the Lagrangian approach.

provides a map from cross-sections of $\bigwedge^2 T^*\mathcal{M} \otimes P \times_G \mathfrak{g}$ to cross-sections of $\bigwedge^2 T^*\mathcal{M} \otimes (\mathcal{M} \times \mathfrak{g})$.

Given that the curvature two form $\Omega^\omega$ is the exterior covariant derivative of the connection one-form $\omega$, the pull-down $F = \sigma^*\Omega^\omega$ of the curvature two-form is related to the pull-down $A = \sigma^*\omega$ of the connection one-form by the equation

$$F = dA + A \wedge A = dA + \frac{1}{2}[A, A].$$

$dA$ is the exterior derivative, and $A \wedge A$ is the exterior tensor product.

Given a coordinate chart $(x_1, \dots, x_n)$, one can express the pull-downs as

$$A = A_\mu(x)\, dx^\mu, \quad A_\mu(x) \in \mathfrak{g},$$

and

$$F = F_{\mu\nu}(x)\, dx^\mu \otimes dx^\nu, \quad F_{\mu\nu}(x) \in \mathfrak{g}.$$

The definition of the curvature two-form entails that, in component terms

$$F_{\mu\nu} = \partial_\mu A_\nu - \partial_\nu A_\mu + \frac{1}{2}[A_\mu, A_\nu].$$

In terms of the pull-downs to the base-space, the Bianchi identities are

$$\partial^\mu F_{\mu\nu} + \left[A^\mu, F_{\mu\nu}\right] = 0,$$

where

$$\partial^\mu = g^{\mu\nu}\partial_\nu, \qquad A^\mu = g^{\mu\nu}A_\nu.$$

Whilst each $A_\mu$ and each $F_{\mu\nu}$ is a Lie-algebra valued field on space–time $\mathcal{M}$, if one chooses a basis $\{E_a \in \mathfrak{g}\colon\ a = 1, \dots, k\}$, $k = \dim \mathfrak{g}$ in the Lie algebra, then one can write expressions for $A$ and $F$ in which the components are real-valued fields rather than Lie-algebra valued fields. One can write each $A_\mu$ as the linear combination

$$A_\mu = A^a_\mu E_a,$$

where each $A^a_\mu$ is a real-valued field on space–time.

Given both a coordinate chart $(x_1, \dots, x_n)$, and a Lie-algebra basis $\{E_a \in \mathfrak{g}\colon\ a = 1, \dots, k\}$, one can write the Lie-algebra valued one-form $A$ as

$$A = A^a_\mu\, dx^\mu \otimes E_a.$$

Similarly, one can write the Lie-algebra valued two-form $F$ as

$$F = F^a_{\mu\nu}\, dx^\mu \otimes dx^\nu \otimes E_a.$$

In these component terms, one can write[8]

$$F^a_{\mu\nu} = \partial_\mu A^a_\nu - \partial_\nu A^a_\mu + C^a_{bc} A^b_\mu A^c_\nu,$$

where $C^a_{bc}$ are the structure constants of the Lie algebra.[9]

The curvature two-form itself transforms under a gauge transformation. In the active point of view, the relationship between a curvature two-form $\Omega$ on $P$ and its gauge-transform $\Omega'$ can be expressed as

$$\begin{aligned}\Omega' &= f^*\Omega \\ &= Ad_{g^{-1}(x)}\,\Omega \\ &= g^{-1}(x)\Omega g(x),\end{aligned}$$

where the $g(x)$ comes from the expression of a gauge transformation as a cross-section $[(\sigma(x), g(x))]$ of $P \times_G G$. The curvature two-form quite literally transforms in the adjoint representation of the gauge group.

In the special case where the gauge group $G$ is Abelian, the curvature two-form is unchanged by a gauge transformation. Hence, in the special case of electromagnetism, all the connection one-forms in a gauge equivalence class correspond to a common curvature two-form. In electromagnetism, a gauge transformation of a connection pull-down $A = \sigma^*\omega$ can be treated as the map

$$A \mapsto A + d\phi,$$

induced by a smooth real-valued function $\phi \in C^\infty(\mathcal{M}, \mathbb{R})$. The smooth function $\phi$ corresponds to the selection of an element $e^{i\phi(x)}$ of the gauge group $U(1)$ at each point, and the right action of this element thereby defines a change in the choice of gauge:

$$\sigma' = R_{e^{i\phi(x)}}\sigma.$$

Under this change in the choice of gauge, one has the distinct pull-down $A' = \sigma'^*\omega = A + d\phi$. Given the properties of the exterior derivative, it follows that

$$F' = dA' = d(A + d\phi) = dA + d(d\phi) = dA = F.$$

Hence, as claimed, for the special case of electromagnetism, all the connection one-forms in a gauge equivalence class correspond to a common curvature two-form.

With the assumption that the set of gauge-equivalence classes corresponds to the set of physically distinct objects, there is a temptation to treat the curvature two-forms, rather than the connection one-forms, as the primary physical objects. However, in

---

[8] A full derivation of the component expression for the curvature two-form $F$ is provided in Appendix H.

[9] The expression for the field strength of a physical gauge field typically contains the coupling constant $g$ of the gauge field as follows:

$$F^a_{\mu\nu} = \partial_\mu A^a_\nu - \partial_\nu A^a_\mu + g\, C^a_{bc} A^b_\mu A^c_\nu.$$

the general, non-Abelian case, the curvature two-forms are not gauge invariant, hence gauge-invariance cannot be used as a criterion to assign primary physical status to the curvature two-forms.

## 3.6. The interaction bundle picture

In the interaction bundle picture (Derdzinski, 1992, pp. 81–83), there is no need to introduce a principal fibre bundle $P$ to define a gauge connection, a choice of gauge, or a gauge transformation. Instead, one deals with a vector bundle $\delta$ equipped with a $G$-structure, the *interaction bundle*. One introduces a bundle $G(\delta)$ of automorphisms of each fibre of $\delta$, and a bundle $\mathfrak{g}(\delta)$ of endomorphisms of each fibre of $\delta$. A cross-section of $G(\delta)$ specifies an automorphism of each fibre of $\delta$, and a cross-section of $\mathfrak{g}(\delta)$ specifies an endomorphism of each fibre of $\delta$. Given the $G$-structure in each fibre of $\delta$, typically a Hermitian inner product, perhaps in tandem with a volume form, an automorphism or endomorphism of each fibre $\delta_x$ is a mapping which preserves this structure.

Each fibre of $G(\delta)$ is a Lie group, and each fibre of $\mathfrak{g}(\delta)$ is a Lie algebra. $G(\delta)$ is said to be a Lie group bundle, and $\mathfrak{g}(\delta)$ is said to be a Lie algebra bundle. Each fibre of $G(\delta)$ is isomorphic to the matrix Lie group $G \subset GL(n, \mathbb{C})$, and each fibre of $\mathfrak{g}(\delta)$ is isomorphic to the matrix Lie algebra $\mathfrak{g} \subset \mathfrak{gl}(n, \mathbb{C})$, but the isomorphisms are not canonical. It is necessary to fix a basis in a fibre $\delta_x$ to establish an isomorphism between $G(\delta)_x$ and $G \subset GL(n, \mathbb{C})$. Similarly, it is necessary to fix a basis in a fibre $\delta_x$ to establish an isomorphism between $\mathfrak{g}(\delta)_x$ and $\mathfrak{g} \subset \mathfrak{gl}(n, \mathbb{C})$.

$G(\delta)$ and $\mathfrak{g}(\delta)$ are both isomorphic to bundles associated to the principal fibre bundle $P$:

$$G(\delta) \cong P \times_G G,$$
$$\mathfrak{g}(\delta) \cong P \times_G \mathfrak{g}.$$

However, there is no need to define either $G(\delta)$ or $\mathfrak{g}(\delta)$ as such associated bundles.

Recall from Section 3.1 that for each interaction bundle $\delta$ equipped with a $G$-structure, there is a corresponding principal fibre bundle $P$, with structure group $G$, whose fibres consist of the bases selected by the $G$-structure. A connection on $\delta$ consists of the selection of a horizontal subspace in each fibre of the tangent bundle $T\delta$, just as a connection on $P$ consists of the selection of a horizontal subspace in each fibre of the tangent bundle $TP$. Recall that a connection upon a principal fibre bundle determines a connection upon any associated bundle, hence a connection on $P$ determines a connection on $\delta \cong P \times_G \mathbb{C}^n$. One can define a $G$-connection on $\delta$ to be a connection on $\delta$ which is induced by a connection upon the principal fibre bundle $P$, where $P$ is the principal fibre bundle that corresponds to the $G$-structure in $\delta$. Alternatively, one can define a $G$-connection on $\delta$ to be a connection which is such that the objects defining the $G$-structure are parallel with respect to the covariant derivative of the connection.

A connection on a vector bundle $\delta$ differs from a connection on a principal fibre bundle $P$ in the sense that only a connection upon the principal fibre bundle can be specified by a Lie-algebra valued one-form upon the total space, $\omega : TP \rightarrow \mathfrak{g}$, a cross-section of $T^*P \otimes (P \times \mathfrak{g})$. As explained in Section 3.3, such a Lie-algebra valued one-form must satisfy conditions with respect to the right action of $G$ upon the fibres of $P$, to constitute a connection one-form. There is no right action of $G$ upon the fibres of $\delta$, and a $G$-connection on $\delta$ does not correspond to a $\mathfrak{g}$-valued one-form, a cross-section of $T^*\delta \otimes (\delta \times \mathfrak{g})$.

The space of connections on $P$, and the space of $G$-connections on $\delta$ are both affine spaces.[10] The translation space of the $G$-connections on $\delta$ is the space of cross-sections of $T^*\mathcal{M} \otimes \mathfrak{g}(\delta)$. If we consider the space of $G$-connections on $\delta$ as the cross-sections of an affine bundle $\mathscr{C}(\delta)$, the translation space bundle of $\mathscr{C}(\delta)$ is $T^*\mathcal{M} \otimes \mathfrak{g}(\delta)$.

A choice of gauge corresponds to a cross-section $\sigma$ of the principal fibre bundle $P$, otherwise thought of as a trivializing $n$-tuple of sections $(\psi_1, \dots, \psi_n)$ of $\delta$ respecting the $G$-structure. E.g. if the $G$-structure consists of an inner product, then the trivializing sections should be orthonormal. A choice of gauge does two things:

- It selects a base connection $\omega_0$, and thereby renders the space of $G$-connections on $\delta$ canonically isomorphic to its translation space, the space of cross-sections of $T^*\mathcal{M} \otimes \mathfrak{g}(\delta)$ (Derdzinski, 1992, p. 91).
- It renders $(P \times_G \mathfrak{g}) \cong \mathfrak{g}(\delta)$ canonically isomorphic with $\mathcal{M} \times \mathfrak{g}$.

In sum, a choice of gauge renders the space of connections canonically isomorphic with $T^*\mathcal{M} \otimes (\mathcal{M} \times \mathfrak{g})$. In other words, a choice of gauge enables one to treat a $G$-connection on $\delta$ as a Lie-algebra valued one-form on the base space $\mathcal{M}$.

To see why a choice of gauge selects a base connection $\omega_0$, think of the choice of gauge $\sigma$ as selecting a principal sub-bundle of $P$ with structure group $\{Id\}$. I.e., a principal fibre bundle for which each fibre consists of a single element, the basis selected by $\sigma = (\psi_1, \dots, \psi_n)$. This principal fibre bundle $P_\sigma$ has a unique connection. The tangent vector space at each point of the total space has the same dimension as the base space. The vertical subspace is the zero vector, and there is only one choice for the horizontal subspace, the entire tangent vector space. Given that the Lie algebra of $\{Id\}$ is $\{Id\}$, the Lie-algebra valued one-form which specifies this unique connection maps the entire tangent vector space to $\{Id\}$. The vertical subspace (the zero vector) is mapped to the Lie-algebra $\{Id\}$, as required, and the horizontal subspace (the whole of the tangent space) must belong to the kernel, and is therefore also mapped to $\{Id\}$, as required.

One can injectively map the sub-bundle $P_\sigma$ back into the principal fibre bundle $P$ corresponding to the $G$-structure in $\delta$. Under the differential map of this injection, the images of the horizontal subspaces on $P_\sigma$ provide horizontal subspaces $H_{\sigma(x)}$ on $P$ at each point in the codomain of the cross-section $\sigma$. Given $H_{\sigma(x)}$ one can then use

---

[10] Recall that an affine space is a set which is acted upon transitively and effectively by the additive group structure of a vector space, the so-called 'translation space.'

the right action of $G$ to define horizontal subspaces at all the other points of the fibres of $P$. One defines

$$H_{\sigma(x)g} = R_{g*}\big(H_{\sigma(x)}\big),$$

and this determines a connection upon $P$. This base connection on $P$ then induces a base $G$-connection on $\delta$.

Given a choice of gauge $\sigma$ which selects a base connection $\omega_0$ on $\delta$, there is a canonical isomorphism between the affine space of connections on $\delta$ and the translation space $\Gamma(T^*\mathcal{M} \otimes \mathfrak{g}(\delta))$. To each connection $\omega$, there is a unique $\tau \in \Gamma(T^*\mathcal{M} \otimes \mathfrak{g}(\delta))$ such that

$$\omega = \omega_0 + \tau.$$

The space $\Gamma(T^*\mathcal{M} \otimes \mathfrak{g}(\delta))$, with the group structure derived from its vector space structure, acts simply transitively upon the affine space of connections. Hence, the selection of a base connection determines a canonical isomorphism between the space of connections and $\Gamma(T^*\mathcal{M} \otimes \mathfrak{g}(\delta))$.

In the interaction bundle $\delta$ picture, a gauge transformation is a cross-section of $G(\delta)$. Hence, a gauge transformation selects, at each point $x$, an automorphism $\alpha_x$ of the fibre $\delta_x$. A gauge transformation is a bundle automorphism which respects the $G$-structure in each fibre. Recall that a cross-section of $P \times_G G$ defines an automorphism of each fibre of $\delta$, hence $G(\delta) \cong P \times_G G$.

A cross-section of $G(\delta)$ also acts upon the Lie algebra bundle of endomorphisms $\mathfrak{g}(\delta)$. At each point $x$, the automorphism $\alpha_x$ acts adjointly, as an inner automorphism upon $\mathfrak{g}(\delta)_x$, mapping an endomorphism $T$ to $\alpha_x T \alpha_x^{-1}$. A gauge transformation changes a cross-section of $\mathfrak{g}(\delta)$, hence it changes a cross-section of $T^*\mathcal{M} \otimes \mathfrak{g}(\delta)$.

As with the principal fibre bundle picture, a gauge transformation can be treated as changing the gauge connection, or as changing the choice of gauge. Treated as an automorphism of each fibre of $\delta$, a gauge transformation maps any chosen basis into another basis, thereby changing the choice of gauge. Treated as an inner automorphism of each fibre of $\mathfrak{g}(\delta)$, a gauge transformation changes the gauge connection.

Chapter 4

# Interactions

## 4.1. Interacting fields

Equipped with an understanding of free matter fields and free gauge fields, we now proceed to consider interacting fields. In particular, we need to consider matter fields interacting with gauge fields.

To recap: in the configuration space approach, a free particle of mass $m$ and spin $s$, represented by a cross-section $\phi$ of a spin-$s$ free-particle bundle $\eta$, must satisfy free field equations (Derdzinski, 1992, p. 84),

$$\mathscr{P}\big(x, \phi(x), (\nabla^\eta \phi)(x), (\nabla^{\eta^2} \phi)(x), \ldots\big) = 0,$$

where $\mathscr{P}$ indicates a polynomial expression, and the mass $m$ is a coefficient in the polynomial expression. $\nabla^\eta$ here is the Levi-Civita connection on $\eta$. Free gauge fields, represented by $G$-connections $\nabla^\delta$ on an interaction bundle $\delta$, must satisfy the free-field Yang–Mills equations (Derdzinski, 1992, p. 84),

$$\operatorname{div} R^{\nabla^\delta} = 0.$$

$R^{\nabla^\delta}$ is the curvature two-form of the connection $\nabla^\delta$. The $G$-connections on $\delta$ correspond to smooth cross-sections of an affine bundle $\mathscr{C}(\delta)$. The space of $G$-connections on $\delta$ which satisfy the free-field Yang–Mills equations correspond to a subspace of this cross-section space.

Interactions can take a number of different forms. For example, whilst a free field of mass $m$ and spin 0 can be represented by the solutions $\phi$ of the Klein–Gordon equation,

$$(\square + m^2)\phi(x) = 0,$$

a *self-interacting* field of mass $m$ and spin 0 can be represented by the solutions of an equation with the form

$$(\square + m^2)\phi(x) = -\mathscr{P}'\big(\phi(x)\big),$$

where $\mathscr{P}$ is a polynomial expression of some degree which specifies the self-interaction potential, and $\mathscr{P}'$ denotes the derivative of the potential.

Secondly, a boson field $\phi$ and a fermion field $\psi$ can interact with each other via a so-called *Yukawa interaction*. Whilst a free boson field can be represented by the solutions of a Klein–Gordon equation $(\Box + m^2)\phi(x) = 0$, and a free fermion field can be represented by the solutions of a Dirac equation $(\gamma^\mu \partial_\mu + m)\psi = 0$, under the Yukawa interaction they must satisfy the coupled pair of equations,

$$(\Box + m^2)\phi(x) = -\tilde{\psi}\psi,$$
$$(\gamma^\mu \partial_\mu + m)\psi = \phi\psi,$$

where $\tilde{\psi} = -i\bar{\psi}\gamma^0$.

Thirdly, a matter field can interact with a gauge field. When such an interaction is 'switched on,' one must deal with pairs $(\psi, \nabla^\delta)$, where $\psi$ is a cross-section of an interacting-particle bundle $\alpha$, and $\nabla^\delta$ is a connection on the corresponding interaction bundle $\delta$ (Derdzinski, 1992, p. 84). One can think of such pairs as cross-sections of the direct sum $\alpha \oplus \mathscr{C}(\delta)$. A pair $(\psi, \nabla^\delta)$ must satisfy coupled field equations, consisting of (i) the interacting field equation upon the cross-sections $\psi$ of $\alpha$, and (ii) the coupled Yang–Mills equation upon the curvature $R^{\nabla^\delta}$ of the connection $\nabla^\delta$ on $\delta$:

$$\mathscr{P}\left(x, \psi(x), \left((\nabla^\eta \otimes \nabla^\delta)\psi\right)(x), \left((\nabla^\eta \otimes \nabla^\delta)^2\psi\right)(x), \ldots\right) = 0,$$
$$\operatorname{div} R^{\nabla^\delta} = C_0 J(\psi).$$

$C_0$ is a constant related to the so-called 'coupling constants' of the gauge field. These specify the strength of the interaction, and are determined by the choice of an adjoint-invariant metric in the Lie algebra $\mathfrak{g}$ (Derdzinski, 1992, pp. 114–115).[1] $J(\psi)$ is a 4-current on $\mathcal{M}$, a cross-section of $T^*\mathcal{M} \otimes \mathfrak{g}(\delta)$. The transition from $\eta$ to $\alpha$, and from the use of $\nabla^\eta$ in the free field equation, to the use of $(\nabla^\eta \otimes \nabla^\delta)$ in the interacting field equation, is often referred to as the 'minimal coupling substitution.'

These coupled equations are non-linear, entailing that the set of all pairs $(\psi, \nabla^\delta)$ which solve the coupled equations does not possess a linear vector space structure. Given a choice of gauge which renders $\mathscr{C}(\delta)$ canonically isomorphic with $T^*\mathcal{M} \otimes$

[1] In the case of a gauge field with a simple gauge group, there is a single degree of freedom in the choice of the adjoint-invariant metric upon the corresponding Lie algebra, hence there is a single coupling constant (see Appendix J for the definition of a simple Lie group). For a gauge field with a more general compact gauge group $G$, there is a coupling constant for every simple Lie algebra and every copy of $\mathfrak{u}(1)$ in a direct sum decomposition of the Lie algebra $\mathfrak{g}$. The gauge group of the electromagnetic field is $U(1)$, hence there is a single electromagnetic coupling constant, determined by $q$, the charge of the electron. The gauge group of the strong force, $SU(3)$, is simple, hence the strong force also has a single coupling constant, $g_s$. In the case of the electroweak force, with gauge group $U(2) \cong SU(2)_L \times U(1)_Y/\mathbb{Z}_2$, there are two coupling constants: $g$, the weak isospin coupling constant, associated with $SU(2)_L$, and $g'$, the weak hypercharge coupling constant, associated with $U(1)_Y$. Alternatively, one can specify the metric on $\mathfrak{u}(2)$ with a combination of the Weinberg angle $\theta_W$ (see Section 4.3), and the charge of the electron $q$. These parameters are related by the expressions $g = q/\sin\theta_W$ and $g' = q/\cos\theta_W$. The coupling constants are free parameters in the standard model, with values that need to be fixed by experiment and observation. Quantum field theory has added to this the notion of 'running coupling constants,' in which the coupling constants are functions of the energy at which an interaction takes place. The unification of different forces requires that their respective coupling constants converge at some energy scale.

$(\mathcal{M} \times \mathfrak{g})$, the set of all pairs $(\psi, \nabla^\delta)$ which solve the coupled equations constitutes a non-linear subset of $\Gamma(\alpha) \times \Gamma(T^*\mathcal{M} \otimes (\mathcal{M} \times \mathfrak{g}))$. The interacting field equation imposed upon the cross-sections $\psi$ of $\alpha$ is linear, but contains $\nabla^\delta$ within its very definition. If one fixes $\nabla^\delta$, then the space of cross-sections $\psi$ of $\alpha$ which solve the interacting field equation is a linear vector space, but this should not be considered as the state space for the interacting particle. Each different $\psi$ entails a different current $J(\psi)$, and a different $\nabla^\delta$ solving the coupled Yang–Mills equation div $R^{\nabla^\delta} = C_0 J(\psi)$ with respect to this current. This feeds back to the definition of the covariant derivative in the interacting-particle equation, hence one cannot treat the solutions $\psi$ and $\nabla^\delta$ separately.

## 4.2. Interaction symmetries

Whilst a free-particle corresponds to an irreducible representation of the local space–time symmetry group, a particle interacting with a gauge force field 'transforms under' both the local space–time symmetry group, and the infinite-dimensional group of gauge transformations. We shall try to explain and clarify this point from the perspective of Derdzinski's interacting-particle bundles.

Recall that in the case of the standard model over curved space–time, a gauge group $G$ does not, in general, determine a unique interaction bundle $\delta$. Hence, in general, a spin-$s$ particle interacting with a group-$G$ gauge field does not have a unique interacting-particle bundle, even if one assumes the simplest type of interacting-particle bundle $\eta \otimes \delta$. Instead, there might exist a family of interaction bundles $\delta_i$, and a corresponding family of interacting-particle bundles $\eta \otimes \delta_i$.

However, because the interaction bundles and spinor bundles over Minkowski space–time are trivial, and because a typical interacting-particle bundle is constructed from a tensor product and direct sum combination of interaction bundles and spinor bundles, it follows that, in the case of the standard model over Minkowski space–time at least, these interacting-particle bundles are trivial bundles.

Now, an interacting-particle bundle $\alpha$ inherits the gauge transformations possessed by the interaction bundle $\delta$. Moreover, the distinction between local and global gauge transformations which applies to the interaction bundle also applies to the interacting-particle bundle. As pointed out in Section 3.4, the distinction only has meaning for a product bundle, or within the domain of a local trivialization.

Given that each interaction bundle over Minkowski space–time is isomorphic to a product bundle, $\delta \cong \mathcal{M} \times \mathbb{C}^n$, each trivialization (i.e., choice of gauge), establishes an isomorphism between each fibre $\delta_x$ and the typical fibre $\mathbb{C}^n$. Hence, as noted in Section 3.4, given a representation of $G$ on $\mathbb{C}^n$, each trivialization induces a representation of $G$ upon the space of cross-sections $\Gamma(\delta)$. This, in turn, induces a representation of $G$ upon the space of cross-sections $\Gamma(\alpha)$ of the spin-$s$ interacting-particle bundle $\alpha = \eta \otimes \delta$. In this sense, one can say that a spin-$s$ interacting particle, represented by the cross-sections $\psi$ of $\alpha = \eta \otimes \delta$, transforms according to a representation of $G$. However, the action of $G$ here corresponds only to a global gauge transformation. The more general case of a local gauge transformation corresponds to a cross-section of $G(\delta)$, which selects a choice of gauge and an element of the matrix group $G$ at each point, and thereby specifies an automorphism of $\delta$. The (infinite-dimensional) group of all such automorphisms $\mathcal{G} = \Gamma(G(\delta))$ acts upon the space of sections $\Gamma(\delta)$, thence it acts upon the space of sections $\Gamma(\alpha)$. Hence, one can say that a spin-$s$ interacting particle transforms under the action of this infinite-dimensional group.

Note that $\mathcal{G}$ is an infinite-dimensional Lie group, and, as an infinite-dimensional manifold, it does not possess a locally compact topology. Hence, *a fortiori*, $\mathcal{G}$ is not a compact group. An interacting particle does not transform under an infinite-dimensional representation of a compact Lie group; rather, it transforms under the action of an infinite-dimensional Lie group upon an infinite-dimensional space.

Whilst a free particle in our universe corresponds to a unitary, irreducible representation of $SL(2,\mathbb{C}) \ltimes \mathbb{R}^{3,1}$, a particle interacting with a gauge field of gauge group $G$ does *not* correspond to a unitary, irreducible representation of $(SL(2,\mathbb{C}) \ltimes \mathbb{R}^{3,1}) \times G$. One could find, and classify, all the unitary, irreducible representations of $(SL(2,\mathbb{C}) \ltimes \mathbb{R}^{3,1}) \times G$, as an extension of the Wigner classification. Indeed, all the unitary, irreducible representations of compact groups are finite-dimensional, so one could set about taking all the tensor products of the unitary, irreducible, infinite-dimensional representations of $SL(2,\mathbb{C}) \ltimes \mathbb{R}^{3,1}$ with the unitary, irreducible, *finite*-dimensional representations of $G$, to obtain all the unitary, irreducible representations of $(SL(2,\mathbb{C}) \ltimes \mathbb{R}^{3,1}) \times G$.[2] These linear vector space representations, however, do not correspond with the state spaces of interacting particles, which are non-linear.

This non-linearity in the state space for an interacting particle, justifies the 'two-step' configuration space approach to specifying the state space of a particle. The two-step approach begins with the introduction of a vector bundle, and then specifies that the particle corresponds to a special set of cross-sections of this bundle. If one, alternatively, takes an exclusively Hilbert space approach to free particles, based upon the unitary, irreducible representations of $SL(2,\mathbb{C}) \ltimes \mathbb{R}^{3,1}$, then the transition to interacting particles is difficult to understand, given that they are not unitary, irreducible representations of anything.

To clarify further, an interacting particle $\psi$ in our universe does not transform under a representation of $SL(2,\mathbb{C}) \times G$, or a representation of $(SL(2,\mathbb{C}) \ltimes \mathbb{R}^{3,1}) \times G$. Rather, it transforms under a group action of $SL(2,\mathbb{C}) \ltimes \mathbb{R}^{3,1}$, and a group action of $\mathcal{G} = \Gamma(G(\delta))$. However, there *is* a representation of $SL(2,\mathbb{C}) \times G$ upon the typical fibre of the interacting-particle bundle $\eta \otimes \delta$. The typical fibre of the spin-$s$ free-particle bundle $\eta$ will carry a spin-$s$ representation of $SL(2,\mathbb{C})$, and the typical fibre of the interaction bundle $\delta$ will carry a representation of the gauge group $G$, so the

[2] Private communication with Heinrich Saller.

typical fibre of the tensor product $\eta \otimes \delta$ must carry a representation of the product group $SL(2, \mathbb{C}) \times G$.

Similarly, a gauge field connection pull-down $A$ in our universe does not transform under a representation of $SL(2, \mathbb{C}) \times G$ or a representation of $(SL(2, \mathbb{C}) \ltimes \mathbb{R}^{3,1}) \times G$. Rather, it transforms under a representation/group action of $SL(2, \mathbb{C}) \ltimes \mathbb{R}^{3,1}$, and a representation/group action of $\mathcal{G} = \Gamma(G(\delta))$. However, there *is* a representation of $SL(2, \mathbb{C}) \times G$ upon $\mathbb{R}^{3,1} \otimes \mathfrak{g}$, the typical fibre of the translation space bundle $T^*\mathcal{M} \otimes \mathfrak{g}(\delta)$.

A crucial difference between an interacting particle and a gauge field is that the representation of $SL(2, \mathbb{C}) \times G$ upon the typical fibre of the interacting-particle bundle $\eta \otimes \delta$ uses the standard representation of $G$, whilst the representation of $SL(2, \mathbb{C}) \times G$ upon $\mathbb{R}^{3,1} \otimes \mathfrak{g}$ uses the adjoint representation of $G$.[3]

Note also that there are two different group actions of $\mathcal{G}$ here. In the case of an interacting particle, there is an action of $\mathcal{G}$ upon $\Gamma(\delta)$, whilst in the case of the gauge field, there is an action of $\mathcal{G}$ upon $\Gamma(\mathfrak{g}(\delta))$.

Given the representation of $SL(2, \mathbb{C}) \times G$ upon $\mathbb{R}^{3,1} \otimes \mathfrak{g}$, the selection of a basis in $\mathfrak{g}$, or the restriction of the representation to $SL(2, \mathbb{C}) \times Id$, enables one to decompose this representation as a direct sum

$$\bigoplus^{\dim \mathfrak{g}} \mathbb{R}^{3,1}.$$

I.e., one decomposes the representation into a direct sum of $\dim \mathfrak{g}$ copies of the representation of $SL(2, \mathbb{C})$ on $\mathbb{R}^{3,1}$. This type of symmetry breaking is basically the way in which one obtains the distinct gauge bosons that correspond to a gauge field. Recall that a choice of gauge renders $\mathscr{C}(\delta)$, the affine bundle housing the $G$-connections on $\delta$, canonically isomorphic with $T^*\mathcal{M} \otimes (\mathcal{M} \times \mathfrak{g})$. The selection of a basis in $\mathfrak{g}$ then enables one to decompose $\mathcal{M} \times \mathfrak{g}$ as the direct sum

$$\bigoplus^{\dim \mathfrak{g}} (\mathcal{M} \times \mathbb{R}^1),$$

and thereby enables one to decompose $T^*\mathcal{M} \otimes (\mathcal{M} \times \mathfrak{g})$ as the direct sum (Derdzinski, 1992, p. 91):

$$\bigoplus^{\dim \mathfrak{g}} T^*\mathcal{M}.$$

[3] Each group element $g \in G$ corresponds, by its conjugation action, to an automorphism $f^g : G \to G$ of the group:

$$f^g(h) = ghg^{-1}.$$

This automorphism has a differential map $f^g_* : TG \to TG$. In the adjoint representation of a group $G$, each $g \in G$ is mapped to $f^g_*|_e$, the restriction of this differential map to the tangent space at the identity, $T_eG \cong \mathfrak{g}$.

Given that $S_0^1 T^*\mathcal{M} = T^*\mathcal{M}$, this is the configuration space bundle for $\dim \mathfrak{g}$ 'real vector bosons,' neutral particles of spin 1 and parity $-1$.[4] Recall that a spin-$s$ configuration space bundle possesses, upon its typical fibre, either a complex, finite-dimensional, irreducible representation of $SL(2,\mathbb{C})$ from the $\mathscr{D}^{s_1,s_2}$ family, for $s = s_1 + s_2$, or a direct sum of such representations. Given that $T^*\mathcal{M}$ is a real vector bundle, it cannot possess upon its typical fibre a member of the $\mathscr{D}^{s_1,s_2}$ family of complex representations, but it does possess the real representation of $SL(2,\mathbb{C})$ which complexifies to the $\mathscr{D}^{1/2,1/2}$ representation. In this sense, $T^*\mathcal{M}$ is a spin-1 configuration space bundle.

The differential equations for a spin 1, parity $-1$ bundle (Derdzinski, 1992, p. 19), consist of the Klein–Gordon equation,

$$\Box\psi = m^2\psi,$$

and the divergence condition,

$$\operatorname{div}\psi = 0.$$

Under a choice of gauge, and the selection of a Lie algebra basis, the cross-sections of the affine bundle $\mathscr{C}(\delta) \cong \bigoplus^{\dim \mathfrak{g}} T^*\mathcal{M}$, which satisfy the free-field Yang–Mills equations, correspond to the mass 0 solutions of these two equations. This is easiest to see in the case of electromagnetism, where a choice of gauge selects an isomorphism $\mathscr{C}(\lambda) \cong T^*\mathcal{M}$ which maps a connection $\nabla$ to a real vector potential $A$. With the Lorentz choice of gauge, the Maxwell equations upon a real vector potential,

$$\Box A = 0, \qquad \operatorname{div} A = 0,$$

clearly correspond to the differential equations for a spin 1, parity $-1$ particle of mass 0. Hence, under a choice of gauge, from the space of $U(1)$ connections satisfying the free-field Maxwell equations, one can construct a space which is the inverse Fourier transform of the space of single photon spates $\Gamma_{L^2}(E^+_{0,1})$ in the Wigner representation (see Appendix G).

In our universe the 'gauge bosons,' or 'interaction carriers' of a gauge field are the spin 1, mass 0, Wigner-representations of $SL(2,\mathbb{C}) \ltimes \mathbb{R}^{3,1}$, which inverse Fourier transform into spaces constructed from mass 0 cross-sections of spin 1 bundles such as $T^*\mathcal{M}$. These spin 1 bundles belong to a decomposition such as $\bigoplus^{\dim \mathfrak{g}} T^*\mathcal{M}$ of the affine bundle $\mathscr{C}(\delta)$ housing the $G$-connections on $\delta$. A choice of gauge renders the affine bundle $\mathscr{C}(\delta)$ isomorphic to the translation space bundle $T^*\mathcal{M} \otimes \mathfrak{g}(\delta)$, and a choice of Lie algebra basis then enables one to decompose the translation space bundle into separate interaction carrier bundles. This type of decomposition is said to be obtained by 'formal' symmetry breaking because it doesn't correspond to a phys-

[4] See Section 2.7 for a definition of $S_0^k T^*\mathcal{M}$.

ical process. We will presently see that for gauge fields which undergo spontaneous symmetry breaking, the decomposition changes slightly from $\bigoplus^{\dim \mathfrak{g}} T^*\mathcal{M}$. Either way, one can refer to the bundle $T^*\mathcal{M} \otimes \mathfrak{g}(\delta)$ as the interaction carrier bundle for the gauge field.

The dimension of the gauge group clearly corresponds to the number of summands in a decomposition such as $\bigoplus^{\dim \mathfrak{g}} T^*\mathcal{M}$, hence the dimension of the gauge group equals the number of interaction carriers associated with the gauge field in question. In the case of the strong force, with $G = SU(3)$, one has $\dim SU(3) = 8$, therefore one has 8 strong force interaction carriers; namely, the gluons, electrically neutral bosons of zero mass and spin 1. In the case of the electroweak force, with $G = U(2)$, one has $\dim U(2) = 4$, therefore one has 4 interaction carriers: the photon $\gamma$, the $W^{\pm}$ particles, and the $Z^0$ particle.

Note that whilst the interaction carriers can be defined by irreducible representations of $SL(2,\mathbb{C}) \ltimes \mathbb{R}^{3,1}$ alone in the Wigner representation, cross-sections of the bundle $T^*\mathcal{M} \otimes \mathfrak{g}(\delta)$ transform under both $SL(2,\mathbb{C}) \ltimes \mathbb{R}^{3,1}$ and $\mathcal{G}$. This tallies with the fact that the space of single-photon states in the Wigner representation is the Fourier transform of a space of $U(1)$-connections in the configuration representation *modulo gauge transformations* (see Appendix G). Gauge bosons in the Wigner representation do not transform under the group of gauge transformations. Note also that it is only under symmetry breaking that $T^*\mathcal{M} \otimes \mathfrak{g}(\delta)$ breaks into a direct sum of bundles housing the inverse Fourier transforms of the Wigner representations.

Mark that there is some distortion of meaning when people say that the interaction carriers of a gauge field 'belong to' the adjoint representation of the gauge group $G$. In the configuration representation, the interaction carriers of a gauge field belong to an infinite-dimensional representation of $\mathcal{G}$, which is certainly not the same thing as the finite-dimensional adjoint representation of $G$. To reiterate, it is the representation of $SL(2,\mathbb{C}) \times G$ upon the typical fibre of $T^*\mathcal{M} \otimes \mathfrak{g}(\delta)$ which uses the finite-dimensional adjoint representation of $G$, tensored with a finite-dimensional representation of $SL(2,\mathbb{C})$ on $\mathbb{R}^{3,1}$.

Thus, in the case of the strong force, the gluons belong to an infinite-dimensional representation of $\mathcal{G} = \Gamma(SU(\rho))$. However, the representation of $SL(2,\mathbb{C}) \times SU(3)$ upon the typical fibre of the translation bundle $T^*\mathcal{M} \otimes \mathfrak{su}(\rho)$ *does* use the eight-dimensional adjoint representation of $SU(3)$, tensored with a finite-dimensional representation of $SL(2,\mathbb{C})$ on $\mathbb{R}^{3,1}$. In the case of the unified electroweak force, the interaction carriers belong to an infinite-dimensional representation of $\mathcal{G} = \Gamma(U(\iota))$. One has a representation of $SL(2,\mathbb{C}) \times U(2)$ upon the typical fibre of the translation bundle $T^*\mathcal{M} \otimes \mathfrak{u}(\iota)$, and this representation *does* use the four-dimensional adjoint representation of $U(2)$.

## 4.3. The electroweak gauge connection bundle

There is a significant difference between the strong force and the electroweak force. Whilst the affine bundle housing the strong force connections decomposes as

$$\mathscr{C}(\rho) \cong \bigoplus^{\dim \mathfrak{su}(3)} T^*\mathcal{M},$$

under *formal* symmetry breaking, the affine bundle housing the electroweak connections decomposes as

$$\mathscr{C}(\iota) \cong \mathscr{C}(\lambda) \oplus (T^*\mathcal{M} \otimes \lambda) \oplus T^*\mathcal{M},$$

under *spontaneous* symmetry breaking (Derdzinski, 1992, pp. 104–111).

Under formal symmetry breaking, the affine bundle $\mathscr{C}(\iota)$ decomposes into $T^*\mathcal{M} \oplus T^*\mathcal{M} \oplus T^*\mathcal{M} \oplus T^*\mathcal{M}$. A choice of gauge, i.e., a cross-section of $P_\iota$, has the dual effect of rendering $\mathscr{C}(\iota)$ canonically isomorphic with the translation bundle $T^*\mathcal{M} \otimes \mathfrak{u}(\iota)$, and rendering the bundle of skew-adjoint endomorphisms $\mathfrak{u}(\iota)$ canonically isomorphic with the product bundle $\mathcal{M} \times \mathfrak{u}(2)$. A choice of gauge therefore renders $\mathscr{C}(\iota)$ canonically isomorphic with $T^*\mathcal{M} \otimes (\mathcal{M} \times \mathfrak{u}(2))$. A choice of basis in the Lie algebra $\mathfrak{u}(2)$ renders $\mathcal{M} \times \mathfrak{u}(2)$ canonically isomorphic with $\oplus^4(\mathcal{M} \times \mathbb{R}^1)$. In turn, this renders $T^*\mathcal{M} \otimes (\mathcal{M} \times \mathfrak{u}(2))$ canonically isomorphic with $\oplus^4 T^*\mathcal{M}$.

Under spontaneous symmetry breaking, one obtains a different decomposition of $\mathscr{C}(\iota)$. Instead of using a choice of gauge and a choice of Lie algebra basis, one uses a fibre metric in $\mathfrak{u}(\iota)$ and a constant length cross-section $\psi_0$ of $\iota$ to select the decomposition (see Section 4.4 for the interpretation of this cross-section as a Higgs field).

Each fibre of the bundle of skew-adjoint endomorphisms $\mathfrak{u}(\iota)$ can be equipped with a positive-definite metric $\langle\, , \rangle_{p_0,q_0}$, the choice of which is determined by two positive real numbers $p_0, q_0$. These two numbers are related to the value of the Weinberg angle $\theta_W$, a free parameter in the standard model, by $\tan^2\theta_W = p_0/q_0$. The value of the Weinberg angle, supposedly fixed during spontaneous symmetry breaking in the early universe, is experimentally determined as (Derdzinski, 1992, p. 111):

$$\sin^2\theta_W = 0.234 \pm 0.013.$$

Given a choice of metric, a choice of $\psi_0$ decomposes each fibre of $\iota$ into $\mathbb{C}\psi_0 \oplus \psi_0^\perp$, where $\mathbb{C}\psi_0$ is the set of all complex scalar multiples of $\psi_0$, and $\psi_0^\perp$ is the set of vectors orthogonal to $\psi_0$. This, in turn, selects a sub-bundle $W(\iota)$ of $\mathfrak{u}(\iota)$ consisting of endomorphisms $a$ in each fibre which are such that $a\psi_0 \in \psi_0^\perp$ and $a(\psi_0^\perp) \subset \mathbb{C}\psi_0$. Each fibre of $W(\iota)$ is a 2-dimensional real vector space. The combined choice of a fibre metric and $\psi_0$ determines a decomposition of $\mathfrak{u}(\iota)$ into $W(\iota) \oplus W^\perp(\iota)$.

If the fibre metric $\langle\, , \rangle_{p_0,q_0}$ in $\mathfrak{u}(\iota)$ is fixed, and the choice of $\psi_0$ is fixed, the direct sum decomposition of $\mathfrak{u}(\iota)$ into $W(\iota) \oplus W^\perp(\iota)$ is orthogonal, but no finer decomposition is determined. To obtain a decomposition which is consistent with the four observed interaction carriers of the *broken* electroweak force, one must use empirical considerations to select an orthogonal decomposition of $W^{\perp}(\iota)$ into a pair of real line bundles. One defines $\gamma(\iota)$ as the endomorphisms in each fibre of $\iota$ which are such that $a\psi_0 = 0$, and one defines $Z(\iota) = \gamma^{\perp}(\iota)$. This obtains the following orthogonal direct sum decomposition of $\mathfrak{u}(\iota)$:

$$\mathfrak{u}(\iota) = \gamma(\iota) \oplus W(\iota) \oplus Z(\iota).$$

It follows that the translation bundle $T^*\mathcal{M} \otimes \mathfrak{u}(\iota)$ decomposes as

$$T^*\mathcal{M} \otimes \mathfrak{u}(\iota) = \big(T^*\mathcal{M} \otimes \gamma(\iota)\big) \oplus \big(T^*\mathcal{M} \otimes W(\iota)\big) \oplus \big(T^*\mathcal{M} \otimes Z(\iota)\big).$$

Now, the cross-section $\psi_0$ selects an affine sub-bundle $\mathscr{C}_{\psi_0}(\iota)$ consisting of all the $U(2)$-connections on $\iota$ which make $\psi_0$ parallel. This affine bundle has $T^*\mathcal{M} \otimes \gamma(\iota)$ as its translation space bundle. Because the translation bundle of the affine bundle $\mathscr{C}_{\psi_0}(\iota)$ is $T^*\mathcal{M} \otimes \gamma(\iota)$, the translation bundle of the following affine bundle

$$\mathscr{C}_{\psi_0}(\iota) \oplus \big(T^*\mathcal{M} \otimes W(\iota)\big) \oplus \big(T^*\mathcal{M} \otimes Z(\iota)\big),$$

is

$$\big(T^*\mathcal{M} \otimes \gamma(\iota)\big) \oplus \big(T^*\mathcal{M} \otimes W(\iota)\big) \oplus \big(T^*\mathcal{M} \otimes Z(\iota)\big).$$

This is simply the translation bundle $T^*\mathcal{M} \otimes \mathfrak{u}(\iota)$ of $\mathscr{C}(\iota)$ under the orthogonal direct sum decomposition obtained above. Hence, the affine bundle $\mathscr{C}_{\psi_0}(\iota) \oplus (T^*\mathcal{M} \otimes W(\iota)) \oplus (T^*\mathcal{M} \otimes Z(\iota))$ and the affine bundle $\mathscr{C}(\iota)$ possess the same translation bundle. Given that an affine bundle can be rendered isomorphic with its translation bundle, affine bundles with isomorphic translation bundles must be isomorphic affine bundles. Hence, we have obtained an orthogonal affine bundle decomposition:

$$\mathscr{C}(\iota) \cong \mathscr{C}_{\psi_0}(\iota) \oplus \big(T^*\mathcal{M} \otimes W(\iota)\big) \oplus \big(T^*\mathcal{M} \otimes Z(\iota)\big).$$

To complete this spontaneous symmetry breaking decomposition, one must note some further isomorphisms. One has the affine bundle isomorphism $\mathscr{C}_{\psi_0}(\iota) \cong \mathscr{C}(\lambda)$, obtained by defining $\lambda = \psi_0^{\perp}$, and by restricting to $\lambda = \psi_0^{\perp}$ those connections on $\iota$ which make $\psi_0$ parallel. One also has the isomorphisms $W(\iota) \cong \lambda$ and $Z(\iota) \cong (\mathcal{M} \times \mathbb{R})$, from which one obtains the final decomposition:

$$\mathscr{C}(\iota) \cong \mathscr{C}(\lambda) \oplus (T^*\mathcal{M} \otimes \lambda) \oplus T^*\mathcal{M}.$$

The $T^*\mathcal{M}$ summand corresponds to the $Z^0$ particle, a strictly neutral, spin 1 particle, but $T^*\mathcal{M} \otimes \lambda$ is the interacting-particle bundle for $W^{\pm}$, a spin 1 particle with the charge of an electron/positron. $T^*\mathcal{M} \otimes \lambda$ is isomorphic to the complexification $T^*\mathcal{M} \otimes \mathbb{C}$, which possesses the $(1/2, 1/2)$ spin-1 finite-dimensional irreducible representation of $SL(2, \mathbb{C})$ upon its typical fibre $\mathbb{R}^4 \otimes \mathbb{C} \cong \mathbb{C}^4 \cong \mathbb{C}^2 \otimes \mathbb{C}^2$. As before, the affine bundle $\mathscr{C}(\lambda) \cong T^*\mathcal{M}$ represents the photon $\gamma$.

## 4.4. The Higgs field

Mathematically, spontaneous symmetry breaking (SSB) is simply a special case of bundle reduction. Given the electroweak interaction bundle $\iota$, equipped with a $U(2)$-structure, SSB obtains the electromagnetism bundle $\lambda$ as a sub-bundle equipped with a $U(1)$-structure. Given the electroweak principal fibre bundle $P_\iota$, consisting of all the orthonormal bases in all the fibres of $\iota$, SSB obtains the electromagnetism principal fibre bundle $P_\lambda$ as a principal fibre sub-bundle of $P_\iota$. There are many ways in which $\lambda$ can be embedded in $\iota$, and there are many ways in which $P_\lambda$ can be embedded in $P_\iota$. Correspondingly, one says that there are many ways in which one can reduce the $U(2)$-symmetry to a $U(1)$ symmetry. Spontaneous symmetry breaking selects one particular embedding, one particular bundle reduction.

One can select a particular reduction of $\iota$ by selecting a constant length cross-section $\phi$ of $\iota$, and by defining $\lambda = \phi^\perp$. As alluded to in Section 4.3, physicists explain spontaneous symmetry breaking as a physical process by proposing the existence of something called the Higgs field. They propose that the Higgs field does not have a single possible vacuum state, represented by the zero cross-section, but many possible vacuum states, each represented by a non-zero cross-section. They also propose that nature has randomly selected one particular Higgs field vacuum state, and that this vacuum state is the cross-section $\phi$ which breaks the electroweak symmetry.

The free-particle bundle of the Higgs field is $\eta = \mathcal{M} \times \mathbb{R}$ (Derdzinski, 1992, pp. 186–187). When the electroweak force is 'switched on,' the interacting Higgs field bundle is $\eta \otimes \iota = \iota$. Thus, any state of the electroweak-interacting Higgs field corresponds to a cross-section of the electroweak interaction bundle $\iota$ itself. In particular, a vacuum state of the electroweak-interacting Higgs field corresponds to a cross-section of the electroweak interaction bundle $\iota$. Hence, it is the choice of $\eta = \mathcal{M} \times \mathbb{R}$ as the free-particle bundle for the Higgs field which enables certain states of the Higgs field to determine bundle reductions of $\iota$. The choice of $\eta = \mathcal{M} \times \mathbb{R}$ as the free-particle bundle entails that the particles associated with the Higgs field are neutral particles of zero spin. Note, however, that the Higgs boson has yet to be empirically detected.

For a *vacuum* state of the electroweak-interacting Higgs field to perform the role of symmetry breaker, it is necessary that there are non-zero vacuum states, and this requires the Lagrangian of the electroweak-interacting Higgs field to assume a specific type of form. In particular, it is necessary to treat the Higgs field as a self-interacting field in which the potential $\mathcal{V} : \iota \to \mathbb{R}$ assumes the following form (Derdzinski, 1992, p. 186, Eq. (11.19)):

$$\mathcal{V}(\phi) = f\big(\langle\phi, \phi\rangle\big) = a\langle\phi, \phi\rangle^2 - b\langle\phi, \phi\rangle,$$

with $a, b > 0$. This corresponds to the requirement that the Higgs field satisfy the following non-linear equation:

$$\Box\phi + 2b\phi = 4a\phi^3.$$

The function $f : [0, \infty) \to \mathbb{R}$ has a non-zero minimum at $z_0 = b/2a$. Hence, in each fibre $\iota_x$, the set of vectors which are such that $\langle \phi, \phi \rangle = b/2a$ will minimize the potential. The specific expression for $f$, determined by fixing values for $a$ and $b$, will determine the particular value of the minimum $b/2a$, but the form of $f$ is determined by the constraint that $f$ must possess a unique minimum point in its domain. The values of $a$ and $b$ are free parameters in the standard model, which have to be fixed by experiment and observation.

If a potential is defined upon a vector bundle $\eta$ of unspecified fibre dimension, then one can obtain a direct sum decomposition of $\eta$ in which each summand corresponds to a particle of a specific mass (Derdzinski, 1992, pp. 181–182). If the potential has the form specified above, then there are three possible cases (Derdzinski, 1992, pp. 184–185):

1. $z_0 f''(z_0) = 0$. In this case, there is one summand, corresponding to a massless particle unless $z_0 = 0$.
2. $z_0 f''(z_0) \neq 0$, and the real fibre dimension of $\eta$ is greater than one. In this case, there are two summands, one of which corresponds to a massive particle, and one of which corresponds to a massless particle.
3. $z_0 f''(z_0) \neq 0$, and $\eta$ is a real line bundle. In this case there is one summand, corresponding to a massive particle.

The first two cases predict massless particles called Goldstone bosons, and are therefore neglected. The third case is the case of the Higgs field, with $\eta = \mathcal{M} \times \mathbb{R}$.

A potential function on an interacting-particle bundle is invariant under the action of the group of gauge transformations. Hence, the electroweak-interacting Higgs field potential $\mathcal{V} : \iota \to \mathbb{R}$ is invariant under the action of $\mathcal{G}(\iota)$. A gauge transformation $g(x) \in \mathcal{G}(\iota)$ selects an element of $U(\iota_x) \cong U(2)$ in each fibre $\iota_x$. If $\phi_0(x)$ is a cross-section of $\iota$ which minimizes the potential in each fibre, then $g(x)\phi_0(x)$ must also minimize the potential in each fibre. In each fibre $\iota_x$, the set of vectors which are such that $\langle \phi, \phi \rangle = b/2a$ will minimize the potential. Given any vector $v_0 \in \iota_x$ which minimizes the potential, then all the points of $U(\iota_x)v_0$, the orbit of $v_0$ under the action of $U(\iota_x)$, will also minimize the potential. By definition, the automorphism group $U(\iota_x)$ preserves the inner product, hence if $\langle v_0, v_0 \rangle = b/2a$, then $\langle gv_0, gv_0 \rangle = b/2a$, for all $g \in U(\iota_x)$. The isotropy group $U_0(\iota_x)$ of a specific $v_0$, the sub-group of $U(\iota_x) \cong U(2)$ which leaves $v_0$ fixed, is isomorphic to $U(1)$, hence the vacuum orbit in each fibre is isomorphic to $U(\iota_x)/U_0(\iota_x) \cong U(2)/U(1)$. Treating $\iota_x$ as a 4-dimensional real space, the vacuum orbit is a 3-sphere of radius $\|v_0\| = \sqrt{\langle v_0, v_0 \rangle} = \sqrt{b/2a}$.

Under the Higgs mechanism, then, symmetry breaking in $\iota$ is determined by a cross-section $\phi_0$ of $\iota$ of constant length $\sqrt{b/2a}$. Such a cross-section picks out an element from the vacuum orbit in each fibre. One then defines the electromagnetism sub-bundle to be $\lambda = \phi_0^\perp$. The set of all the unit vectors in all the fibres of $\phi_0^\perp$ provides $\phi_0^\perp$ with a $U(1)$-structure, corresponding to the selection of $P_\lambda$, a $U(1)$-principal fibre sub-bundle of $P_\iota$.

A general Higgs field $\phi$ is often represented in the physics literature as a two component field

$$\phi = \begin{pmatrix} \phi^+ \\ \phi^0 \end{pmatrix}.$$

One can treat $\phi^0$ as the component of $\phi$ belonging the one-dimensional subspace spanned by $\phi_0$, and one can treat $\phi^+$ as the component of $\phi$ in the subspace $\lambda = \phi_0^\perp$. The particles corresponding to $\phi^0$ are Higgs bosons of zero electric charge, and the particles corresponding to $\phi^+$ are Higgs bosons of unit electric charge. The constant length $\sqrt{b/2a}$ of a vacuum state cross-section in the first-quantized theory is considered to equal the vacuum state expectation value of the second-quantization of $\phi^0$. Thus, it is written that $\langle \phi^0 \rangle = v/\sqrt{2}$, where $v = \sqrt{b/a}$, and where $\langle \phi^0 \rangle := \langle \Omega_\phi | \hat{\phi}^0 \Omega_\phi \rangle$, the expectation value of the field operator $\hat{\phi}^0$ in the Higgs field vacuum state $\Omega_\phi$.

As we shall see in Section 4.6, the Higgs mechanism is also considered to be responsible for the non-zero masses of the quarks, the leptons, and the interaction carriers of the weak force (see Derdzinski, 1992, Section 11.5).

The electroweak-unified standard model represents the left-handed electron and (left-handed) neutrino to be merely different states of a single particle type. Whilst weak interactions enable a left-handed electron to transform into a left-handed neutrino, and vice versa, under electroweak unification no transformation of particle type is involved. It is, supposedly, only because of spontaneous symmetry breaking that the electron acquires a mass much greater than that of the neutrino, and prior to symmetry breaking, there is no difference between the charge of the left-handed electron and neutrino. It is, purportedly, the interactions between the electron and the Higgs boson which bestow the electron with mass. This, claims Smolin, entails that the mass of the electron is not an *intrinsic* property, merely a relationship the electron has with Higgs bosons (1997, p. 54). Under electroweak unification, there is no *intrinsic* difference between left-handed electrons and neutrinos. If, however, prior to symmetry breaking, $e_L$ and $\nu_e$ are simply two states of the same type of particle, which possess the same intrinsic properties, then in virtue of what do they interact in different ways with the Higgs boson? What distinguishes them as states and determines that they will interact in different ways with the Higgs boson, and thereby become different particle types under symmetry breaking? If it is not their intrinsic properties, then is it their extrinsic properties? If so, what types of extrinsic properties? This appears to be unexplained, or, at best, obscure.

## 4.5. Minimal coupling

Recall that when a matter field interacts with a gauge field, one must deal with pairs $(\psi, \nabla^\delta)$, where $\psi$ is a cross-section of an interacting-particle bundle $\alpha$, and $\nabla^\delta$ is a connection on the corresponding interaction bundle $\delta$ (Derdzinski, 1992, p. 84). The move from $\eta$ to $\alpha$, and the move from the use of $\nabla^\eta$ in the free field equation, to the use of $(\nabla^\eta \otimes \nabla^\delta)$ in the interacting field equation, is often referred to as the 'minimal coupling substitution.'

Recall that an *interacting* particle of mass $m$ and spin $s$ is represented by a mass-$m$ solution of a $\nabla^\delta$-dependent differential equation imposed upon the cross-sections of a spin-$s$ interacting-particle bundle $\alpha$. The connection $\nabla^\delta$ is a connection upon the interaction bundle $\delta$. A spin-$s$ interacting-particle bundle $\alpha$ is a construction from a spin-$s$ free-particle bundle $\eta$, and an interaction bundle $\delta$. In the simplest case, if the free-particle bundle is $\eta$, and the interaction bundle is $\delta$, then the interacting-particle bundle will be the tensor product $\alpha = \eta \otimes \delta$.

To make this notion more concrete, let us consider the case of an electron with the electromagnetic force 'switched on.' The free field bundle is the Dirac spinor bundle $\sigma$, the interaction bundle is the complex line bundle $\lambda$, and the interacting-particle bundle is $\sigma \otimes \lambda$. The minimal coupling substitution is the move from $\nabla^\sigma$ to $\nabla^\sigma \otimes \nabla^\lambda$, where $\nabla^\lambda$ is a $U(1)$-connection on $\lambda$. Recall that the free field equation for an electron is the Dirac equation, defined as

$$(\mathcal{D} + m_e c/\hbar)\psi = 0, \quad \psi \in \Gamma(\sigma),$$

where $m_e$ is the mass of the electron, and the Dirac operator on the spinor bundle $\sigma$ is defined by

$$\mathcal{D}\psi = c\big(\nabla^\sigma \psi\big), \quad \psi \in \Gamma(\sigma),$$

$c$ denoting Clifford multiplication.

To obtain the interacting field equation for an electron, we need to obtain a 'twisted' Dirac operator $\mathcal{D}^T$, defined upon cross-sections of $\sigma \otimes \lambda$. To achieve this, first we substitute $\nabla^\sigma \otimes \nabla^\lambda$ in the place of $\nabla^\sigma$, then we substitute $c \otimes Id_\lambda$ for the Clifford multiplication, to obtain:

$$\mathcal{D}^T\psi = (c \otimes Id_\lambda)\big(\nabla^\sigma \otimes \nabla^\lambda\big)\psi, \quad \psi \in \Gamma(\sigma \otimes \lambda).$$

On this occasion, the covariant derivative $(\nabla^\sigma \otimes \nabla^\lambda)$ is a map

$$\nabla^\sigma \otimes \nabla^\lambda : \Gamma(\sigma \otimes \lambda) \rightarrow \Gamma(T^*\mathcal{M} \otimes \sigma \otimes \lambda),$$

and the Clifford multiplication becomes a map

$$c \otimes Id_\lambda : \Gamma(T^*\mathcal{M} \otimes \sigma \otimes \lambda) \rightarrow \Gamma(\sigma \otimes \lambda),$$

hence the twisted Dirac operator is a map

$$\mathcal{D}^T : \Gamma(\sigma \otimes \lambda) \rightarrow \Gamma(\sigma \otimes \lambda).$$

The interacting field equation for an electron can then be defined as the twisted Dirac equation,

$$\big(\mathcal{D}^T + m_e c/\hbar\big)\psi = 0, \quad \psi \in \Gamma(\sigma \otimes \lambda).$$

To obtain the more familiar expression of this equation, first note that each nowhere-zero unit cross-section $\xi$ of $\lambda$ enables one to treat a $U(1)$-connection on $\lambda$ as a Lie-algebra valued one-form $A$ on the base space $\mathcal{M}$. Because $\lambda$ is a complex line bundle, a *nowhere-zero* cross-section of the bundle also selects a basis in each fibre. Recall that, in general, a choice of gauge is a cross-section of the principal fibre bundle corresponding to the $G$-structure in the interaction bundle. In the case of a complex line bundle, the general linear frame bundle consists of all the non-zero vectors in each fibre, and in the case of a complex line bundle equipped with a $U(1)$-structure, the corresponding principal fibre bundle is the orthonormal frame bundle, consisting of all the unit vectors in each fibre of $\lambda$. Hence, a unit cross-section $\xi$ of $\lambda$ provides a choice of gauge, and establishes a canonical isomorphism between the affine bundle $\mathscr{C}(\lambda)$ and the translation-space bundle $T^*\mathcal{M} \otimes (\mathcal{M} \times \mathfrak{u}(1))$. In addition, each unit cross-section $\xi$ of $\lambda$ establishes a correspondence $\phi \mapsto \phi \otimes \xi$ between the cross-sections $\phi$ of $\sigma$, and the cross-sections $\psi$ of $\sigma \otimes \lambda$.

The covariant derivative $\nabla^\lambda_v$ with respect to a vector field $v$ is such that (Derdzinski, 1992, p. 74),

$$\nabla^\lambda_v \xi = \frac{-iq}{\hbar} A(v)\xi,$$

where $q$ is the charge of the electron. Using a rule for the tensor product of covariant derivatives, it follows that (Derdzinski, 1992, p. 76),

$$\begin{aligned}
\left(\nabla^\sigma \otimes \nabla^\lambda\right)_v (\phi \otimes \xi) &= \nabla^\sigma_v \phi \otimes \xi + \phi \otimes \left(\nabla^\lambda_v \xi\right) \\
&= \nabla^\sigma_v \phi \otimes \xi + \phi \otimes (-iq/\hbar) A(v)\xi \\
&= \nabla^\sigma_v \phi \otimes \xi + (-iq/\hbar) A(v)\phi \otimes \xi \\
&= \left(\nabla^\sigma_v \phi - (iq/\hbar) A(v)\phi\right) \otimes \xi \\
&= \left(\nabla^\sigma_v - (iq/\hbar) A(v)\right)\phi \otimes \xi.
\end{aligned}$$

As a special case,

$$\left(\nabla^\sigma \otimes \nabla^\lambda\right)_{\partial/\partial x^\mu} (\phi \otimes \xi) = \left(\nabla^\sigma_{\partial/\partial x^\mu} - (iq/\hbar) A_\mu\right)\phi \otimes \xi,$$

and in the case of a flat spinor connection $\nabla^\sigma$,

$$\left(\nabla^\sigma \otimes \nabla^\lambda\right)_{\partial/\partial x^\mu} (\phi \otimes \xi) = \left(\partial/\partial x^\mu - (iq/\hbar) A_\mu\right)\phi \otimes \xi.$$

Hence, in the case of a flat spinor connection, and recalling from Section 2.8 the use of gamma matrices in Clifford multiplication, the twisted Dirac operator can be written as:

$$\begin{aligned}
\mathcal{D}^T(\phi \otimes \xi) &= (c \otimes Id_\lambda)\left(\nabla^\sigma \otimes \nabla^\lambda\right)\phi \otimes \xi \\
&= \gamma^\mu \left(\partial/\partial x^\mu - (iq/\hbar) A_\mu\right)\phi \otimes \xi.
\end{aligned}$$

Having fixed a cross-section $\xi$ of $\lambda$, one can treat the interacting field equation for an electron as the equation

$$\big(\gamma^{\mu}\big(\partial/\partial x^{\mu} - (iq/\hbar)A_{\mu}\big) + (m_e c/\hbar)\big)\phi = 0, \quad \phi \in \Gamma(\sigma),$$

which is the familiar form of the expression.

Note that in the physics literature, the covariant derivative $(\nabla^{\sigma} \otimes \nabla^{\lambda})_{\partial/\partial x^{\mu}}$ is often written as

$$\nabla_{\mu} = \partial_{\mu} - (iq/\hbar)A_{\mu},$$

where $\partial_{\mu}$ is an abbreviation for $\partial/\partial x^{\mu}$.

In the general case of an interacting-particle bundle $\eta \otimes \delta$, a gauge transformation $a(x) \in \Gamma(G(\delta))$ changes a cross-section $\psi = \phi \otimes \xi$ into $\psi' = \phi \otimes a(x)\xi$, and changes the covariant derivative from $\nabla^{\eta} \otimes \nabla^{\delta}$ to $\nabla^{\eta} \otimes \nabla'^{\delta}$. The latter has the effect of changing the differential equation which cross-sections of $\eta \otimes \delta$ must obey. The coupled field equations are invariant under gauge transformations in the sense that if $(\nabla^{\delta}, \psi)$ solve the coupled equations, then $(\nabla'^{\delta}, \psi')$ also solve the coupled equations. One can therefore speak of gauge equivalence classes of pairs $(\nabla^{\delta}, \psi)$ (Derdzinski, 1992, p. 92).

In the special case of an electron with the electromagnetic field 'switched on,' a gauge transformation is specified by a function $a(x) = e^{iq\theta(x)}$, where $q$ is again the charge of the electron. Such a gauge transformation changes a cross-section $\psi = \phi \otimes \xi$ of the interacting electron bundle $\sigma \otimes \lambda$ into $\psi' = \phi \otimes e^{iq\theta(x)}\xi = e^{iq\theta(x)}\phi \otimes \xi$, and changes the covariant derivative from $\nabla^{\sigma} \otimes \nabla^{\lambda}$ to $\nabla^{\sigma} \otimes \nabla'^{\lambda}$. This change in covariant derivative corresponds to a change from

$$\nabla_{\mu} = \partial_{\mu} - (iq/\hbar)A_{\mu},$$

to

$$\nabla'_{\mu} = \partial_{\mu} - (iq/\hbar)A'_{\mu},$$

where

$$A'_{\mu} = A_{\mu} - \partial_{\mu}\theta(x).$$

The change in covariant derivative from $\nabla^{\sigma} \otimes \nabla^{\lambda}$ to $\nabla^{\sigma} \otimes \nabla'^{\lambda}$ has the effect of changing the twisted Dirac operator $\mathcal{D}^{T}$ into $\mathcal{D}^{T'}$, thereby changing the twisted Dirac equation. The twisted Dirac equation is invariant under a gauge transformation in the sense that if

$$\mathcal{D}^{T}(\psi) = (m_e c/\hbar)\psi,$$

then

$$\mathcal{D}^{T'}(\psi') = (m_e c/\hbar)\psi'.$$

Moving away from the special case of electromagnetism, we can obtain general expressions for the covariant derivative under the minimal coupling substitution, and general expressions for the effect of a gauge transformation upon this covariant derivative. In general, under the minimal coupling substitution, the covariant derivative $\nabla^\eta \otimes \nabla^\delta$ is a map

$$\nabla^\eta \otimes \nabla^\delta : \Gamma(\eta \otimes \delta) \rightarrow \Gamma(T^*\mathcal{M} \otimes \eta \otimes \delta).$$

Assuming a flat Levi-Civita connection $\nabla^\eta$, one obtains the following general expression:

$$\left(\nabla^\eta \otimes \nabla^\delta\right)_{\partial/\partial x^\mu}(\phi \otimes \xi) = (\partial_\mu - igW_\mu)\phi \otimes \xi.$$

$W_\mu$ is the pull-down of a general gauge connection, $\phi \otimes \xi$ is a simple tensor cross-section of the interacting-particle bundle $\eta \otimes \delta$, and $g$ is a general coupling constant.[5] $W_\mu$ is a Lie-algebra valued one-form, just like $A_\mu$, but in the general case the Lie-algebra is multi-dimensional. As a consequence, $W_\mu$, unlike $A_\mu$, cannot necessarily be treated as a simple multiplication operator upon cross-sections of $\eta \otimes \delta$. Instead, one requires a representation of the Lie algebra $\mathfrak{g}$ upon each fibre of the interacting-particle bundle $\eta \otimes \delta$. With such a representation, $W_\mu$ can be treated as a linear operator on each fibre. I.e., $W_\mu$ can be treated as an endomorphism of each fibre. There is a representation of $\mathfrak{g}$ on the typical fibre of the associated bundle $\delta$, and a choice of gauge induces a representation of $\mathfrak{g}$ on each fibre of $\delta$, which, in turn, induces a representation on each fibre of the interacting-particle bundle $\eta \otimes \delta$.

If one chooses a basis $\{E_a\}$ for the Lie algebra $\mathfrak{g}$, then the expression for the covariant derivative can be written as

$$\left(\nabla^\eta \otimes \nabla^\delta\right)_{\partial/\partial x^\mu}(\phi \otimes \xi) = \left(\partial_\mu - igW_\mu^a E_a\right)\phi \otimes \xi.$$

Again, the representation on each fibre of $\eta \otimes \delta$ is implicit in this expression.

In the case of a general gauge group, if a choice of gauge has been selected, then a gauge transformation can be specified by a $G$-valued function on the base space:

$$a(x) = e^{ig(\chi^a(x)E_a)}.$$

This expression selects an element $\chi^a(x)E_a$ of the Lie algebra $\mathfrak{g}$ at each point, then uses the Lie exponential map to obtain an element of the Lie group $G$ at each point.[6] In general, this function selects an element of a multi-dimensional Lie group at each point, rather than a complex number of unit modulus.

[5] For simplicity of notation, we have suppressed $\hbar$ from this expression, and do so hereafter.

[6] Given that the Lie exponential map is surjective for a compact, connected Lie group (Simon, 1996, p. 166), there is no need to take a product of such functions to obtain an arbitrary element of the gauge group at each point.

Under such a gauge transformation, a particle cross-section $\psi$ changes into

$$\psi' = e^{ig(\chi^a(x)E_a)}\psi,$$

the gauge field pull-down $W_\mu^a$ changes into

$$W_\mu'^a = W_\mu^a - \partial_\mu \chi^a(x) - gC_{bc}^a \chi^b(x) W_\mu^c,$$

where $C_{bc}^a$ are the structure constants of the Lie algebra, and the covariant derivative $\nabla_\mu$ changes into

$$\nabla'_\mu = \partial_\mu - igW_\mu'^a E_a.$$

In terms of the specifics of the standard model, the covariant derivative for the electroweak gauge field is as follows:

$$\nabla_\mu^{\text{EW}} = \partial_\mu - ig\frac{\sigma_a}{2}W_\mu^a - ig'\frac{Y}{2}B_\mu.$$

$W_\mu$ here is the weak isospin $SU(2)_L$ gauge field and $B_\mu$ is the weak hypercharge $U(1)_Y$ gauge field (see Section 5.6 for an explanation of the role of these groups in the electroweak interaction). $\{\sigma^a\colon\ a = 1, 2, 3\}$ denotes the Pauli matrices and $\{i\sigma^a/2\colon\ a = 1, 2, 3\}$ provide a basis of the Lie algebra $\mathfrak{su}(2)$. $Y$ is the hypercharge generator of the one-dimensional Lie algebra $\mathfrak{u}(1)$. $g$ is the $SU(2)_L$ coupling constant and $g'$ is the $U(1)_Y$ coupling constant.

The covariant derivative of the entire standard model is:

$$\nabla_\mu^{\text{SM}} = \partial_\mu - ig_s\frac{\lambda_i}{2}G_\mu^i - ig\frac{\sigma_a}{2}W_\mu^a - ig'\frac{Y}{2}B_\mu.$$

$G_\mu$ here is the strong $SU(3)$ gauge field, $g_s$ is the coupling constant of the strong force, $\{\lambda_i\colon\ i = 1, \dots, 8\}$ are the self-adjoint, trace-free, complex $3 \times 3$ Gell-Mann matrices, and $\{i\lambda_i/2\colon\ i = 1, \dots, 8\}$ provide a basis of $\mathfrak{su}(3)$, the Lie algebra of *skew-adjoint*, trace-free, complex $3 \times 3$ matrices. The Gell-Mann matrices are as follows:

$$\lambda_1 = \begin{pmatrix} 0 & 1 & 0 \\ 1 & 0 & 0 \\ 0 & 0 & 0 \end{pmatrix}, \quad \lambda_2 = \begin{pmatrix} 0 & -i & 0 \\ i & 0 & 0 \\ 0 & 0 & 0 \end{pmatrix}, \quad \lambda_3 = \begin{pmatrix} 1 & 0 & 0 \\ 0 & -1 & 0 \\ 0 & 0 & 0 \end{pmatrix},$$

$$\lambda_4 = \begin{pmatrix} 0 & 0 & 1 \\ 0 & 0 & 0 \\ 1 & 0 & 0 \end{pmatrix}, \quad \lambda_5 = \begin{pmatrix} 0 & 0 & -i \\ 0 & 0 & 0 \\ i & 0 & 0 \end{pmatrix},$$

$$\lambda_6 = \begin{pmatrix} 0 & 0 & 0 \\ 0 & 0 & 1 \\ 0 & 1 & 0 \end{pmatrix}, \quad \lambda_7 = \begin{pmatrix} 0 & 0 & 0 \\ 0 & 0 & -i \\ 0 & i & 0 \end{pmatrix}, \quad \lambda_8 = \frac{1}{\sqrt{3}}\begin{pmatrix} 1 & 0 & 0 \\ 0 & 1 & 0 \\ 0 & 0 & -2 \end{pmatrix}.$$

## 4.6. The Lagrangian approach

The free field equations for a matter field, the free field Yang–Mills equations for a gauge field, and the coupled field equations for a matter field interacting with a gauge field, can all be derived from a Lagrangian (Derdzinski, 1992, Chapter 10). A Lagrangian in the vector bundle $\eta$ for a free matter field is a map from cross-sections $\psi$ to real-valued functions $\mathcal{L}(\psi) : \mathcal{M} \rightarrow \mathbb{R}$. A Lagrangian in the affine bundle $\mathscr{C}(\delta)$ for a free gauge field is a map from connections $\nabla$ to real-valued functions $\mathcal{L}(\nabla) : \mathcal{M} \rightarrow \mathbb{R}$. A Lagrangian in a bundle such as $(\eta \otimes \delta) \oplus \mathscr{C}(\delta)$ for a matter field interacting with a gauge field, is a map from each pair $(\psi, \nabla)$ to a real-valued function $\mathcal{L}(\psi, \nabla) : \mathcal{M} \rightarrow \mathbb{R}$. In each case, the corresponding field equations are derived from the principle that the solutions of the field equations should be extrema of the 'action,' $A = \int_{\mathcal{M}} \mathcal{L}\, d^4x$, the space–time integral of the Lagrangian. The solutions are extrema of the action in the sense that the 'variation' of the action vanishes, $\delta A = 0$. The equations so obtained are said to be the Euler–Lagrange equations, and they are said to be obtained by a variational principle.

Just as the field equations for a free gauge field, or the coupled field equations for a matter field interacting with a gauge field, are invariant under a gauge transformation, so the corresponding Lagrangians are also invariant under a gauge transformation. Thus, if $\nabla'$ is the gauge transformation of $\nabla$, then $\mathcal{L}(\nabla') = \mathcal{L}(\nabla)$. Similarly, if $(\psi', \nabla')$ are the gauge transforms of $(\psi, \nabla)$, then $\mathcal{L}(\psi', \nabla') = \mathcal{L}(\psi, \nabla)$. This is true for Abelian and non-Abelian gauge fields, despite the fact, alluded to in Section 3.5, that the curvature $R^{\nabla}$ of a non-Abelian gauge field changes under a gauge transformation. The reason is that the Lagrangians can be expressed in terms of inner products in the fibres of $\eta$, $\delta$, $\mathfrak{g}(\delta)$ and $T^*\mathcal{M}$. A gauge transformation $a(x) \in \mathcal{G} = \Gamma(G(\delta))$ provides an automorphism of each fibre of $\delta$, and the adjoint of the gauge transformation $Ad(a(x))$ provides an automorphism of each fibre of $\mathfrak{g}(\delta)$, hence, by definition, the respective inner products are preserved. Given a gauge transformation $a(x)$, $\nabla'\psi' = a(x)\nabla\psi$, and although $\nabla\psi \neq \nabla'\psi'$, it is true that

$$\langle \nabla'\psi', \nabla'\psi' \rangle = \langle a(x)\nabla\psi, a(x)\nabla\psi \rangle = \langle \nabla\psi, \nabla\psi \rangle,$$

and it is also true that

$$\langle R^{\nabla'}, R^{\nabla'} \rangle = \langle Ad(a(x)) R^{\nabla}, Ad(a(x)) R^{\nabla} \rangle = \langle R^{\nabla}, R^{\nabla} \rangle.$$

Lagrangians play a prominent role in most of the theoretical physics literature on the standard model. This is primarily because of their utility in second-quantized quantum field theory for obtaining interaction Hamiltonian density operators, and thence for obtaining scattering operators, as explained in Section 2.11. Expounding the numerical recipes and algorithms used by physicists to make the second-quantized standard model account for empirical particle phenomena is not the purpose of this text, hence Lagrangians take something of a back-seat here.

However, as an illustration of the Lagrangian technique, and as a further insight into the electroweak interaction, let us consider the Weinberg–Salam electroweak Lagrangian density for the first lepton generation (the electron and electron-neutrino $(e, \nu_e)$), and the first anti-lepton generation (the positron and positron-neutrino $(e^+, \bar{\nu}_e)$).

Begin by noting that, although the correct gauge group of the electroweak force is $U(2)$, physicists typically consider the electroweak gauge group to be $SU(2)_L \times U(1)_Y$, where $SU(2)_L$ is the gauge group of the 'weak isospin' gauge field, and $U(1)_Y$ is the gauge group of the 'weak hypercharge' gauge field.[7]

This electroweak Lagrangian is constructed from the following elements:

- The lepton fields, which can be partitioned into weak isospin $SU(2)_L$ 'doublets' and 'singlets.' In the context of the standard model, when physicists refer to a pair of particles as a doublet, or, more generally, when they refer to a collection of particles as a multiplet, they mean at least one of two things: (i) the particles in question can be transformed into each other by some interaction process, and (ii) the states of the particles in question belong to the same irreducible representation of a symmetry group. Singlets cannot be transformed into anything else via the interaction in question. The first lepton generation contains the right-handed electron $e_R$ as a weak isospin singlet, and the weak isospin doublet

$$\psi_L = \begin{pmatrix} \nu_L \\ e_L \end{pmatrix},$$

consisting of the (left-handed) electron-neutrino $\nu_L$ and the left-handed electron $e_L$. (For ease of notation here we have dropped the '$e$' subscript from the electron-neutrino.)
- The Higgs field $\phi$, represented as a weak isospin doublet

$$\phi = \begin{pmatrix} \phi^+ \\ \phi^0 \end{pmatrix}.$$

The anti-particle doublet to this is $\tilde{\phi} = i\sigma^2\bar{\phi}$,

$$\tilde{\phi} = \begin{pmatrix} \bar{\phi}^0 \\ -\bar{\phi}^+ \end{pmatrix},$$

where $\bar{\phi}$ here is simply the complex conjugate. In vector bundle terms the Higgs field is a cross-section of the electroweak interaction bundle $\iota$.
- The gauge field strength $W_{\mu\nu}$ of the weak isospin $SU(2)_L$ gauge field.
- The gauge field strength $B_{\mu\nu}$ of the weak hypercharge $U(1)_Y$ gauge field.

---

[7] See Section 5.6 for an explanation of the role of these groups in the electroweak interaction.

The Lagrangian constructed from these elements can be written as (Ticciati, 1999, p. 456)[8]:

$$\begin{aligned}\mathcal{L}_{\text{EW}} = &-\frac{1}{4}W_{\mu\nu}W^{\mu\nu} - \frac{1}{4}B_{\mu\nu}B^{\mu\nu} + \bar{\psi}_L i\gamma^\mu \nabla_\mu \psi_L \\ &+ \bar{e}_R i\sigma^\mu \nabla_\mu e_R + \overline{(\nabla_\mu \phi)}(\nabla^\mu \phi) \\ &- \bar{\lambda}_e \bar{\psi}_L \phi\, e_R - \lambda_e \bar{e}_R \tilde{\phi} \psi_L + a(\bar{\phi}\phi)^2 - b(\bar{\phi}\phi).\end{aligned}$$

The first thing to note here is a potentially misleading notational convention used by physicists[9]: $\bar{\psi}_L$ denotes not the complex conjugate, or the anti-particle doublet, but $\psi_L^* \gamma^0$, sometimes called the 'adjoint spinor,' or the 'Dirac adjoint.' If one is thinking in terms of a component expression, where $\psi_L$ is a column vector, then $\psi_L^*$ is the adjoint matrix — the transpose of the complex conjugate — and therefore a row vector. In component-independent terms, one can think of it as a dual vector with respect to an inner product in a spinor space. Similarly, $\bar{e}_R$ denotes $e_R^* \gamma^0$.

The terms in the Lagrangian can be grouped together into the following parts: a pure gauge part $-\frac{1}{4}W_{\mu\nu}W^{\mu\nu} - \frac{1}{4}B_{\mu\nu}B^{\mu\nu}$; the 'kinetic terms' for the first lepton/anti-lepton generation $\bar{\psi}_L i\gamma^\mu \nabla_\mu \psi_L + \bar{e}_R i\sigma^\mu \nabla_\mu e_R$; the Higgs field kinetic term $\overline{(\nabla_\mu \phi)}(\nabla^\mu \phi)$; the lepton–Higgs interaction term, referred to as the Yukawa coupling $-\bar{\lambda}_e \bar{\psi}_L \phi e_R - \lambda_e \bar{e}_R \tilde{\phi} \psi_L$, where $\lambda_e$ is the Yukawa coupling constant for the electron; and the Higgs self-interaction term $a(\bar{\phi}\phi)^2 - b(\bar{\phi}\phi)$. Note that the two Yukawa terms,

$$\begin{aligned}\bar{\lambda}_e \bar{\psi}_L \phi e_R &= \bar{\lambda}_e \left[\bar{\nu}_L \phi^+ + \bar{e}_L \phi^0\right] e_R, \\ \lambda_e \bar{e}_R \tilde{\phi} \psi_L &= \lambda_e \bar{e}_R \left[\bar{\phi}^0 \nu_L - \bar{\phi}^+ e_L\right],\end{aligned}$$

are considered to be Hermitian conjugates of each other.

The way in which physicists write this Lagrangian conceals the fact that the terms involve fibre-wise inner products on vector bundle cross-sections. Thus, one could write the Lagrangian as

$$\begin{aligned}\mathcal{L}_{\text{EW}} = &-\frac{1}{4}\langle W, W\rangle - \frac{1}{4}\langle B, B\rangle + \bar{\psi}_L i\gamma^\mu \nabla_\mu \psi_L + \bar{e}_R i\sigma^\mu \nabla_\mu e_R \\ &+ \big\langle (\nabla_\mu \phi), (\nabla_\mu \phi)\big\rangle - \bar{\lambda}_e \bar{\psi}_L \phi e_R - \lambda_e \bar{e}_R \tilde{\phi} \psi_L + a\langle \phi, \phi\rangle^2 - b\langle \phi, \phi\rangle.\end{aligned}$$

Note that $\sigma^\mu \nabla_\mu$ is the gauge covariant Dirac operator $\mathcal{D}^\nabla$ in a Weyl spinor bundle, and $\gamma^\mu \nabla_\mu$ is the gauge covariant Dirac operator in a Dirac spinor bundle, where, in this case, $\nabla$ is the electroweak gauge covariant derivative. (The 'twisted Dirac operator' of Section 4.5 is a gauge covariant Dirac operator in the case where $\nabla$ is the electromagnetic gauge covariant derivative.)

---

[8] Assuming, for simplicity, zero neutrino mass.

[9] Private communication with Maria Herrero.

To cast the electroweak Lagrangian into a form in which the physical implications can be directly read from it, one must first transform the gauge fields $W^a_\mu$ and $B_\mu$ into a form in which they can be interpreted as representing the photon and the $W^\pm$ and $Z$ gauge bosons. Doing so changes the expressions for the covariant derivatives in the kinetic terms of the Lagrangian, and this changes the kinetic terms into a form which admits a direct physical interpretation.

We begin by defining

$$W^+_\mu = \frac{W^1_\mu - iW^2_\mu}{\sqrt{2}} \quad \text{and} \quad W^-_\mu = \frac{W^1_\mu + iW^2_\mu}{\sqrt{2}}.$$

Next we define a photon field, which we shall denote as $A_\mu$, and a $Z$ gauge boson field $Z_\mu$:

$$A_\mu = \cos\theta_W B_\mu + \sin\theta_W W^3_\mu,$$
$$Z_\mu = -\sin\theta_W B_\mu + \cos\theta_W W^3_\mu,$$

where $\theta_W$ is the Weinberg angle. Substituting new expressions for the covariant derivative into the fermionic kinetic terms $\mathcal{L}_f = \bar{\psi}_L i\gamma^\mu \nabla_\mu \psi_L + \bar{e}_R i\sigma^\mu \nabla_\mu e_R$, one obtains an expression which indicates more explicitly the couplings between the first lepton/anti-lepton generation and the gauge bosons of the electroweak force (Ticciati, 1999, p. 460):

$$\begin{aligned}\mathcal{L}_f &= \mathcal{L}_{cc} + \mathcal{L}_{nc} + \mathcal{L}_{em}\\ &= \frac{q}{\sqrt{2}\sin\theta_W}\left(\bar{\nu}_L\sigma^\mu W^+_\mu e_L + \bar{e}_L\sigma^\mu W^-_\mu \nu_L\right)\\ &\quad + \frac{q}{\sin 2\theta_W}\left(\bar{\nu}_L\sigma^\mu Z_\mu \nu_L - \cos 2\theta_W \bar{e}_L\sigma^\mu Z_\mu e_L + 2\sin^2\theta_W \bar{e}_R\sigma^\mu Z_\mu e_R\right)\\ &\quad - q\left(\bar{e}_L\sigma^\mu A_\mu e_L + \bar{e}_R\sigma^\mu A_\mu e_R\right).\end{aligned}$$

The charge of the electron is denoted here as $q$, and the weak isospin coupling constant $g$ can be substituted into this expression via

$$g = \frac{q}{\sin\theta_W}.$$

As the notation indicates, the fermionic kinetic part of the electroweak Lagrangian can itself be split into three parts. Each part contains so-called ‘currents.’ The first part contains the weak charged-currents, $j^\mu = \bar{\nu}_L\sigma^\mu e_L$ and $\bar{j}^\mu = \bar{e}_L\sigma^\mu \nu_L$,

$$\mathcal{L}_{cc} = \frac{g}{\sqrt{2}}\left(j^\mu W^+_\mu + \bar{j}^\mu W^-_\mu\right).$$

The second part contains the weak neutral-current $j^\mu_{nc}$,

$$\mathcal{L}_{nc} = \frac{g}{\cos\theta_W} j^\mu_{nc} Z_\mu,$$

and the third part contains the electromagnetic current $j^\mu_{em} = -(\bar{e}_L \sigma^\mu e_L + \bar{e}_R \sigma^\mu e_R)$, $\mathcal{L}_{em} = q j^\mu_{em} A_\mu$.

Recall from Section 2.11, that under second-quantization one substitutes creation and annihilation operators into the expressions above to obtain field operators. Recall also that the expression for a field operator contains annihilation operators for the particle, and creation operators for the anti-particle, and the expression for the anti-field operator contains annihilation operators for the anti-particle and creation operators for the particle. Note that in the case of a spinor field, whilst the adjoint spinor $\bar{\psi} = \psi^* \gamma^0$ does not represent the anti-particle in the first-quantized theory, physicists do, at least heuristically, treat the second-quantization $\hat{\bar{\psi}}$ of this field as the anti-field operator.

Thus, the quantization of $e_L$ contains annihilation operators for left-handed electrons and creation operators for right-handed positrons. The quantization of $\bar{\nu}_e$ contains annihilation operators for right-handed positron-neutrinos and creation operators for left-handed electron-neutrinos. (Note that we have reverted to the '$e$' subscript to indicate the electron-neutrino.) Hence, one can say that the product in the weak charged-current $j^\mu = \bar{\nu}_e \sigma^\mu e_L$ contains (i) annihilation operators for left-handed electrons, and creation operators for left-handed electron neutrinos, plus (ii) annihilation operators for right-handed positron neutrinos and creation operators for right-handed positrons. The current $j^\mu$ couples with $W^+_\mu$, and under second quantization $W^+_\mu$ contains annihilation operators for $W^+$ bosons and creation operators for $W^-$ bosons. Hence, the $j^\mu W^+_\mu$ term is associated with weak-force interactions that involve either (i) $e_L \to \nu_e + W^-$ or (ii) $\bar{\nu}_e \to e^+_R + W^-$ transitions. $e_L$ possesses an electric charge of $-1$ and $\nu_e$ possesses zero electric charge, hence the transition $e_L \to \nu_e$ increases electric charge by $+1$, and needs to be balanced by the emission of the $W^-$, which possesses an electric charge of $-1$. Anti-particles, of course, possess opposite charges to their particle counterparts, hence $\bar{\nu}_e$ possesses zero electric charge and $e^+_R$ possesses an electric charge of $+1$, thus the transition $\bar{\nu}_e \to e^+_R$ also increases electric charge by $+1$, and is similarly balanced by the emission of a $W^-$.

Applying the same logic, the other weak charged-current $\bar{j}^\mu = e^+_R \sigma^\mu \nu_e$ contains (i) annihilation operators for left-handed electron-neutrinos, and creation operators for left-handed electrons, plus (ii) annihilation operators for right-handed positrons and creation operators for right-handed positron-neutrinos. The current $\bar{j}^\mu$ couples with $W^-_\mu$, which under second quantization contains annihilation operators for $W^-$ bosons and creation operators for $W^+$ bosons. Hence, the $\bar{j}^\mu W^-_\mu$ term is associated with weak force interactions that involve either (i) $\nu_e \to e_L + W^+$ or (ii) $e^+_R \to \bar{\nu}_e + W^+$ transitions. A transition such as $\nu_e \to e_L$ involves the creation of $-1$ electric charge, and is balanced by the creation of the $W^+$ particle, which possesses an electric charge of $+1$.

Thus, in summary, the left-handed leptons in the first lepton generation $(\nu_e, e_L)$ and the right-handed anti-leptons $(e^+_R, \bar{\nu}_e)$, are capable of transforming into each other under weak force processes. This permits interactions, called weak charged-current events, in which two weakly interacting particles respectively gain or lose a unit of electric charge by exchanging a charged $W$ gauge boson. Such interactions were predicted prior to the proposal of the unified electroweak theory, and were represented using the so-called V–A theory of weak interactions, which, in turn, was a development of Fermi's 1934 theory of weak interactions. The unified electroweak theory went beyond the V–A theory in predicting weak *neutral*-current events, interactions between weakly interacting particles mediated by the chargeless $Z$ boson. Weak charged-current events in neutrino-neutron scattering such as

$$\nu_\mu + n \rightarrow \mu^- + p,$$

and weak charged-current events in neutrino-electron scattering such as

$$\nu_e + e^- \rightarrow \nu_e + e^-,$$

were represented by the V–A theory, and, in the case of such neutrino-neutron scattering, were experimentally observed in the 1960s. Weak neutral-current events in neutrino-neutron scattering were not represented by the V–A theory, but were predicted by the electroweak theory. This includes reactions such as

$$\nu_\mu + N \rightarrow \nu_\mu + X,$$

where $N$ denotes a nucleon (either a proton $p$ or a neutron $n$), and $X$ denotes the hadron or spray of hadrons which result from the decay of the nucleon after the interaction. Weak neutral-current events in neutrino-electron scattering were also not represented by the V–A theory. For example, the following reaction

$$\nu_\mu + e^- \rightarrow \nu_\mu + e^-,$$

between a *muon*-neutrino and an electron can only be mediated by the $Z$ boson (Ryder, 1986, pp. 235–236).

Following neutrino experiments at CERN in the 1960s, it was claimed that the upper limit on the number of possible weak neutral-current events was well below that predicted by the electroweak theory (Pickering, 1984, p. 185). However, after theoretical developments in the early 1970s which favoured the electroweak theory, additional pressure was placed upon the experimenters, and the prediction of weak neutral-currents was duly verified in 1973 by the new giant bubble chamber Gargamelle installed at CERN (Pickering, 1984, p. 187). The proton synchotron at CERN was used to produce a secondary beam[10] of neutrinos, and this beam of neutrinos was fired into the bubble chamber. A bubble chamber acts as both the target for accelerated particles, and the detector for the experimentalists. In contrast to a cloud

[10] A secondary beam is produced by firing a primary beam, in this case a proton beam, at, say, a metal target, from which a shower of secondary particles emerges. Secondary beams can then be selected from this shower of particles (Pickering, 1984, p. 24).

chamber, a bubble chamber contains not air, but a liquid, which is initially at a temperature just below its boiling point. The pressure on the liquid is then relaxed, which reduces its boiling point, and the liquid is said to be superheated. If a charged, ionizing particle passes through such a chamber, bubbles of gas form around the ions, and the particle tracks so created can be photographed (Ne'eman and Kirsh, 1996, p. 25). In particular, Gargamelle held 18 tonnes of liquid Freon in such a superheated state. The beam of high-energy neutrinos fired into Gargamelle underwent weak interactions with the nucleons and electrons in the Freon liquid. Although the neutrinos cannot be observed in the bubble chamber photographs, the interactions they cause are observable. For example, weak charged neutrino-nucleon events, such as those specified above, can be inferred from the creation of the muon $\mu^-$, and weak neutral neutrino-nucleon events can be inferred from the creation of the hadron spray.

In 1973, one instance of a weak neutral-current event between a neutrino and an atomic electron was discovered from over 700 000 photographs, and 100 weak neutral-current interactions between neutrinos and nucleons were found from over 290 000 photographs (Pickering, 1984, p. 187 and p. 204, Note 6). However, as Pickering explains, the weak neutral-current events could have been discovered from the CERN neutrino experiments conducted in the 1960s, and were only neglected because the experimenters interpreted their data in a different way. The same spray of hadrons produced by a weak neutral interaction between a neutrino and nucleon, was also produced by the decay of a neutron which issued forth from a charged-current interaction between a neutrino and a neutron outside the Freon tank, in the 1000 tonnes of ancillary equipment (Pickering, 1984, p. 190). In the CERN neutrino experiments of the 1960s, this neutron background was known about, but the experimenters merely used a practical rule-of-thumb to exclude it, neglecting all events below a certain energy level. E.C.M. Young attempted to calculate the neutron background in 1967, and upon doing so, found numerous events that could not be accounted for by the neutron background, which, in retrospect, were probably weak neutral events (Pickering, 1984, p. 191). Young's results, however, were largely ignored until the search for weak neutral events was given added theoretical impetus. Thereafter, estimates of the neutron background were calculated, and the most detailed results, published in 1975, used a computer simulation to model both the generation of neutrons in the ancillary equipment of Gargamelle, and their propagation into the main tank (Pickering, 1984, p. 192).

Pickering attempts to reduce the change of interpretative procedure to a sociological phenomenon, asserting that "the 1960s order, in which a particular set of interpretative procedures pointed to the non-existence of the neutral current, was displaced in the 1970s by a new order, in which a new set of interpretative procedures made the neutral current manifest. Each set of procedures was in principle questionable, and yet the HEP [High Energy Physics] community chose to accept first one and then the other" (Pickering, 1984, p. 193). The two sets of interpretative procedures were not, however, relevantly similar: whilst the first relied upon a rule-of-thumb, the second relied upon calculations derived from theoretical models. Whilst Pickering (1984, p. 192) correctly points out that even calculations involve assumptions, it is nevertheless true that calculations derived from *good* theoretical models are superior to a practical rule-of-thumb. The change of interpretative procedure employed by the HEP community was a reflection of increased understanding, not merely a sociological fashion.

The detection of weak neutral-current events should not be equated with the discovery of the $Z$ boson itself. The $W^{\pm}$ and $Z$ gauge bosons were first discovered by the Sp$\bar{\text{p}}$S proton–anti-proton collider at CERN in 1982/1983. The creation of these weak gauge bosons in the aftermath of a proton–anti-proton collision, was inferred from the observation of their decay products. The evidence for the charged gauge bosons $W^{\pm}$ was provided by the reactions

$$p + \bar{p} \to W^{+} \to e^{+} + \nu_e,$$
$$p + \bar{p} \to W^{-} \to e^{-} + \bar{\nu}_e,$$

in which the lepton pair carry considerable momentum in a direction transverse to the $p\bar{p}$ beam axis (Ne'eman and Kirsh, 1996, p. 259). The evidence for the $Z$ bosons was provided by reactions

$$p + \bar{p} \to Z \to e^{+} + e^{-},$$
$$p + \bar{p} \to Z \to \mu^{+} + \mu^{-},$$

in which, once again, the decay products carry considerable transverse momentum (Ne'eman and Kirsh, 1996, p. 260). Subsequently, between 1989 and 1995, 20 million $Z$ bosons were detected by the LEP I electron–positron collider at CERN (Okun, 1996).

Whilst this section has only addressed the electroweak Lagrangian for the first lepton and anti-lepton generation, the electroweak Lagrangian can, of course, be extended to include all three lepton and anti-lepton generations, and can also be extended to include quarks as well as leptons. It is to the subject of quarks that we turn in the next section.

## 4.7. Quark colours

A quark interacting via the strong force is represented to have an 'internal' degree of freedom, called colour. This degree of freedom enables a quark to have three colour states, often dubbed red, green and blue. An anti-quark can have anti-red, anti-green and anti-blue colour states. Gluons are continually exchanged between the quarks in a hadron, and of the eight gluons which mediate the strong force, six are considered to be colour-changing gluons in the sense that, when a pair of quarks exchange such a gluon, the colours of those quarks change in tandem (Ne'eman and Kirsh, 1996, pp. 223–224).

In fibre-bundle terms, the existence of colour is merely a manifestation of the fact that a quark interacting via the strong force is represented by the tensor product $\sigma \otimes \rho$ of the Dirac spinor bundle $\sigma$ and a strong interaction bundle $\rho$, a complex vector bundle of fibre dimension three. Colour is an internal degree of freedom in the sense that it pertains to the fibres of the interacting bundle $\sigma \otimes \rho$.

To respect the Pauli exclusion principle for fermions, the states of a bound multi-quark system must be totally anti-symmetric with respect to the colour degree of freedom. Hence, if we let $\psi_r$ denote a red quark state, $\psi_g$ a green quark state, and $\psi_b$ a blue quark state, then the space of fibre-states of a composite quark system is spanned by anti-symmetric tensor products of such vectors. For example, in the case of a three-quark system, the physically permissable states are those such as

$$\begin{aligned}\psi_{[rgb]} &= \frac{1}{\sqrt{6}}\psi_r \wedge \psi_g \wedge \psi_b \\ &= \frac{1}{\sqrt{6}}(\psi_r \otimes \psi_g \otimes \psi_b - \psi_r \otimes \psi_b \otimes \psi_g + \psi_b \otimes \psi_r \otimes \psi_g \\ &\quad - \psi_b \otimes \psi_g \otimes \psi_r + \psi_g \otimes \psi_b \otimes \psi_r - \psi_g \otimes \psi_r \otimes \psi_b).\end{aligned}$$

A hadron therefore cannot possess states in which more than one particle has the same colour. Such states belong to the kernel of the anti-symmetric projection. For example, $\psi_{[rrr]} = 0$. Similarly, a meson must possess states in which the quark and anti-quark possess opposite colours.

Just as one can apply formal symmetry breaking to obtain the eight interaction carriers of the strong force, one can apply formal symmetry breaking to the interacting quark bundles to obtain the three quark 'colours' (Derdzinski, 1992, p. 100). The tensor product of two vector spaces, $W \otimes V$, is isomorphic to the $n$-fold direct sum of $W$ with itself, where $n = \dim V$. This entails that $\sigma \otimes \rho$ is isomorphic to $\sigma \oplus \sigma \oplus \sigma$. The specific choice of isomorphism depends upon the choice of a basis $\{v_j\colon\ j = 1, 2, 3\}$ in each fibre of $\rho$. A choice of gauge for the strong force corresponds to the selection of a basis $\{v_j\colon\ j = 1, 2, 3\}$ in each fibre of $\rho$ which is compatible with the $SU(3)$-structure in each fibre. In other words, a choice of gauge corresponds to a cross-section of the principal bundle $P_\rho$ of all oriented, orthonormal bases in the fibres of $\rho$. With a choice of gauge, then, $\sigma \otimes \rho$ can be decomposed into a direct sum $\bigoplus_{j=1}^{j=3} \sigma_j$ of sub-bundles $\sigma_j \subset \sigma \otimes \rho$, each of which is isomorphic with $\sigma$. At each point $x$, the fibre of each direct summand $\sigma_j$ is the span of the set of simple tensors $\{w \otimes v_j\colon\ w \in \sigma_x\}$.

Given a choice of basis $\{w_i\}$ in $\sigma_x$, any element of $\sigma_x \otimes \rho_x$ can be expressed as a linear combination

$$\sum_i \sum_j c_{ij} w_i \otimes v_j = \sum_j \left( \sum_i c_{ij} w_i \right) \otimes v_j.$$

Given that the fibre of each direct summand $\sigma_j$ over $x$ is the span of the set of simple tensors $\{w \otimes v_j\}$, this vector maps to

$$\left(\sum_i c_{i1} w_i, \sum_i c_{i2} w_i, \sum_i c_{i3} w_i\right) \in \left(\bigoplus_{j=1}^{j=3} \sigma_j\right)_x .$$

The three summands of the direct sum $\sigma \oplus \sigma \oplus \sigma$, are the three so-called 'colour sectors' of a quark flavour. The general state of any quark flavour is a linear combination of quark states in each colour sector.

Whilst the three internal degrees of freedom provided by colour make the theory empirically adequate, colour itself appears to be unobservable. Bound systems of quarks are required to be colourless, and neither free quarks nor free gluons have ever been observed. The colour possessed by a quark is considered to be the charge which is the source of the strong force, hence the strong force between colourless hadrons is merely a residual force. The strong force between quarks does not decrease in proportion to the quark separation, leading to the phenomenon of 'quark confinement.' The gluons which mediate strong force interactions, themselves possess a combination of a colour and an anti-colour charge, and therefore interact with each other via the strong force. This is one reason used to explain why the strong force does not decrease in proportion to separation (Ne'eman and Kirsh, 1996, p. 223). However, quark confinement does not entail that quarks are, by definition, nothing else other than a part of something else, rather than particles in their own right. Quark confinement simply entails that, once bound into mesons and baryons, quarks cannot be freed from their confinement. According to particle physics cosmology, the universe originally contained, amongst other things, free quarks and free anti-quarks, but after about a billionth of a second, the energy density dropped to the critical level for a phase transition from the quark state to a state dominated by nucleons. At this density, the inter-quark forces became sufficiently strong to bind those particles which escaped quark–anti-quark annihilation into baryons and mesons (Barrow and Tipler, 1986, pp. 370–371).

The experimental evidence for quarks is strong. For example, the evidence from numerous scattering experiments indicates that the charge and mass in a hadron are not uniformly distributed. Moreover, in proton–proton scattering, jets of particles which issue perpendicular to the direction of the proton beams can only be explained as "the result of a head-on collision between a quark in one proton and a quark in the other proton" (Ne'eman and Kirsh, 1996, p. 219). Furthermore, whilst muons can be produced in pion–proton scattering at a certain energy, the observation of muons is a rare event in proton–proton scattering at the same energy; the muon production can be explained by the annihilation of an anti-quark in the pion and a quark in the proton, and no such annihilation can take place in collisions between protons (Ne'eman and Kirsh, 1996, p. 220).

There is also strong experimental evidence for the existence of gluons. In high-energy electron–positron annihilation events, it was noted that the result of the annihilation is often the creation of a new particle–anti-particle pair. On other occasions, however, the result is two jets of hadrons moving in opposite directions. It is believed that when a quark–anti-quark pair are created from the annihilation event, the pair of separating particles produce a train of additional quarks and anti-quarks, which ultimately form the two hadron jets (Ne'eman and Kirsh, 1996, p. 219). At sufficiently high energies, it was predicted that at least one of the quarks created would emit a gluon, and this gluon would give rise to a third jet. The first three-jet event, found at the PETRA collider in 1979, was therefore considered to be evidence for the existence of gluons (Ne'eman and Kirsh, 1996, p. 224).

Each different choice of gauge (each different cross-section of the principal bundle $P_\rho$) selects a different decomposition of $\sigma \otimes \rho$ into three colour sectors. The selection of such a cross-section provides the formal symmetry breaking. Before the selection of a decomposition, there is a representation of $SL(2, \mathbb{C}) \times SU(3)$ upon the typical fibre of $\sigma \otimes \rho$. Afterwards, the only element of $SU(3)$ which preserves the selected basis in each fibre of $\rho$ is the identity $Id$. Hence, after selecting a decomposition, there is a (reducible) representation of $SL(2, \mathbb{C}) \times Id \cong SL(2, \mathbb{C})$ on the typical fibre of $(\sigma \otimes \rho) \cong \sigma \oplus \sigma \oplus \sigma$. One can say that the direct sum decomposition is obtained by restricting the representation of $SL(2, \mathbb{C}) \times SU(3)$ to a representation of $SL(2, \mathbb{C})$. One can also say that the $SU(3)$-symmetry has been broken. To reiterate, this symmetry breaking is referred to as 'formal' because it doesn't correspond to a physical process.

A direct sum of multiple free-particle bundles can be thought of as the free-particle bundle which represents the generalisation of the individual free-particles. In this sense, a quark with the strong force switched on, can be thought of as a generalisation of three quarks. Using the metaphorical language of quark colours, a quark with the strong force switched on can be thought of as a generalisation of a red quark, a green quark, and a blue quark. However, whilst the tensor product bundle $\sigma \otimes \rho$ represents an *interacting*-quark, the individual spinor bundles in the direct sum decomposition are bundles which only represent *free*-quarks.

## 4.8. Quark mixing and the Cabibbo angle

In Section 4.6 we considered the electroweak Lagrangian for the first lepton and anti-lepton generation, and alluded to the fact that this Lagrangian can be easily extended to include each generation of leptons and anti-leptons. The electroweak Lagrangian can also be extended to include quarks, but in this case a complication called quark 'mixing' enters the fray. By convention, this mixing is placed in the down-type quarks $(d, s, b)$. The identity of the down-type quark states in the strong force Lagrangian is distinct from the identity of the down-type quark states in the electroweak Lagrangian. Whilst the strong force Lagrangian contains the down-type quark-states $(d, s, b)$, the electroweak Lagrangian contains the down-type quark states $(d', s', b')$. Whilst the quark states $(d, s, b)$ are mass eigenstates, each one of the primed quark states is a superposition of the mass eigenstates. This so-called mixing is specified by the Cabibbo–Kobayashi–Maskawa matrix $V_{\text{CKM}} \in U(3)$ in the sense that

$$\begin{pmatrix} d' \\ s' \\ b' \end{pmatrix} = V_{\text{CKM}} \begin{pmatrix} d \\ s \\ b \end{pmatrix}.$$

Recall from Section 4.6 that the kinetic fermionic terms of the electroweak Lagrangian contain things called weak charged-currents. In that section we expressed these currents in terms of only the first generation of leptons. If we extend the electroweak model to include not only each generation of leptons, but each generation of quarks, then the part of the electroweak Lagrangian containing the weak charged-currents,

$$\mathcal{L}_{cc} = \frac{g}{\sqrt{2}}(j^\mu W_\mu^+ + \bar{j}^\mu W_\mu^-),$$

is such that the currents themselves are expressed as

$$j^\mu = \bar{u}_L \sigma^\mu d'_L + \bar{c}_L \sigma^\mu s'_L + \bar{t}_L \sigma^\mu b'_L + \bar{\nu}_e \sigma^\mu e_L + \bar{\nu}_\mu \sigma^\mu \mu_L + \bar{\nu}_\tau \sigma^\mu \tau_L,$$

and

$$\bar{j}^\mu = \bar{d}'_L \sigma^\mu u_L + \bar{s}'_L \sigma^\mu c_L + \bar{b}'_L \sigma^\mu t_L + \bar{e}_L \sigma^\mu \nu_e + \bar{\mu}_L \sigma^\mu \nu_\mu + \bar{\tau}_L \sigma^\mu \nu_\tau.$$

To repeat what was said in Section 4.6, under second-quantization, one substitutes creation and annihilation operators into the expressions above to obtain field operators. Under second-quantization, the expression for a field operator contains annihilation operators for the particle, and creation operators for the anti-particle, and the expression for the anti-field operator contains annihilation operators for the anti-particle and creation operators for the particle.

Take the expression $\bar{u}_L \sigma^\mu d'_L$ for example. Given that $d'_L$ is a superposition of $(d_L, s_L, b_L)$ quark states, the quantization of $d'_L$ contains annihilation operators for $(d_L, s_L, b_L)$ quarks, and creation operators for the right-handed anti-down quarks $(\bar{d}_R, \bar{s}_R, \bar{b}_R)$. The quantization of $\bar{u}_L$ contains annihilation operators for right-handed anti-up quarks and creation operators for left-handed up-quarks. Hence, one can say that the product $\bar{u}_L \sigma^\mu d'_L$ contains (i) annihilation operators for $(d_L, s_L, b_L)$ quarks, and creation operators for $u_L$ quarks, and (ii) annihilation operators for $\bar{u}_R$ quarks and creation operators for $(\bar{d}_R, \bar{s}_R, \bar{b}_R)$ quarks. The current $j^\mu$ couples with $W_\mu^+$, and under second quantization $W_\mu^+$ contains creation operators for $W^-$ bosons (along with annihilation operators for $W^+$ bosons). Hence, $\bar{u}_L \sigma^\mu d'_L$ is associated with weak-force interactions that involve transitions such as

$$d_L \to u_L + W^-,$$
$$s_L \to u_L + W^-,$$
$$b_L \to u_L + W^-,$$

and transitions such as

$$\bar{u}_R \to \bar{d}_R + W^-,$$
$$\bar{u}_R \to \bar{s}_R + W^-,$$
$$\bar{u}_R \to \bar{b}_R + W^-.$$

Mixing thereby permits transmutations between quarks in different generations. $d_L$, $s_L$ and $b_L$ possess an electric charge of $-1/3$ and $u_L$ possesses an electric charge of $+2/3$, hence the transition $(d_L, s_L, b_L) \to u_L$ increases electric charge by $+1$, and needs to be balanced by the emission of the $W^-$, which possesses an electric charge of $-1$. Anti-particles, of course, possess opposite charges to their particle counterparts, hence $\bar{u}_R$ possesses an electric charge of $-2/3$ and $(\bar{d}_R, \bar{s}_R, \bar{b}_R)$ possess an electric charge of $+1/3$, hence the transition $\bar{u}_R \to (\bar{d}_R, \bar{s}_R, \bar{b}_R)$ also increases electric charge by $+1$, and is similarly balanced by the emission of a $W^-$.

Note that the $j^\mu$ and $\bar{j}^\mu$ are left-handed currents. Weak processes cannot create or annihilate right-handed quarks and leptons, or left-handed anti-quarks and anti-leptons. These particles form weak isospin singlets.

A mixture of quark flavours can be represented by forming a generalised particle bundle for the quark flavours involved. One example is a mixture of the $d$ and $s$ quarks.[11] The $d$ quark can be represented by one copy $\sigma_d$ of the Dirac spinor bundle $\sigma$, and the $s$ quark can be represented by another copy $\sigma_s$. The generalised particle bundle which represents any mixture of these two quarks is the direct sum $\sigma_d \oplus \sigma_s$. A specific mixture of the two quark flavours is a bundle $\sigma_{d'}$ isomorphic to $\sigma$, which is a summand in an alternative decomposition $\sigma_{d'} \oplus \sigma_{s'}$ of the generalised particle bundle. Where $V$ is an arbitrary fibre of $\sigma$, one can specify such a sub-bundle by selecting, first, the subspace

$$V'_{a_1,a_2} = \{(a_1 v, a_2 v) \in V \oplus V\colon\ v \in V\},$$

with $|a_1|^2 + |a_2|^2 = 1$. This determines an orthogonal complement $V''_{\bar{a}_2,-\bar{a}_1}$. One can then use the requirement that the decomposition be invariant under parallel transport, defined by the spinor connection on $\sigma \oplus \sigma$, to determine a decomposition $\sigma_{d'} \oplus \sigma_{s'}$. The constants $a_1, a_2$ are independent of $x$, determine the mixture of the $d$ quark and $s$ quark states, and can be expressed in terms of the Cabibbo angle $\theta_C$:

$$\tan\theta_C = |a_2/a_1|, \quad 0 \leqslant \theta_C \leqslant \frac{\pi}{2}.$$

Experiment indicates that $\sin\theta_C = 0.231 \pm 0.003$ (Derdzinski, 1992, p. 159). Note that whilst the Weinberg angle of the electroweak theory is considered to have been fixed during spontaneous symmetry breaking, the Cabibbo angle concerns the relation between the electroweak and strong forces, and the standard model itself makes no commitment on how these two forces might be incorporated in a Grand Unified Theory (Penrose, 2004, p. 651).

[11] Here the exposition closely follows Derdzinski (1992, p. 153 and p. 159).

The mixing of the $d$ and $s$ quarks in weak interactions was generalised by Kobayashi and Maskawa to the mixture of the $\{d, s, b\}$ quarks already referred to. The bundle which represents the generalisation of these three quark flavours is $\sigma_d \oplus \sigma_s \oplus \sigma_b$, and different mixtures of these flavours correspond to different orthogonal decompositions $\sigma_{d'} \oplus \sigma_{s'} \oplus \sigma_{b'}$. Each such decomposition is defined by the Cabibbo–Kobayashi–Maskawa matrix $V_{\text{CKM}}$, which can be specified by four parameters, $\{\theta_1, \theta_2, \theta_3, u\}$, called the Cabibbo–Kobayashi–Maskawa parameters. The first three parameters $\theta_1, \theta_2, \theta_3$ are angular parameters with values in $[0, \frac{\pi}{2}]$. The fourth parameter is a phase factor $u = e^{i\delta}$ (Derdzinski, 1992, p. 160). The three angular parameters have values fixed by experiment.

This notion of quark mixing is considered to be a consequence of the interaction of quarks with the electroweak Higgs bosons. If, as current evidence indicates, the neutrinos possess mass, then there is a corresponding notion of lepton mixing, and the CKM matrix has a lepton counterpart called the Maki–Nakagawa–Sakata matrix. This matrix also requires four parameters for its specification, hence the standard model acquires four more parameters whose values must be established by experiment rather than theory.

Quark mixing entails that the identity of quarks interacting via the strong force alone, are distinct from the identity of quarks interacting via the strong and the electroweak force. A down quark $d$ interacting via the strong force alone can be represented as a cross-section $\psi$ of the interacting-particle bundle $\rho \otimes \sigma_d$. In other words, $\psi \in \Gamma(\rho \otimes \sigma_d)$. In contrast, a down quark $d'$ interacting via the electroweak force can be represented as a cross-section of the interacting-particle bundle $\iota \otimes \sigma_{d'}$, and a down quark interacting via the strong and electroweak forces can be represented as a cross-section $\psi'$ of the interacting-particle bundle $\rho \otimes \iota \otimes \sigma_{d'}$. In other words, $\psi' \in \Gamma(\rho \otimes \iota \otimes \sigma_{d'})$.

## 4.9. The standard model Lagrangian

From a somewhat Olympian perspective, all the information about the standard model of particle physics, considered as a classical field theory, is specified by the standard model Lagrangian (Nachtmann, 1990, pp. 369–370; Ticciati, 1999, p. 680):

$$\begin{aligned}\mathcal{L}_{sm} = &-\frac{1}{4}G_{\mu\nu}G^{\mu\nu} - \frac{1}{4}W_{\mu\nu}W^{\mu\nu} - \frac{1}{4}B_{\mu\nu}B^{\mu\nu} \\ &+ \bar{\psi}^i_L i\gamma^\mu \nabla^{\text{EW}}_\mu \psi^i_L + \bar{\psi}^i_R i\sigma^\mu \nabla^{\text{EW}}_\mu \psi^i_R \\ &+ \bar{\chi}^i_L i\gamma^\mu \nabla^{\text{SM}}_\mu \chi^i_L + \bar{U}^i_R i\sigma^\mu \nabla^{\text{SM}}_\mu U^i_R + \bar{D}^i_R i\sigma^\mu \nabla^{\text{SM}}_\mu D^i_R \\ &- Y^u_{ij}\bar{\chi}^i_L\tilde{\phi}U^j_R - Y^d_{ij}\bar{\chi}^i_L\phi D^j_R - Y^e_{ij}\bar{\psi}^i_L\phi\psi^j_R - \text{h.c.} \\ &+ \overline{\left(\nabla^{\text{EW}}_\mu\phi\right)}\left(\nabla^\mu_{\text{EW}}\phi\right) + a(\bar{\phi}\phi)^2 - b(\bar{\phi}\phi).\end{aligned}$$

$G_{\mu\nu}$ is the gauge field strength of the strong $SU(3)$ gauge field, $W_{\mu\nu}$ is the gauge field strength of the weak isospin $SU(2)_L$ gauge field, and $B_{\mu\nu}$ is the gauge field strength of the weak hypercharge $U(1)_Y$ gauge field. The fermionic doublets and singlets are denoted as follows:

$$(\psi_L^1, \psi_L^2, \psi_L^3) = \left( \begin{pmatrix} \nu_e \\ e_L \end{pmatrix}, \begin{pmatrix} \nu_\mu \\ \mu_L \end{pmatrix}, \begin{pmatrix} \nu_\tau \\ \tau_L \end{pmatrix} \right),$$

$$(\chi_L^1, \chi_L^2, \chi_L^3) = \left( \begin{pmatrix} u_L \\ d'_L \end{pmatrix}, \begin{pmatrix} c_L \\ s'_L \end{pmatrix}, \begin{pmatrix} t_L \\ b'_L \end{pmatrix} \right),$$

$$(\psi_R^1, \psi_R^2, \psi_R^3) = (e_R, \mu_R, \tau_R),$$
$$(U_R^1, U_R^2, U_R^3) = (u_R, c_R, t_R),$$
$$(D_R^1, D_R^2, D_R^3) = (d'_R, s'_R, b'_R).$$

The abbreviation 'h.c.' indicates the hermitian conjugates of the preceding Yukawa terms. $Y_{ij}^u$, $Y_{ij}^d$ and $Y_{ij}^e$ are the Yukawa matrices, which specify the strength of the interaction between the fermions and the Higgs bosons. The Higgs field $\phi$ is represented as a weak isospin doublet

$$\phi = \begin{pmatrix} \phi^+ \\ \phi^0 \end{pmatrix},$$

with the anti-particle doublet $\tilde{\phi} = i\sigma^2\bar{\phi}$, such that

$$\tilde{\phi} = \begin{pmatrix} \bar{\phi}^0 \\ -\bar{\phi}^+ \end{pmatrix}.$$

Note that because the leptons do not interact via the strong force, the covariant derivative used in their kinetic terms is the electroweak covariant derivative,

$$\nabla_\mu^{\text{EW}} = \partial_\mu - ig\frac{\sigma_a}{2}W_\mu^a - ig'\frac{Y}{2}B_\mu,$$

whilst the covariant derivative used in the kinetic quark terms is that of the entire standard model (see the end of Section 4.5),

$$\nabla_\mu^{\text{SM}} = \partial_\mu - ig_s\frac{\lambda_i}{2}G_\mu^i - ig\frac{\sigma_a}{2}W_\mu^a - ig'\frac{Y}{2}B_\mu.$$

The standard model Lagrangian is an economical way of specifying a non-linear interacting field equation and coupled Yang–Mills equation for a generic elementary matter field, with all the non-gravitational force fields switched on. The solution space of these equations is a non-linear space, and if we accept the proposal of Section 2.11, that the second-quantization of interacting fields should be a functor from a category of such non-linear spaces into another category of non-linear spaces, then such a functor would supply all the information needed to specify the non-gravitational structure of a standard model universe. In the final chapter, we will approach the same structure specified by the standard model Lagrangian, but from a fibre bundle perspective.

## 4.10. Gauge fields in other universes

At first sight, the gauge force fields capable of existing in a universe appear to be independent of the space–time dimension and signature. Given a space–time $\mathcal{M}$ of arbitrary dimension and signature, a gauge field in such a universe is specified by the selection of a compact connected Lie group $G$. With the exception of regions in which the gravitational field is very strong, a gauge field in such a universe can be represented by a connection upon a principal $G$-bundle or interaction bundle over the Minkowski space–time $\mathbb{R}^{p,q}$ of the relevant dimension and signature.

The structure theorem of compact connected Lie groups (see Section 5.1), enables one to decompose any such group into a quotient of a direct product of simple, simply connected compact groups, and a compact Abelian group. These groups can be given an exhaustive, non-repetitious listing, hence one apparently obtains a classification of all the possible gauge fields which can exist in a universe.

This approach is, however, deceptive. Recall from Section 4.2 that the interaction carriers ('gauge bosons') of a gauge force field correspond to (integer spin) unitary irreducible representations of the local space–time symmetry group. A physically legitimate gauge field must be such that the space of $G$-connections satisfying the free Yang–Mills equations, modulo the gauge transformations, decompose into a direct sum of unitary irreducible representations of the local space–time symmetry group. Given a choice of gauge, and given a choice of basis in the Lie algebra of $G$, the space of $G$-connections will always decompose as

$$\bigoplus^{\dim G} T^*\mathcal{M}.$$

In the special case of a universe with three spatial dimensions and one time dimension, the cotangent bundle $T^*\mathcal{M}$ possesses upon its typical fibre, isomorphic to $\mathbb{R}^{3,1}$, a spin-1 irreducible representation of $Spin(3,1) \cong SL(2,\mathbb{C})$. In the case of universes with an arbitrary number of space and time dimensions, a space of $G$-connections will still decompose as $\bigoplus^{\dim G} T^*\mathcal{M}$, but the typical fibre of the cotangent bundle $T^*\mathcal{M}$ will be isomorphic to $\mathbb{R}^{p,q}$, and will possess an irreducible representation of $\widetilde{SO}_0(p,q)$. As already pointed out in Section 2.6, bundles equipped with representations of $\widetilde{SO}_0(p,q)$ upon their typical fibres cannot, in general, be interpreted as spin-$s$ free-particle bundles. In particular, $T^*\mathcal{M}$ cannot be interpreted as a spin-1 free-particle bundle in the case of an arbitrary space–time $\mathcal{M}$. In those cases where there is only one temporal dimension, $q = 1$, and an arbitrary number of spatial dimensions $p$, the cotangent bundle $T^*\mathcal{M}$ will always possess the irreducible representation of $\widetilde{SO}_0(p,1) = Spin(p+1)$ with dominant weights $(1,0,\dots,0)$, but this doesn't correspond to a spin-1 representation.

## 4.11. Composite systems

A bound and stable collection of particles can be referred to as a composite system. In such a bound state, the particles involved tend to neutralize each other's ability to interact with the environment, so bunches of more than one interacting particle can be described by the cross-sections of a free-particle bundle (Derdzinski, 1992, p. 86–88).

Consider a collection of $n$ particles, represented individually by cross-sections of the vector bundles $\alpha_1, \ldots, \alpha_n$. If the collection is not necessarily considered to form a bound system, it is represented collectively by cross-sections of the Cartesian product vector bundle $\alpha_1 \times \cdots \times \alpha_n$ (Derdzinski, 1992, p. 22). The base space of this Cartesian product bundle is the Cartesian product $\mathcal{M}_1 \times \cdots \times \mathcal{M}_n$ of the individual base spaces. However, the typical fibre of the Cartesian product vector bundle is understood to be the tensor product $V_1 \otimes \cdots \otimes V_n$ of the individual typical fibres, and *not* the Cartesian product of the individual typical fibres.

To emphasise, the vector bundle for the collection of $n$ particles is not the $n$-fold tensor product bundle $\alpha_1 \otimes \cdots \otimes \alpha_n$, but the $n$-fold Cartesian product $\alpha_1 \times \cdots \times \alpha_n$. This is actually consistent with the quantum theoretical principle that the state space of a $n$-particle system is (a subspace of) the $n$-fold tensor product of the individual particle state spaces. Given that the state space of an interacting particle is not a linear vector space, let us suppose that each of the vector bundles $\alpha_1, \ldots, \alpha_n$ is a free-particle bundle, and that the state space of particle $k$, for $k = 1, \ldots, n$, is a vector subspace of the set of cross-sections $\Gamma(\alpha_k)$. Consider the simple case where each bundle $\alpha_k$ is trivial, and the set of cross-sections is isomorphic to $\mathcal{F}(\mathcal{M}, \mathbb{C}^m)$, the space of $\mathbb{C}^m$-valued functions on $\mathcal{M}$. Two isomorphisms need to be considered here. Firstly,

$$\mathcal{F}(\mathcal{M}, \mathbb{C}^m) \cong \mathcal{F}(\mathcal{M}) \otimes \mathbb{C}^m,$$

and then the fact that

$$\bigotimes^n \mathcal{F}(\mathcal{M}) \cong \mathcal{F}(\mathcal{M}^n).$$

It follows from these isomorphisms that

$$\begin{aligned}
\bigotimes^n \mathcal{F}(\mathcal{M}, \mathbb{C}^m) &\cong \bigotimes^n (\mathcal{F}(\mathcal{M}) \otimes \mathbb{C}^m) \\
&\cong \left(\bigotimes^n \mathcal{F}(\mathcal{M})\right) \otimes \left(\bigotimes^n \mathbb{C}^m\right) \\
&\cong \mathcal{F}(\mathcal{M}^n) \otimes \left(\bigotimes^n \mathbb{C}^m\right) \\
&\cong \mathcal{F}\left(\mathcal{M}^n, \bigotimes^n \mathbb{C}^m\right).
\end{aligned}$$

Given the assumption that $\Gamma(\alpha_k) \cong \mathcal{F}(\mathcal{M}, \mathbb{C}^m)$,

$$\Gamma(\alpha_1) \otimes \cdots \otimes \Gamma(\alpha_n) \cong \bigotimes^n \mathcal{F}(\mathcal{M}, \mathbb{C}^m) \cong \mathcal{F}\left(\mathcal{M}^n, \bigotimes^n \mathbb{C}^m\right).$$

This demonstrates, under the assumptions made here, that the states of the $n$-particle system are represented by cross-sections of a vector bundle which has a Cartesian product $\mathcal{M}^n$ as base space, and a tensor product $\bigotimes^n \mathbb{C}^m$ as typical fibre. In the case where $\Gamma(\alpha_k) \cong \mathcal{F}(\mathcal{M}, \mathbb{C}^m)$,

$$\mathcal{F}\left(\mathcal{M}^n, \bigotimes^n \mathbb{C}^m\right) \cong \Gamma(\alpha_1 \times \cdots \times \alpha_n),$$

hence

$$\Gamma(\alpha_1) \otimes \cdots \otimes \Gamma(\alpha_n) \cong \Gamma(\alpha_1 \times \cdots \times \alpha_n) \neq \Gamma(\alpha_1 \otimes \cdots \otimes \alpha_n).$$

The vector bundle for a collection of $n$ particles which *are* bound together to form a composite system, is the $n$-fold tensor product $\alpha_1 \otimes \cdots \otimes \alpha_n \subset \alpha_1 \times \cdots \times \alpha_n$. In other words, one restricts the base space to the subset of $\mathcal{M}_1 \times \cdots \times \mathcal{M}_n$ consisting of $n$-tuples $(x_1, \ldots, x_n)$ in which $x_1 = x_2 = \cdots = x_n$. This is naturally isomorphic to $\mathcal{M}$ (Derdzinski, 1992, p. 22). From the fact that $\alpha_1 \otimes \cdots \otimes \alpha_n \subset \alpha_1 \times \cdots \times \alpha_n$, it follows that $\Gamma(\alpha_1 \otimes \cdots \otimes \alpha_n) \subset \Gamma(\alpha_1 \times \cdots \times \alpha_n)$. Hence, the fact that the bound states of an $n$-particle system are cross-sections of $\alpha_1 \otimes \cdots \otimes \alpha_n$, remains consistent with the principle that the state space of an $n$-particle system is (a subspace of) the $n$-fold tensor product of the individual particle state spaces. Note, however, that particles require interactions to bind together, so the $\alpha_k$ will be interacting-particle bundles in this case.[12]

A composite system can possess orbital angular momentum about its centre of mass, and, as a consequence, if a composite system has an orbital angular momentum of $\ell\hbar$, its states belong to $S_0^\ell T^*\mathcal{M} \otimes \alpha_1 \otimes \cdots \otimes \alpha_n$ (Derdzinski, 1992, pp. 12–13). (The $S_0^\ell T^*\mathcal{M}$ factor for the orbital angular momentum of a composite system should be distinguished from the $S_0^k T^*\mathcal{M}$ factor for an elementary system with intrinsic spin.) There is then a surjective bundle morphism onto a free-particle bundle $\eta$, which represents the bound states of the composite system as if it were a free elementary particle:

$$S_0^\ell T^*\mathcal{M} \otimes \alpha_1 \otimes \cdots \otimes \alpha_n \to \eta.$$

A *composite* system of spin $s$ and mass $m$ can be represented, in the configuration space approach, by the cross-sections of a spin-$s$ free-particle bundle which provide mass $m$ solutions to a relevant differential equation. The Hilbert space constructed from these cross-sections is, under Fourier transform, the Hilbert space for a spin $s$,

[12] Note that collections of interaction carriers can also form bound states. See Derdzinski (1992, pp. 88–89), for details.

mass $m$ particle in the Wigner approach. Hence, the irreducible unitary representations of the local symmetry group, $SL(2, \mathbb{C}) \ltimes \mathbb{R}^{3,1}$, can be used to represent not only free elementary particles, but also stable and bound collections of elementary particles. A composite system of spin $s$ and mass $m$ can be represented by Wigner's spin $s$, mass $m$, unitary, irreducible representation of $SL(2, \mathbb{C}) \ltimes \mathbb{R}^{3,1}$. One can conclude from this that the unitary, irreducible representations of $SL(2, \mathbb{C}) \ltimes \mathbb{R}^{3,1}$ specify not merely the possible elementary particles in a universe, but all possible stable particles, whether they be elementary or composite. The irreducibility of a representation of $SL(2, \mathbb{C}) \ltimes \mathbb{R}^{3,1}$ does not entail the elementarity of the corresponding particle.

## 4.12. Baryons, mesons and hadron symmetries

The composite particles which participate in strong force interactions are referred to as 'hadrons.' Any hadron will eventually decay into some collection of protons, electrons, neutrinos, photons, and their anti-particles (Sternberg, 1994, p. 276). For example, a neutron will eventually undergo $\beta$-decay, $n \rightarrow p + e + \bar{\nu}_e$, into a proton, an electron, and an anti-electron-neutrino. However, the decay products of a hadron are not, in general, uniquely determined. The different possible decay products are referred to as the 'decay modes.'

Hadrons are divided into mesons and baryons. Mesons are defined as the bosonic hadrons, with integer spin $s = 0, 1, 2, \dots$, and baryons are defined as fermionic hadrons, with half integer spin $s = \frac{1}{2}, \frac{3}{2}, \dots$. Whilst there is some experimental evidence for the existence of hypothetical hadrons called 'glueballs,' which are composed purely of gluons, one can say, almost without qualification, that hadrons are composed of quarks. Mesons consist of bound pairs of quarks and anti-quarks, baryons consist of bound quark triples, and anti-baryons consist of bound anti-quark triples. Protons and neutrons, for example, are baryons.

Some hadrons decay via the strong interaction, while others decay via the weak interaction. The lifetime of those hadrons which decay via the weak interaction is of the order $10^{-10}$ s, whilst the lifetime of those which decay via the strong interaction is much shorter, at the order of $10^{-23}$ s. As a consequence, experimentalists were unable to use particle tracks in bubble chambers, or on photographic emulsions, to detect hadrons which decay via the strong interaction. These hadrons were dubbed 'resonances,' and many, but not all of them, are the excited states of longer lifetime hadrons which can be detected by means of particle tracks (Ne'eman and Kirsh, 1996, pp. 188–189). Those resonances which are the excited states of other hadrons are still classified as distinct particles because, by virtue of being the excited states of composite systems, they possess higher mass and spin than the ground state hadrons. The intrinsic spin of a composite system is the sum of the orbital angular momentum and the intrinsic spin of the components. In a ground state hadron, the orbital angular momentum is zero. The sum of the quark spins in a meson is either 1 or 0, and the sum of the quark spins in a baryon is either $\frac{1}{2}$ or $\frac{3}{2}$. The orbital angular momentum of a composite system is always a natural number multiple $\ell = 0, 1, 2, \ldots$ of $\hbar$, hence the excited meson resonances must have integral spin, and the excited baryon resonances must have half-integral spin (Ne'eman and Kirsh, 1996, pp. 211–212).

Whilst resonances could not be detected by particle tracks, their existence was inferred from the experimental data (Sternberg, 1994, pp. 284–286). For example, associated to any scattering reaction, such as $A+B \rightarrow C+D$, there is an experimental quantity called the cross-section, which is expressed in units of area, and which is related to the probability of the reaction taking place. The reaction cross-section is a function of the incident energy, and one can infer the existence of resonances from blips in the curve which plots cross-section as a function of incident energy. Fermi used this very technique to infer the existence of the $\Delta^{++}$ resonance in 1952 from pion-proton scattering $\pi^+ + p \rightarrow \pi^+ + p$ (Ne'eman and Kirsh, 1996, p. 182). Alternatively, the kinetic energy of certain decay products might have a statistical distribution, and one can infer the existence of intermediate particles from blips in this statistical distribution (Ne'eman and Kirsh, 1996, pp. 185–186).

It would be incorrect to say that whilst long-lifetime hadrons are directly observed, the existence of a resonance is only inferred from experiment. After all, the existence of a non-resonant hadron is itself inferred from the macroscopic phenomenon which is the particle track. However, it is clear that there is a second stage of inference to the existence of resonances: the experimental particle physicist uses the macroscopic experimental data to infer the existence of certain decay products, and thence infers the prior existence of other, intermediate particles. The inferred existence of quarks, as the constituents of hadrons, is yet another step removed from observation. For example, the first evidence for the existence of the $c$ quark came in 1974 from the discovery of a new meson, the $J/\psi$, whose properties were explained by proposing it to be composed of a $c\bar{c}$ quark–anti-quark pair (Ne'eman and Kirsh, 1996, pp. 229–235).[13] Similarly, the existence of the $b$ quark was inferred in 1977 from the discovery at Fermilab of a new resonance, the $\Upsilon$, interpreted as a quark–anti-quark meson of composition $b\bar{b}$ (Sternberg, 1994, p. 300).

In subsequent chapters we will see that the interacting *elementary* particles in each fermion generation can be partitioned into so-called 'multiplets' by finite-dimensional irreducible representations of the standard model gauge group $SU(3) \times SU(2) \times U(1)$. However, *composite* particles can also be partitioned into multiplets by group representations. In particular, the hadrons can be partitioned into hadron multiplets (Derdzinski, 1992, pp. 138–154). Specifically, the set of hadrons are partitioned into multiplets by representations of $SU(n)$, for each $2 \leqslant n \leqslant 6$. These symmetries are referred to as hadron symmetries. The partition is different for each value of $n$.

[13] This meson resonance was independently discovered by a team at the Brookhaven National Laboratory, and another team at the Stanford Linear Accelerator Centre (SLAC). It was called the $J$ by the team at the Brookhaven National Laboratory, and dubbed the $\psi$ by the team at SLAC.

The set of 6 quark flavours can also be partitioned into multiplets by representations of $SU(n)$, for each $2 \leqslant n \leqslant 6$. These symmetries are referred to as flavour symmetries. Once again, the partition is different for each value of $n$. In the case of $n = 3$, it is important to distinguish this flavour symmetry group from the *bona fide* $SU(3)$ gauge group of the strong force, often called the $SU(3)$ colour symmetry group.

As O'Raifeartaigh notes (1986, pp. 66–67), one can take the existence of quarks as the starting point, and one can derive the $SU(n)$ representations for the hadrons, or, alternatively, one can start with the $SU(n)$ representations for hadrons, and one can then infer the existence of quarks. Indeed, the observation that no hadrons belong to the lowest dimensional irreducible representation of $SU(3)$ on $\mathbb{C}^3$, led Gell-Mann and Zweig, independently, to postulate in 1964 the existence of the $u, d, s$ quarks as orthogonal vectors in the $\mathbb{C}^3$ representation, and to explain the spectrum of hadrons as various compositions of these quarks.[14] In this section we shall mostly adopt the first approach.

In general, if a hadron or a quark is represented by a free-particle bundle $\eta$, then one obtains a hadron or quark multiplet by taking the tensor product $\eta \otimes (\mathcal{M} \times W)$, where $W$ is a complex vector space possessing an $SU(n)$-structure, for $2 \leqslant n \leqslant 6$. The number of hadrons or quarks in the multiplet equals the dimension of $W$.

For each value of $n$, the lightest $n$ quarks belong to an $SU(n)$ quark flavour $n$-plet, with the remaining quarks each belonging to an $SU(n)$ singlet. Given that the free-particle bundle of each quark flavour is the Dirac spinor bundle $\sigma$, the $SU(n)$ quark flavour $n$-plet is housed by the vector bundle $\sigma \otimes (\mathcal{M} \times \mathbb{C}^n)$, where $\mathbb{C}^n$ possesses the $SU(n)$-structure which corresponds to the standard representation of $SU(n)$. Given a choice of oriented, orthonormal basis in $\mathbb{C}^n$, this vector bundle can then be decomposed into an $n$-fold direct sum of $\sigma$.

For example, for $n = 3$, the $u, d, s$ quarks belong to the $SU(3)$ quark flavour triplet. The free-particle bundle for this triplet is $\sigma \otimes (\mathcal{M} \times \mathbb{C}^3)$. Given the choice of an oriented, orthonormal basis in $\mathbb{C}^3$, this bundle decomposes into $\sigma \oplus \sigma \oplus \sigma$.

Let us consider the baryons and mesons composed of the $u$, $d$, $s$ quarks and their anti-quarks. Given the strong force interacting-particle bundle for an individual quark, $\sigma \otimes \rho$, a meson is represented by a free-particle bundle $\eta$ satisfying the map

$$S_0^\ell T^*\mathcal{M} \otimes \sigma \otimes \rho \otimes \bar{\sigma} \otimes \bar{\rho} \to \eta,$$

and a baryon is represented by a free-particle bundle $\eta$ satisfying the map

$$S_0^\ell T^*\mathcal{M} \otimes \sigma \otimes \rho \otimes \sigma \otimes \rho \otimes \sigma \otimes \rho \to \eta.$$

The $SU(3)$ meson multiplets are the multiplets of mesons composed of the $u$, $d$, $s$ quarks and anti-quarks. To obtain the vector bundles which represent these meson multiplets, one begins by forming the $u, d, s$ interacting quark multiplet bundle $\sigma \otimes \rho \otimes (\mathcal{M} \times \mathbb{C}^3)$ and the $u, d, s$ interacting anti-quark multiplet bundle $\bar{\sigma} \otimes \bar{\rho} \otimes (\mathcal{M} \times \bar{\mathbb{C}}^3)$.

[14] Zweig proposed to name the constituents of hadrons as 'aces.' Zweig's term has yet to catch-on.

Then one forms the 2-fold tensor product bundle

$$\sigma \otimes \rho \otimes (\mathcal{M} \times \mathbb{C}^3) \otimes \bar{\sigma} \otimes \bar{\rho} \otimes (\mathcal{M} \times \overline{\mathbb{C}^3}).$$

Adding the orbital angular momentum space, and re-arranging, one obtains

$$S_0^\ell T^*\mathcal{M} \otimes \sigma \otimes \rho \otimes \bar{\sigma} \otimes \bar{\rho} \otimes (\mathcal{M} \times \mathbb{C}^3 \otimes \overline{\mathbb{C}^3}).$$

The states of a bound multi-quark system must be totally antisymmetric with respect to the colour degree of freedom, hence the surjective bundle morphism for mesons,

$$S_0^\ell T^*\mathcal{M} \otimes \sigma \otimes \rho \otimes \bar{\sigma} \otimes \bar{\rho} \to \eta,$$

must include the projection $\rho \otimes \bar{\rho} \to \rho \wedge \bar{\rho}$. One then obtains

$$\eta \otimes (\mathcal{M} \times \mathbb{C}^3 \otimes \overline{\mathbb{C}^3}).$$

Next, one looks for a decomposition of the representation of $SU(3)$ on $\mathbb{C}^3 \otimes \overline{\mathbb{C}^3}$ into irreducible direct summands. For each irreducible direct summand $W$, the vector bundle $\eta \otimes (\mathcal{M} \times W)$ represents a meson multiplet, with the dimension of $W$ being the number of mesons in the multiplet. In the case of $\mathbb{C}^3 \otimes \overline{\mathbb{C}^3}$, it decomposes into the direct sum of the 8-dimensional adjoint representation of $SU(3)$, and the 1-dimensional trivial representation. Hence, there are 8 mesons composed of $u$, $d$, $s$ quarks in one $SU(3)$ octet, and one $u$, $d$, $s$ meson in an $SU(3)$ singlet.

Each $SU(n)$ quark or hadron multiplet is contained in a unique $SU(n+1)$ multiplet (Derdzinski, 1992, p. 147), and one consequence of this is that each $SU(3)$ multiplet is a union of $SU(2)$ multiplets. The $SU(2)$ hadron multiplets are called strong isospin multiplets (this type of isospin is distinct from weak isospin). The members of each strong isospin multiplet have the same spin and parity, their masses are almost equal, and their electric charges form a progression of step 1 (Derdzinski, 1992, p. 36). For example, take the following strong isospin meson multiplets: the pion-triplet $\pi$ consists of the $\pi^-$, the $\pi^0$, and the $\pi^+$; the kaon-doublet $K$ consists of the $K^0$ and the $K^+$; the anti-kaon doublet $\bar{K}$ consists of the $K^-$ and the $\bar{K}^0$; and the $\eta$-singlet consists of the $\eta^0$. The union of these strong isospin multiplets, $\pi \cup K \cup \bar{K} \cup \eta$, forms an $SU(3)$ meson octet. The $u$, $d$, $s$ quark composition of the mesons in this octet is as follows: $\pi^-(d\bar{u})$, $\pi^0(u\bar{u})$ or $\pi^0(d\bar{d})$, $\pi^+(u\bar{d})$, $K^-(\bar{u}s)$, $K^0(d\bar{s})$, $\bar{K}^0(\bar{d}s)$, $K^+(u\bar{s})$, and $\eta^0(u\bar{u})$ or $\eta^0(d\bar{d})$ or $\eta^0(s\bar{s})$ (Derdzinski, 1992, p. 100).

Each of the particles in this meson octet possess a constant value for the third component of strong isospin $I_3$, and a constant value for the strong hypercharge $Y$ (to be distinguished from weak hypercharge). Together, these values can be used to plot the members of the meson octet as points on a 2-dimensional plane. So doing reveals that the particles in the meson octet correspond to the weight vectors in the eight-dimensional adjoint representation of $SU(3)$ (Sternberg, 1994, p. 287). $I_3$ and $Y$ can be represented as elements which span a maximal commuting subalgebra in the Lie algebra $\mathfrak{su}(3)$. Recall that a weight vector in a Lie group representation is a simultaneous eigenvector of the operators representing a maximal commuting subalgebra (a 'Cartan' subalgebra) from the Lie algebra of the Lie group. The weights of such a vector are the simultaneous eigenvalues.

The $SU(3)$ baryon multiplets are the multiplets of baryons composed of the $u$, $d$, $s$ quarks. To obtain the vector bundles which represent these baryon multiplets, one begins by forming the $u$, $d$, $s$ interacting quark multiplet bundle $\sigma \otimes \rho \otimes (\mathcal{M} \times \mathbb{C}^3)$, and then one forms the 3-fold tensor product bundle

$$\sigma \otimes \rho \otimes \left(\mathcal{M} \times \mathbb{C}^3\right) \otimes \sigma \otimes \rho \otimes \left(\mathcal{M} \times \mathbb{C}^3\right) \otimes \sigma \otimes \rho \otimes \left(\mathcal{M} \times \mathbb{C}^3\right).$$

Adding the orbital angular momentum space, and re-arranging, one obtains

$$S_0^\ell T^* \mathcal{M} \otimes \sigma \otimes \rho \otimes \sigma \otimes \rho \otimes \sigma \otimes \rho \otimes \left(\mathcal{M} \times \mathbb{C}^3 \otimes \mathbb{C}^3 \otimes \mathbb{C}^3\right).$$

Once again, the states of a bound multi-quark system must be totally antisymmetric with respect to the colour degree of freedom, hence the surjective bundle morphism for baryons,

$$S_0^\ell T^* \mathcal{M} \otimes \sigma \otimes \rho \otimes \sigma \otimes \rho \otimes \sigma \otimes \rho \to \eta,$$

must include the projection $\rho \otimes \rho \otimes \rho \to \rho \wedge \rho \wedge \rho$. One then obtains

$$\eta \otimes \left(\mathcal{M} \times \mathbb{C}^3 \otimes \mathbb{C}^3 \otimes \mathbb{C}^3\right).$$

Next, one looks for a direct sum decomposition of the representation of $SU(3)$ on $\mathbb{C}^3 \otimes \mathbb{C}^3 \otimes \mathbb{C}^3$ into irreducible direct summands. For each irreducible direct summand $W$, the vector bundle $\eta \otimes (\mathcal{M} \times W)$ represents a baryon multiplet, with the dimension of $W$ being the number of baryons in the multiplet. In the case of $\mathbb{C}^3 \otimes \mathbb{C}^3 \otimes \mathbb{C}^3$, it decomposes into the direct sum of the 10-dimensional space $S^3(\mathbb{C}^3)$ of symmetric tensors, two copies of the 8-dimensional adjoint representation, and the 1-dimensional space of antisymmetric tensors $\bigwedge^3(\mathbb{C}^3)$. Hence, there are 10 baryons composed of $u$, $d$, $s$ quarks in one $SU(3)$ decuplet, two $SU(3)$ octets containing $u$, $d$, $s$ baryons, and one $u$, $d$, $s$ baryon in an $SU(3)$ singlet.

Recall that each $SU(3)$ multiplet is a union of $SU(2)$ strong isospin multiplets. Again, the members of each strong isospin baryon multiplet have the same spin and parity, their masses are almost equal, and their electric charges form a progression of step 1. For example, take the following strong isospin baryon multiplets: the nucleon doublet $N$, contains the neutron $N^0 = n$ and the proton $N^+ = p$; the lambda singlet $\Lambda$ contains just the lambda hyperon $\Lambda^0$; the sigma-hyperon triplet $\Sigma$ contains the $\Sigma^-$, the $\Sigma^0$, and the $\Sigma^+$; and the xi-hyperon doublet $\Xi$ contains the $\Xi^-$ and the $\Xi^0$. The union of these strong isospin multiplets, $N \cup \Lambda \cup \Sigma \cup \Xi$, forms an $SU(3)$ baryon octet. The $u$, $d$, $s$ quark compositions of these baryons is as follows: $N^+(uud)$, $N^0(ddu)$, $\Lambda^0(uds)$, $\Sigma^-(dds)$, $\Sigma^0(uds)$, $\Sigma^+(uus)$, $\Xi^-(ssd)$, $\Xi^0(ssu)$ (Derdzinski, 1992, p. 100). Each of the particles in this baryon octet possess a constant value for the third component of strong isospin $I_3$, and a constant value for the strong hypercharge $Y$, and, as with the members of the meson octet, plotting these values reveals that the particles in the baryon octet correspond to the weight vectors in the eight-dimensional adjoint representation of $SU(3)$ (Sternberg, 1994, p. 286). The meson and baryon resonances composed of the $u$, $d$, $s$ quarks can also be organised into multiplets by representations of $SU(3)$ (see Sternberg, 1994, pp. 286–288).

## 4.13. Elementary particles and intrinsic properties

One of the interpretational themes of this text is to address the question: 'What is an elementary particle is the first-quantized standard model?'. In this section we consider the possibility that the elementarity of a free particle can be defined in terms of the local space–time symmetry group and the notion of intrinsic and extrinsic properties. As before, we use the phrase 'free particle' in the knowledge that we are really speaking of free matter fields and free gauge fields, and the particle concept may only be derivative from the notion of such fields.

In particular, two related claims will be critically appraised in this section: (i) that an elementary particle has only one intrinsic state; and (ii) that the intrinsic properties of an elementary particle equal the invariant properties defining the irreducible representation of the local space–time symmetry group.

Let us begin by recalling some key concepts. The state of a physical object is the set of all properties possessed by that object at a moment in time. To echo Section 2.5, let us agree to define an intrinsic property of an object to be a property which the object possesses independently of its relationships to other objects, and let us also agree to define an extrinsic property of an object to be a property which the object possesses depending upon its relationships with other objects. To reiterate, if the value of a quantity possessed by an object can change under a change of reference frame, then the value of that quantity must be an extrinsic property of the object, not an intrinsic property. The value of such a quantity must be a relationship between the object and a reference frame, and under a change of reference frame, that relationship can change. As Fleming and Butterfield comment, "in quantum theory the statevectors do *not* represent the intrinsic physical situation (in the Heisenberg picture: the intrinsic history) of the system. Rather they represent the relationships of the physical situation or history to the reference frame for which they are employed. This is why a change of reference frame entails a change of statevector for the same physical history (a passive transformation)" (1999, p. 112).

Recall that each free particle corresponds to a unitary representation of the local, external (space–time) symmetry group $SL(2, \mathbb{C}) \ltimes \mathbb{R}^{3,1}$. In the 'passive' approach to external symmetries, $SL(2, \mathbb{C}) \ltimes \mathbb{R}^{3,1}$ acts upon the set of (local) inertial reference frames. Each $g \in SL(2, \mathbb{C}) \ltimes \mathbb{R}^{3,1}$ maps a reference frame $\sigma$ to a reference frame $g\sigma$. For each type of free particle, the group element $g$ is represented by a unitary linear operator $U_g$ on a Hilbert space. If $v$ is the state of a system as observed from a reference frame $\sigma$, then $w = U_g v$ will be the state of the system as observed from the reference frame $g\sigma$. If $A$ is a self-adjoint operator representing a physical quantity, then, in general, the expectation value $\langle v, Av \rangle$ will differ from the expectation value $\langle U_g v, AU_g v \rangle$. J.M.G. Fell argues that if $v$ and $w$ are a pair of unit vectors in a Hilbert space such that $w = U_g v$ for some $g \in SL(2, \mathbb{C}) \ltimes \mathbb{R}^{3,1}$, then "in a sense," $v$ and $w$ "(or rather the rays through them) describe the same 'intrinsic state'… for the transition from one state to the other can be exactly duplicated by a change in the standpoint of the observer" (Fell and Doran, 1988, pp. 30–31).

Note that what we have here is a sufficient condition for a property to be an extrinsic property: If a property possessed by an object changes under a change of reference frame, then that property must be an extrinsic property of the object, not an intrinsic property. This does not entail that the only extrinsic properties are those which change under a change of reference frame. Conceivably, an object could stand in some relationship to the rest of the universe, and that relationship could be invariant under a change of reference frame. In such an event, one would have an extrinsic property which is invariant under a change of reference frame. Thus, variation under a change of reference frame may not be a necessary condition for a property to be extrinsic. Conversely, invariance under a change of reference frame may not be a sufficient condition for a property to be intrinsic.

The intrinsic state of a physical object can be defined as the set of all intrinsic properties possessed by that object at a moment in time. When the intrinsic state of an object doesn't change, it means that the intrinsic properties of the object don't change. The extrinsic properties of an object, its relationships with other objects, in particular its relationships with a reference frame, can change even if the intrinsic properties of the object don't change. Hence, the intrinsic state of an object can remain unchanged even though the overall state of the object, taking into account its extrinsic properties, does change.

Let us assume for the purpose of argument that the intrinsic properties of a physical system can be defined to be those which are invariant under the action of the physical symmetry group, and the extrinsic properties can be defined to be those which are not invariant under the action of the physical symmetry group. One might be able to find support for this proposal in Weyl's well-known text *Symmetry* (1952, pp. 127–133). For example, believing at the time that spatial reflections were physical symmetries, Weyl asserts that the physical symmetry group "contains the reflections because no law of nature indicates an *intrinsic* difference between left and right" (1952, p. 129, with my italics). However, one can interpret this as merely the claim that if a property is intrinsic, then it must be invariant under the action of the symmetry group. One could accept this, but reject the claim that if a property is invariant under the action of the symmetry group, then it must be intrinsic.

The issue is slightly confused by Weyl's interest in defining what it is for a property to be objective. Weyl states that "objectivity means invariance with respect to the group of [physical] automorphisms" (1952, p. 132). When Weyl speaks of two congruent squares in the same plane which "may show many differences when one regards their relation to each other" (1952, p. 127), he then states that "if each is taken by itself, any objective statement made about one will hold for the other" (1952, p. 128). If this suggests that Weyl identifies objective properties with intrinsic properties, then by defining objectivity to be invariance under the symmetry group, Weyl means to define intrinsic properties as those which are invariant under the symmetry group.

The difficulty with this is the plausibility of identifying objective properties with intrinsic properties. If one accepts that the extrinsic properties of objects, the relationships between objects, can also be objective, then one would reject the claim that the only objective properties of objects are those which are invariant under the action of the physical symmetry group. For example, the speed of a particle is an objective extrinsic property of a particle, an objective relationship between that particle and a reference frame. The fact that the speed of a particle is not invariant under a change of reference frame does not entail that the speed of a particle is not an objective property.

Let us turn then to elementary particles. Whilst speculative ventures such as string theory and supersymmetry attempt to reduce the number of different types of elementary particle, it is perfectly possible that all the different leptons, quarks and gauge bosons are elementary. There is no contradiction between the notion that there are multiple different types of elementary particle, and the mereological notion of an elementary particle as a thing which has no parts.

There is also no contradiction with the notion of elementarity analysed in this section: Each different type of elementary particle is specified by the values of invariant properties, such as mass, spin and charge. If such invariant properties are considered to be intrinsic properties, then it follows that the different types of elementary particle do not differ by virtue of the relationships they stand in to other objects, but by virtue of the intrinsic properties they possess. In subsequent chapters we will see that the spectrum of interacting elementary particles, and the definition of what an interacting elementary particle is, changes with different gauge groups: If one interprets a gauge transformation to be a change of 'internal' reference frame, then the larger the gauge group, the smaller the set of intrinsically distinct interacting elementary particles, and the smaller the set of genuine intrinsic properties. However, this section is not concerned with the question of whether intrinsically distinct types of elementary particle exist. Rather, the debate in this section concerns the properties of the individual particles of a fixed elementary particle type. Such properties can vary, even though elementary particles have no parts. At the very least, this is because the extrinsic properties of a particle, its relationships to other objects, are variable. However, this still leaves open the question: Can the intrinsic properties of an elementary particle also vary, or does an individual elementary particle have only one intrinsic state, only one set of intrinsic properties?

Wigner proposed that an *elementary* particle corresponds to an *irreducible* representation of the physical symmetry group, and if it is also assumed that the intrinsic properties of an object are those which are invariant under the action of that group, then it follows that the intrinsic properties of an elementary particle are those which are invariant under an irreducible representation of the physical symmetry group. However, this is distinct from the proposal that an elementary particle has only one intrinsic state, most clearly expressed by J.M.G. Fell[15] (Fell and Doran, 1988, pp. 29–32).

Fell adopts Wigner's notion that the irreducibility of a representation is the defining characteristic of an elementary particle representation, and argues that the group action is "essentially" transitive upon the state space of such a representation. He argues, therefore, that an elementary particle has only one 'intrinsic' state: "It can never undergo any intrinsic change. Any change which it *appears* to undergo (change in position, velocity, etc.) can be 'cancelled out' by an appropriate change in the frame of reference of the observer. Such a material system is called an *elementary system* or an *elementary particle*. The word 'elementary' reflects our preconception that, if a physical system undergoes an intrinsic change, it must be that the system is 'composite,' and that the change consists in some rearrangement of the 'elementary parts' " (Fell and Doran, 1988, p. 31). Fell implies that when an elementary system is observed to undergo a change within some reference frame $\sigma$, it is the particle's relationship to the reference frame $\sigma$ which changes, not any of the particle's intrinsic, non-relational properties.

The notion that an elementary particle has only one intrinsic state can be found elsewhere in the physics and philosophy of physics literature, albeit not necessarily in such an explicit form. For example, French and Rickles (2003) equate state-independent properties such as mass, charge and spin, with intrinsic properties (p. 221), and equate state-dependent properties with non-intrinsic properties (p. 226). French states that "two or more electrons, say, possess all intrinsic properties in common" (2006, Section 4), on the basis that intrinsic properties are equated with the state-independent properties that define a particle type. Similarly, Castellani states that "in physics, the main invariants are quantities like mass (rest-mass), spin, charge — those properties of physical particles which are usually known as 'intrinsic properties' " (1998, p. 10). If one equates intrinsic properties with state-independent properties, then this entails that there can only be one intrinsic state. There is, indeed, a distinction between the state-independent properties of a particle, which define the particle type, and the state-dependent properties of a particle, which are variable for a fixed type of particle; and there may, indeed, be some intrinsic properties, such as mass, charge and spin, which are state-independent, but this does not entail that all intrinsic properties are state-independent.

Note that both composite particles and elementary particles can undergo transmutations into other particle types, and thereby change their intrinsic properties, but, at first sight at least, this is not the same thing as changing the intrinsic state of a fixed particle type. For example, a neutron can change into a proton via $\beta$-decay, thereby losing mass and gaining charge, both of which are putative intrinsic properties. As

[15] Private communication with R.S. Doran.

explained in Section 4.8, under weak interaction processes, a $d_L$ quark can change into a $u_L$ quark, accompanied by the emission of a $W^-$ gauge boson. The quark composition of a neutron is $(ddu)$, and under the $\beta$-decay of a neutron, one of the down quarks in the neutron changes into a $u$ quark. The resulting composite system $(uud)$ is a proton. The $W^-$ gauge boson then decays into a left-handed electron $e_L$ and an anti-electron-neutrino $\bar{\nu}_e$, the conventional products of $\beta$-decay. Electric charge is a conserved quantity as well as an intrinsic property, hence the electron carries the unit of negative charge which cancels the unit increase in the charge of the nucleon. This example demonstrates that elementary particles, despite the fact that they have no parts, can undergo decay processes. A left-handed down quark, of charge $-1/3$, can decay into a left-handed up-quark, of charge $+2/3$ and lower mass, by emitting a $W^-$ gauge boson. Whilst elementary particles cannot shed parts, they can undergo transmutations into different elementary particle types, defined by different intrinsic properties.

Note also that, as explained in Section 4.11, the unitary, irreducible representations of $SL(2, \mathbb{C}) \ltimes \mathbb{R}^{3,1}$ in the one-particle, first-quantized theory can be used to represent stable, composite particles as well as elementary particles.[16] Hence, the irreducibility of a representation of $SL(2, \mathbb{C}) \ltimes \mathbb{R}^{3,1}$ does not entail the elementarity of the corresponding particle. Moreover, if irreducibility entails only one intrinsic state, then stable, composite systems would also have only one intrinsic state.

In addition, note that the Fock spaces used to represent elementary physical systems in the second-quantized theory do not possess irreducible representations of $SL(2, \mathbb{C}) \ltimes \mathbb{R}^{3,1}$. Given that Fock spaces are multi-particle spaces, often said to represent aggregates of elementary particles (as excitation modes of an underlying quantum field), there is not necessarily any inconsistency here with the notion that elementary particles correspond to irreducible representations.

However, the argument that irreducibility itself entails a single intrinsic state is flawed. The argument only has plausibility if one thinks in terms of classical particle mechanics. In quantum theory, there is no reason why the irreducibility of a particle representation should entail that there is only one intrinsic state. The extended, field-like aspects of particle-states in quantum theory (i.e., the fact that they are wave-functions), make for an infinite-dimensional state space. This is true in non-relativistic quantum mechanics, first-quantized relativistic quantum theory and second-quantized relativistic quantum theory. Because the state space of a free elementary particle is represented in the first-quantized theory by an *infinite-dimensional*[17] irreducible representation of $SL(2, \mathbb{C}) \ltimes \mathbb{R}^{3,1}$, the finite-dimensional space–time symmetry group cannot act transitively upon such a state

[16] It is well known that stable, composite systems correspond to irreducible representations, and, in fact, Newton and Wigner (1949) distinguish between 'elementary systems' and 'elementary particles,' the former being the broader class, defined by an irreducible representation, and including stable, composite systems.

[17] Assuming space is represented to be a continuum, the range of possible values of a position or momentum observable will also be a continuum. Such an observable must be represented in quantum

space. $SL(2,\mathbb{C}) \ltimes \mathbb{R}^{3,1}$ is ten-dimensional, hence the orbits of its action on a state-space are, at most, ten-dimensional. Given that the state spaces are infinite-dimensional, this entails that there is an uncountable infinity of orbits of the symmetry group $SL(2,\mathbb{C}) \ltimes \mathbb{R}^{3,1}$. If one accepts that intrinsic properties are those invariant under the symmetry group, this entails that each different orbit corresponds to a different intrinsic state. There are many changes in the state of an elementary particle which cannot be cancelled out by a change in observational standpoint. In fact, there are an uncountable infinity of such changes! This is essentially because the state of an elementary particle (in the first-quantized, one-particle theory), is represented by a field-like object, a cross-section of a vector bundle, and the value of the cross-section can change in an independent fashion at different points of space–time. A change of reference frame, in the special relativistic sense mandated by $SL(2,\mathbb{C}) \ltimes \mathbb{R}^{3,1}$, is a more rigid, global transformation. $SL(2,\mathbb{C})$ acts transitively[18] upon the set of one-dimensional subspaces in the typical fibre of a free-particle bundle $\eta$, but the transformation

$$\psi(x) \mapsto \psi'(x) = \mathscr{D}^{s_1,s_2}(A) \cdot \psi\big(\Lambda^{-1}(x-a)\big),$$

permits only a global $SL$-symmetry of the fibres, and a shift in the field values assigned to coordinate quadruples. The idea that an elementary particle has only one intrinsic state is destroyed by the infinite-dimensional nature of particle representations in quantum theory.

One needs to carefully distinguish the false notion that the space–time symmetry group acts transitively upon the quantum state space of an elementary system, from the correct notion that any vector in such a state space is 'cyclic' with respect to the action of the space–time symmetry group. For example, Fleming and Butterfield correctly state that for the state space of an elementary system, equipped with an irreducible representation of the space–time symmetry group, all the states "are to be obtainable by superposing the images, under elements of the spacetime symmetry group, of any state" (1999, p. 114). In other words, if one takes the orbit of the action of the space–time symmetry group upon an arbitrary vector, and if one then takes the (closure of the) complex linear span of all the elements in the orbit, then one does indeed obtain the entire state space. The vector chosen is said to be a cyclic vector, and the representation is said to be cyclic. In fact, any irreducible representation of a locally compact group must be cyclic. In the case of an irreducible representation of the space–time symmetry group, the orbit of a single state takes one through a sufficient number of orthogonal states to span the entire infinite-dimensional state space. However, this does not entail that there is only one intrinsic state! The mathematical operation of taking a linear combination of a set of states doesn't correspond to a change of physical reference frame.

theory by a self-adjoint operator with a continuous spectrum, and such an operator requires an infinite-dimensional Hilbert space.

[18] Private communication with Shlomo Sternberg.

Kuhlmann asserts that "the physical justification for linking up irreducible representations with elementary systems is the requirement that 'there must be no relativistically invariant distinction between the various states of the system' (Newton and Wigner, 1949)... the state space of an elementary system must not contain any relativistically invariant subspaces... The requirement that a state space has to be relativistically invariant means that starting from any of its states it must be possible to get to all the other states by superposition of those states which result from relativistic transformations of the state one started with" (2006, Section 5.1.1). It is indeed true, by definition, that under an irreducible representation of the space–time symmetry group, there can be no non-trivial subspace which is invariant under the action of the symmetry group. However, for the reasons explained above, this does not entail that there can be no 'relativistically invariant distinction between the various states of the system.' There can indeed be such a distinction, defined by the different orbits of the symmetry group. Note also that Kuhlmann conflates an irreducible group representation with a cyclic representation; irreducibility is not the same thing as cyclicity.

It might be argued that the quantum state space is only infinite-dimensional because it contains all the possible superpositions of states. In the sense that the quantum state space is the span of the symmetry group orbit of a single intrinsic state, it is true to say that every state is merely a superposition of states with the same intrinsic properties. However, if one interprets quantum theory to provide a maximal description of nature, rather than merely providing an expression of incomplete knowledge, then each such linear combination of states, must itself be an equally legitimate state, and a state which, in general, lies within a different orbit of the symmetry group action, and therefore defines a different intrinsic state.

Mathematically, it is quite possible to introduce an infinite-dimensional group of external symmetries. Each fibre of a free-particle bundle $\eta$ is equipped with an $SL(2, \mathbb{C})$ structure, and there is an automorphism bundle $SL(\eta)$, consisting of all the automorphisms in each fibre of $\eta$. The typical fibre of $SL(\eta)$ is isomorphic to $SL(2, \mathbb{C})$. The space of cross-sections $\mathcal{E} = \Gamma(SL(\eta))$ is the group of vertical bundle automorphisms of $\eta$. $\mathcal{E}$ provides an infinite-dimensional group which acts upon the cross-sections of the free-particle bundle $\eta$. Given a cross-section $\psi(x)$ of $\eta$, and an element $a(x)$ of $\mathcal{E}$, the cross-section is simply mapped to $a(x)\psi(x)$. It seems reasonable to call $\mathcal{E} = \Gamma(SL(\eta))$ a group of external (space–time) symmetries because it provides a double cover of $\Gamma(SO_0(T\mathcal{M}))$, the infinite-dimensional group of local oriented Lorentz transformations. This is the group of vertical automorphisms of the oriented Lorentz frame bundle. The latter consists of all the orthonormal bases $\{e_\mu\colon \mu = 0, 1, 2, 3\}$ of the tangent spaces at all the points of the manifold $\mathcal{M}$, such that each $e_0$ is a future-pointing, timelike vector, and such that each $\{e_i\colon i = 1, 2, 3\}$ is a right-handed triple of spacelike vectors. This principal fibre bundle has the restricted Lorentz group $SO_0(3, 1)$ as its structure group. A cross-section of the automorphism bundle $SO_0(T\mathcal{M})$ selects a linear isometry of the tangent space at each point, and thereby maps an oriented Lorentz frame at each point into another oriented Lorentz frame.

To reiterate, whilst $SL(2, \mathbb{C}) \ltimes \mathbb{R}^{3,1}$ does act upon cross-sections of $\eta$ as well as the base space, it does not act transitively upon the space of cross-sections. Given that $SL(2, \mathbb{C})$ acts transitively upon the set of one-dimensional subspaces in the typical fibre of $\eta$, and given that the elements in the infinite-dimensional group $\mathcal{E} = \Gamma(SL(\eta))$ provide a locally variable choice of $SL$-symmetry, $\mathcal{E}$ does act transitively upon the space of cross-sections. Hence, if $\mathcal{E}$ were a physical symmetry group, then a free elementary particle (in the first-quantized theory), would only have one intrinsic state.

Recall, however, that a free-particle bundle $\eta$ houses many different particle species. The various $\mathscr{H}_{m,s}$ which are constructed out of cross-sections of $\eta$ are not invariant under the action of the infinite-dimensional group $\mathcal{E}$. Whilst one particular particle may be represented by the space constructed from the mass $m$, positive-energy solutions of a differential equation in $\eta$, the group $\mathcal{E}$ is more than capable of mapping such cross-sections into objects which solve that differential equation for a different mass value, or which don't solve the equation at all. The automorphism group of each $\mathscr{H}_{m,s}$ is the unitary group $\mathscr{U}(\mathscr{H}_{m,s})$, into which $SL(2, \mathbb{C}) \ltimes \mathbb{R}^{3,1}$ is mapped. $\mathcal{E}$ is not a group of automorphisms of any $\mathscr{H}_{m,s}$, even if it is the group of vertical automorphisms of $\eta$.

In Fell's argument for the uniqueness of the intrinsic state, he includes changes of velocity (i.e., accelerations), amongst the things which can be cancelled out by a change of reference frame. This implies that he is not merely thinking of the transformations between inertial reference frames provided by $SL(2, \mathbb{C}) \ltimes \mathbb{R}^{3,1}$, but general coordinate transformations. This leads us to consider interacting particles, given that the latter are capable of undergoing acceleration.

In the case of an interacting, first-quantized, elementary fermion, one forms, in the simplest case, an interacting-particle bundle $\eta \otimes \delta$, and in contrast with the free-particle case, one *does* use the infinite-dimensional group of vertical automorphisms of $\delta$ as a physical symmetry group. This, then, is a significant difference between external symmetries and internal symmetries. The internal symmetry group is the *infinite-dimensional* group of cross-sections $\mathcal{G} = \Gamma(G(\delta))$ of an automorphism bundle $G(\delta)$. This entails that any change in the internal degrees of freedom of an interacting particle, even if the change occurs in an independent fashion at different points in space–time, can be cancelled out by an internal symmetry (i.e., a gauge transformation). This allows the group of internal symmetries to act transitively upon the infinite-dimensional space of internal states of an interacting particle. The gauge groups $SU(3)$, $U(2)$, $SU(2)$, and $U(1)$ act transitively[19] upon the set of one-dimensional subspaces in the typical fibres of the relevant interaction bundles, and because an internal symmetry is, in each case, a locally varying cross-section of the corresponding $G(\delta)$, the infinite-dimensional group $\mathcal{G} = \Gamma(G(\delta))$ acts transitively upon the space of internal states of an interacting elementary particle. It is the external degrees of freedom which prevent an elementary particle, free or interacting, from having only one intrinsic state.

---

[19] Private communication with Shlomo Sternberg.

As a brief digression, let us consider the possible implications of the argument in this section for Mach's principle. This is not to endorse Mach's principle, merely to establish the logical relationship between it and the argument in this section. In particular, consider the following two sub-principles of Mach's principle:

- The rest mass of an object is a relationship, and in particular, an interaction, between that object and all the other objects in the universe.
- The acceleration or non-acceleration of an object is a relationship between that object and all the other masses in the universe.

If intrinsic properties are defined as those invariant under the local space–time symmetry group, then given that the latter only corresponds to transformations between local inertial reference frames, it follows that whether or not a particle has non-zero acceleration is an intrinsic property of the particle. A transition from zero acceleration to non-zero acceleration, or vice versa, is an intrinsic change of state. This is clearly inconsistent with the second sub-principle of Mach's principle, which, in effect, holds that acceleration is an extrinsic property, a relationship between a local system and the rest of the universe. A general coordinate transformation can change non-zero acceleration into zero acceleration, and vice versa, hence one could retain the second sub-principle of Mach's principle if one defined intrinsic properties to be those which are invariant under general coordinate transformations. More generally, however, one can reject the claim that intrinsic properties include all those which are invariant under a group of transformations between reference frames. Equivalently, one can reject the claim that extrinsic properties are only those which vary under a change of reference frame. As stated earlier in the section, if a property changes under a change of reference frame, then that is a sufficient condition for the property to be an extrinsic property, but one can reject the claim that variance under a change of reference frame is a necessary condition for a property to be extrinsic. If there is a relationship between an object and the rest of the universe which is invariant under a change of reference frame, then that relationship would be both an extrinsic property of the object, and a property invariant under a change of reference frame.

The first sub-principle of Mach's principle holds that the rest mass of an object is also an extrinsic property of the object. No change of reference frame, whether it belongs to the local space–time symmetry group, or whether it belongs to the group of general coordinate transformations, can change the observed rest mass of an object. Hence, if one defined intrinsic properties to be those invariant under a change of reference frame, rest-mass would be an intrinsic property. To retain the first sub-principle of Mach's principle it would be necessary to reject the claim that extrinsic properties are only those which vary under a change of reference frame. If the rest-mass of an object is a relationship to the rest of the objects in the universe, and if this relationship is invariant under a change of reference frame, then the rest-mass would be an extrinsic property which is invariant under a change of reference frame.

To return to the main line of argument, Fell claims that a composite object can possess different intrinsic properties at different times, but he also appears to hold what philosophers would call an 'endurantist' notion of the persistence of an object through time. The endurantist position holds that the same whole object is capable of possessing a property at one time, and not possessing that property at another time. One popular endurantist account of change holds that properties which are capable of being possessed by an object at one time, and not being possessed at another time, are properties which are possessed in relation to certain times (Weatherson, 2002, Section 1.1). As Hawley puts it, "objects change by standing in different relations to different times" (Hawley, 2004, Section 3). If moments of time correspond to the state of other objects in the universe, then one might argue that properties which are possessed by an object in relation to certain times, must be extrinsic properties. Under this argument, then, the endurantist position entails that all properties capable of change must be extrinsic properties. Under the endurantist view, one might have to concede that the changing properties of all objects, composite or elementary, are extrinsic properties.

To render the notion of variable intrinsic properties consistent with endurantism may require one of the following lines of attack: Firstly, one might argue that an object can possess an 'internal' clock, hence the claim that an object can only possess a changing property in relation to certain times does not entail that such properties are only possessed by an object depending upon its relationships with other objects. Secondly, one might argue that a changing property can be an intrinsic property even if the times at which it is possessed by an object are relationships between that object and other objects in the universe. The intrinsic-ness of a property, one might argue, is not affected by the relationships which are necessary to define the times at which it is possessed.

There is an alternative to endurantism, dubbed the 'perdurantist' view, which holds that an object has temporal parts, and different temporal parts can possess different properties. In particular, under the perdurantist view the different temporal parts can possess different intrinsic properties. On the perdurantist view, the persistence of an object through time is analogous to the extension of an object in space, and the different temporal parts can possess different properties just as much as the different spatial parts of an object can possess different properties (Hawley, 2004, Section 1).

In perdurantism, the ascription of a property to an object at a particular time corresponds to the ascription of a property to a temporal part of a 4-dimensional object. The proposition '$x$ possesses $F$ at time $t$' means that '$x$' is a 4-dimensional object which has a temporal part '$t$' possessing the property '$F$.' Quentin Smith describes the notion of temporal parts in these terms: "If an object $x$ is a whole of temporal parts, then $x$ is composed of distinct particulars, each of which exists at one instant only, such that whatever property $x$ is said to have at a certain time is [possessed by] the particular (temporal part) that exists at that time" (Smith, 1995, p. 84). With the notion of temporal parts, an object can be defined to undergo change if "one temporal part of $x$ possesses a certain property $F$ at one time and... another temporal part of $x$ does not possess $F$ at another time" (Smith, 1995, p. 84). Smith contrasts the 'temporal parts' notion of change with the endurantist notion of change that "the particular that possesses the property at one time is identical with the particular that does not possess the property at another time" (Smith, 1995, p. 84).

If the arguments in this section are correct, then any object, composite or elementary, can possess different intrinsic properties at different times.[20] If endurantism cannot be rendered consistent with the notion of variable intrinsic properties, then this, and Fell's claim that only a composite object can possess different intrinsic properties at different times, is inconsistent with endurantism, but consistent with perdurantism.

[20] This conclusion also applies in string theory, as explained in Appendix I.

Chapter 5

# Standard Model Gauge Groups and Representations

To recap, whilst a *free* elementary particle in our universe corresponds to an infinite-dimensional, irreducible unitary representation of the 'external' space–time symmetry group $SL(2, \mathbb{C}) \ltimes \mathbb{R}^{3,1}$, an *interacting* elementary particle transforms under the external symmetry group $SL(2, \mathbb{C}) \ltimes \mathbb{R}^{3,1}$, *and* an infinite-dimensional group of gauge transformations $\mathcal{G}$. The latter is associated with a compact, connected Lie group $G$, the 'gauge group' or 'internal symmetry group' of the interaction(s) in question. The spectrum of interacting elementary particles which are capable of existing, and the mathematical definition of what an interacting elementary particle is, changes with different gauge groups. The standard model has a gauge group $G = SU(3) \times SU(2) \times U(1)$ which incorporates the strong force, and unifies the electromagnetic and weak interactions. This gauge group defines which interacting elementary particles are consistent with the electroweak-unified standard model: they are those which belong to interacting-particle bundles or interaction carrier bundles possessing a finite-dimensional irreducible representation of $SL(2, \mathbb{C}) \times SU(3) \times SU(2) \times U(1)$ upon their typical fibres. However, because our universe has undergone electroweak symmetry breaking, the gauge group has broken from $SU(3) \times SU(2) \times U(1)$ into $SU(3) \times U(1)_Q$, and the interacting elementary particles with which we are most familiar, actually correspond to bundles that possess a finite-dimensional representation of $SL(2, \mathbb{C}) \times SU(3) \times U(1)_Q$ upon their typical fibres.[1]

The spectrum of interacting elementary particles, and the definition of what an interacting elementary particle is, changes again in a Grand Unified Theory (GUT), where the gauge group unifies the strong and electroweak interactions. For example, in the $Spin(10)$ GUT, an interacting elementary *fermion* corresponds to an interacting-particle bundle which possesses a finite-dimensional irreducible representation of $SL(2, \mathbb{C}) \times Spin(10)$ upon its typical fibre. For the collection of distinct elementary particles in each fermion generation of today's universe, there is only one corresponding elementary fermion in the $Spin(10)$ GUT. Such an elementary fermion is represented by an interacting-particle bundle $\sigma_L \otimes \zeta$ (Derdzinski, 1992, p. 127).

[1] $U(1)_Q$ here is the electromagnetic gauge group, which will be defined in Section 5.6.

The interaction bundle $\zeta$ is a complex vector bundle of fibre dimension 16, which possesses the irreducible spinorial representation of $Spin(10)$ upon its typical fibre, whilst $\sigma_L$ is the left-handed Weyl spinor bundle, possessing the $(1/2, 0)$ irreducible representation of $SL(2, \mathbb{C})$ upon its typical fibre. $\sigma_L \otimes \zeta$ is a complex vector bundle of fibre dimension 32, which is capable of representing an entire fermion generation, such as $(e, \nu_e, u, d)$, as a single elementary fermion. The fermions which are considered to possess distinct identities after GUT symmetry breaking and electroweak symmetry breaking, are merely considered to be different states of a single type of fermion in the $Spin(10)$ GUT. As such, changes between lepton states and quark states are possible, and GUTs predict the existence of additional gauge bosons, collectively referred to as $X$ bosons, which mediate transmutations between quarks and leptons. This leads GUTs to predict that protons have a long, but finite lifetime, and eventually decay by reactions such as $p \rightarrow e^+ + \pi^0$. In such a reaction, an $X$ boson is exchanged between the $d$ quark and a $u$ quark in the proton $(uud)$, changing the $d$ quark into a positron $e^+$, and the $u$ quark into an anti-quark $\bar{u}$. The proton thereby decays into a lepton $e^+$ and a neutral pion $(\bar{u}u)$. After symmetry breaking, different types of elementary fermion are identifiable; quarks and leptons become elementary fermions with distinct identities.

According to particle physics cosmology, there was a period of time in the early universe when the energy density was sufficiently high that the strong and electroweak forces were unified, and massless $X$ bosons mediated transmutations between the quark states and lepton states of elementary fermions. GUTs imply the existence of a second set of Higgs fields, and suggest that after the energy density of the universe dropped below a critical level, the GUT symmetry was spontaneously broken by the GUT Higgs fields, the $X$ bosons acquired a heavy mass, the quarks and leptons became distinct elementary particles, and the strong and electroweak forces became different in both strength, and the spectrum of particles they act upon.

Supersymmetry attempts to take this unification process a step further, postulating a supergroup as the gauge group, so that bosons and fermions can be treated as merely different states of the same type of elementary particle.[2] As Smolin comments, "the idea of unification requires that the different kinds of elementary particles, such as quarks, electrons or neutrinos, arise as different manifestations of one single, most elementary particle" (1997, p. 65).

Considerations such as these mitigate against the notion that there are intrinsically distinct types of elementary particle. If one interprets a gauge transformation as a change of 'internal' reference frame, and if one deems that properties which change under a gauge transformation cannot be intrinsic properties, then the introduction of larger groups of gauge symmetries has the effect of re-casting intrinsic properties as extrinsic properties. Given that GUTs possess gauge symmetries which transform

[2] In a slightly different vein, the 'preon' theories first proposed in the 1970s suggested that quarks and leptons, and perhaps even the spin 1 gauge bosons and Higgs bosons, are composite systems, composed of two or three preons (Bilson-Thompson *et al.*, 2006).

quarks into leptons, and vice versa, the properties which define quarks and leptons cannot be considered to be intrinsic properties in GUTs. The larger the gauge group, the smaller the set of intrinsically distinct elementary particle types, and the smaller the set of intrinsic properties. Although the proton decay predicted by GUTs has never been observed, and the current lack of empirical verification for GUTs and supersymmetry blunts the edge of this argument somewhat, the electroweak theory is empirically correct, and, as we shall explain in this chapter, it suggests that before spontaneous symmetry breaking, left-handed electron states and left-handed neutrino states were simply manifestations of one elementary particle type.

Understanding, then, the philosophical context, we now proceed to expound the gauge group of the standard model and its irreducible representations.

## 5.1. The structure of compact groups

If one starts with the assumption that the gauge group $G$ of the standard model is a compact connected Lie group, then one can employ the structure theorem for compact connected Lie groups (Simon, 1996, p. 155; Hofmann and Morris, 1998, pp. 204–207), to understand the structure of the standard model gauge group, and to clearly see what the alternative gauge groups are. The structure theorem entails that any compact, connected Lie group $G$ is isomorphic to a quotient of a finite direct product,[3]

$$G \cong L_1 \times L_2 \times \cdots \times L_r \times \mathbb{T}^p / D,$$

where each $L_i$ is a compact, simple, and simply connected Lie group, $\mathbb{T}^p$ is a $p$-dimensional torus, and $D$ is a finite central subgroup of $L_1 \times L_2 \times \cdots \times L_r \times \mathbb{T}^p$.

The only compact, simple, and simply connected Lie groups are the special unitary groups $SU(n)$, $n \geqslant 2$, the symplectic groups $Sp(n)$, $n \geqslant 2$, the spin groups $Spin(2n+1)$, $n \geqslant 3$, the spin groups $Spin(2n)$, $n \geqslant 4$, and the five exceptional Lie groups $E_6$, $E_7$, $E_8$, $F_4$, and $G_2$ (Simon, 1996, p. 151). The reason that the list of spin groups begins at $Spin(7)$ is that $Spin(3) \cong SU(2)$, $Spin(4) \cong SU(2) \times SU(2)$ (a non-simple group anyway), $Spin(5) \cong Sp(2)$, and $Spin(6) \cong SU(4)$ (Simon, 1996, p. 152). The list of symplectic groups begins at $Sp(2)$ because $Sp(1) \cong SU(2)$ (Simon, 1996, p. 144).

Given this exhaustive, non-repetitious list of compact, simple, and simply connected Lie groups, it follows that each one of the $L_i$ in the structural decomposition of a compact connected Lie group is a copy of one of the groups in this list. Multiple copies of the same group are, of course, permitted.

Of the infinite number of possible compact connected Lie groups available, the standard model of the gauge force fields in our universe corresponds to the case where

$$L_1 \times L_2 \times \cdots \times L_r = SU(3) \times SU(2),$$

[3] See Appendix J for a more lengthy discussion of this theorem.

and where $p = 1$. The gauge group $G$ of the standard model is such that

$$G \cong \big(SU(3) \times SU(2) \times U(1)\big)/D,$$

where $D$ is a finite central subgroup of $SU(3) \times SU(2) \times U(1)$.

A collection of finite-dimensional, irreducible representations of $SU(3) \times SU(2) \times U(1)$ define the interacting elementary particle 'multiplets' in our universe. In terms of the structural decomposition of the standard model gauge group in our universe, the finite central subgroup $D$ is the subgroup which acts trivially in all of these representations. We will see later that $D \cong \mathbb{Z}_6$, hence the standard model gauge group in our universe is

$$G \cong \big(SU(3) \times SU(2) \times U(1)\big)/\mathbb{Z}_6.$$

The gauge group of the standard model admits a non-trivial structural decomposition precisely because the strong and electroweak gauge fields are not unified in the standard model. The defining characteristic of a gauge field Grand Unified Theory (GUT), is that it postulates a *simple*, connected and compact Lie group as the gauge group of our universe. Such a theory unifies the strong and electroweak forces by postulating that the simple GUT gauge group contains $(SU(3) \times SU(2) \times U(1))/\mathbb{Z}_6 \cong S(U(3) \times U(2))$ as a subgroup. In effect, GUTs suggest that the non-trivial structural decomposition of the standard model gauge group is the result of the spontaneous breaking of the strong-electroweak symmetry.

A universe with non-unified gauge fields different to our own, would have a standard model in which $SU(3) \times SU(2)$ is replaced by a different finite product $L_1 \times L_2 \times \cdots \times L_r$, where each $L_i$ is a copy of one of the infinite number of compact, simple, simply connected Lie groups available. One could use special unitary groups of higher dimension, spin groups, symplectic groups, or even copies of the exceptional groups. One could also replace $U(1) = \mathbb{T}$ with an alternative compact Abelian Lie group $\mathbb{T}^p$.

## 5.2. Standard model irreducible representations

In the standard model of the particle world in our universe, a select collection of finite-dimensional irreducible representations of $SL(2, \mathbb{C}) \times SU(3) \times SU(2) \times U(1)$ are said to define the elementary particle multiplets. A particle multiplet in our universe can be represented by an interacting-particle bundle $\alpha$ or interaction carrier bundle $T^*\mathcal{M} \otimes \mathfrak{g}(\delta)$ which possesses a finite-dimensional irreducible representation of $SL(2, \mathbb{C}) \times SU(3) \times SU(2) \times U(1)$ upon its typical fibre.

We shall consider the relationship between standard model interacting-particle bundles and the finite-dimensional irreducible representations of $SU(3) \times SU(2) \times U(1)$ in more detail in Chapter 6. This chapter will be concerned primarily with the latter issue, the question of multiplet structure.

In general, given a product group $G \times H$, a finite-dimensional, irreducible representation $r : G \rightarrow Aut\, V_1$, and a finite-dimensional, irreducible representation $s : H \rightarrow Aut\, V_2$, the tensor product representation $r \otimes s : G \times H \rightarrow Aut(V_1 \otimes V_2)$ is also a finite-dimensional, irreducible representation. Furthermore, every finite-dimensional, irreducible representation of $G \times H$ is equivalent to such a tensor product representation (Sternberg, 1994, p. 371). Hence, extending this to a three-fold group product, every finite-dimensional, irreducible representation of $SU(3) \times SU(2) \times U(1)$ is a tensor product of finite-dimensional, irreducible representations of the component groups. We therefore need to study the finite-dimensional, irreducible representations of each component group, $SU(3)$, $SU(2)$, and $U(1)$.

In general, the finite-dimensional, irreducible representations of $SU(n)$, for $n \geqslant 2$, are parameterized by the elements of the Cartesian product

$$\left(\frac{1}{n}\mathbb{Z}_+\right)^{n-1}.$$

In other words, each finite-dimensional, irreducible representation of $SU(n)$ is parameterized by a sequence of rational numbers $(s_1, \dots, s_{n-1})$ called the 'spins' of the representation (Derdzinski, 1992, pp. 132–134). Hence, in the special case of $SU(3)$, the finite-dimensional, irreducible representations are parameterized by the elements of the Cartesian product

$$\left(\frac{1}{3}\mathbb{Z}_+\right) \times \left(\frac{1}{3}\mathbb{Z}_+\right),$$

and in the special case of $SU(2)$, the finite-dimensional, irreducible representations are parameterized by the elements of

$$\frac{1}{2}\mathbb{Z}_+.$$

In other words, the finite-dimensional, irreducible representations of $SU(3)$ are parameterized by pairs $(s_1, s_2)$, each of which is a non-negative integral multiple of $1/3$, and the finite-dimensional, irreducible representations of $SU(2)$ are parameterized by single numbers $s$, each of which is a non-negative integral multiple of $1/2$.

The finite-dimensional, irreducible representations of the unitary group $U(1)$ are indexed by the integers $\mathbb{Z}$. For any integer $n \in \mathbb{Z}$, the index $n$ representation maps $e^{i\theta} \in U(1)$ to $e^{in\theta}$, acting upon $\mathbb{C}^1$ by complex multiplication.[4] However, the finite-dimensional, irreducible representations of $U(1)$ can also be parameterized by rational numbers in $\frac{1}{n}\mathbb{Z}$, according to convenience. For the representation of weak hypercharge and electric charge, parameterizing the representations of $U(1)$ by $\frac{1}{3}\mathbb{Z}$ will be particularly convenient. To do so, one simply deems that, say, the weak hypercharge $y$ representation corresponds to the index $n = 3y$ representation of $U(1)$.[5]

[4] A more detailed explication of this can be found in Appendix K.

[5] However, see Appendix L for an alternative approach to dealing with such representations.

Given that every finite-dimensional, irreducible representation of $SU(3) \times SU(2) \times U(1)$ is a tensor product of finite-dimensional, irreducible representations of the component groups, it follows that the equivalence classes of finite-dimensional, irreducible representations of $SU(3) \times SU(2) \times U(1)$ will be indexed by some subset of the Cartesian product

$$\left(\frac{1}{3}\mathbb{Z}_+\right) \times \left(\frac{1}{3}\mathbb{Z}_+\right) \times \left(\frac{1}{2}\mathbb{Z}_+\right) \times \frac{1}{n}\mathbb{Z}.$$

Excluding the gauge boson representations, which utilise the adjoint representations, the standard model in our universe only uses finite-dimensional, irreducible representations of $SU(3) \times SU(2) \times U(1)$ which are tensor products of either the standard representation or trivial representation of $SU(3)$ with finite-dimensional, irreducible representations of $SU(2) \times U(1)$. Furthermore, those finite-dimensional, irreducible representations of $SU(2) \times U(1)$ are themselves tensor products of either the standard representation or trivial representation of $SU(2)$ with finite-dimensional, irreducible representations of $U(1)$.

The standard representation of $SU(3)$ is merely one of the countable infinity of finite-dimensional, irreducible representations of $SU(3)$, parameterized by $(\frac{1}{3}\mathbb{Z}_+) \times (\frac{1}{3}\mathbb{Z}_+)$. The standard representation of $SU(3)$ is indexed as the $(1/3, 0)$ representation. Similarly, the standard representation of $SU(2)$ is merely one of the countable infinity of finite-dimensional, irreducible representations of $SU(2)$, parameterized by $(\frac{1}{2}\mathbb{Z}_+)$. The standard representation of $SU(2)$ is indexed as the $s = 1/2$ representation.

Because only the trivial and standard representations of $SU(3)$ and $SU(2)$ are used to specify the elementary fermion multiplets in the standard model, it is practical, and notationally much simpler, to denote the representations of $SU(3) \times SU(2) \times U(1)$ with the dimension, rather than the 'spins,' of the $SU(3)$ and $SU(2)$ representations. Thus, for the trivial representation, a simple '1' can be used, and for the standard representation of $SU(n)$, a simple '$n$' can be used.

Let us take the first fermion generation as an example. Using the notation defined above, the particles, or, more accurately, the parts of the state spaces of the particles, are partitioned into multiplets by the following finite-dimensional irreducible representations of $SU(3) \times SU(2) \times U(1)$ (Baez, 1998, 1999; Schücker, 1997, pp. 30–31)[6]:

- The neutrino and the 'left-handed' part of the state-space of the electron $(\nu_L, e_L)$, transform according to the $(1, 2, -1)$ irreducible representation of $SU(3) \times SU(2) \times U(1)$. I.e., the tensor product of the trivial representation of $SU(3)$ with the 2-dimensional standard representation of $SU(2)$ with the 1-dimensional representation of $U(1)$ with hypercharge $-1$.

[6] We assume here, for pedagogical purposes, that the neutrino is massless. Also note that in the final chapter, we shall have cause to question the uniqueness of these multiplet representations.

- The left-handed part of the state-spaces of the up quark and down quark $(u_L, d_L)$ transform according to the $(3, 2, 1/3)$ representation. I.e., the tensor product of the standard representation of $SU(3)$ with the 2-dimensional standard representation of $SU(2)$ with the 1-dimensional representation of $U(1)$ with hypercharge $1/3$.
- The right-handed part of the state-space of the electron $e_R$ transforms according to the $(1, 1, -2)$ representation.
- The right-handed part of the state-space of the up quark $u_R$ transforms according to the $(3, 1, 4/3)$ representation.
- The right-handed part of the state-space of the down quark $d_R$ transforms according to the $(3, 1, -2/3)$ representation.

There is a tacit understanding here that each of these representations is tensored with an irreducible, finite-dimensional representation of $SL(2, \mathbb{C})$. In the case of the left-handed multiplets, the representation of the gauge group is tensored with the standard representation of $SL(2, \mathbb{C})$, and in the case of the right-handed multiplets, the representation of the gauge group is tensored with the conjugate representation of $SL(2, \mathbb{C})$.

Universes with a gauge group other than $SU(3) \times SU(2) \times U(1)$, or a quotient thereof, will possess different force fields to those that exist in our own universe, and will possess different sets of possible interacting elementary particles, interaction carriers, and elementary particle multiplets. Given a universe with a group of non-translational local space–time symmetries $SO_0(p, q)$, and a gauge group $L_1 \times \cdots \times L_r \times \mathbb{T}^p$, the set of finite-dimensional irreducible representations of $\widetilde{SO_0}(p, q) \times L_1 \times \cdots \times L_r \times \mathbb{T}^p$ define the set of possible interacting elementary particles, interaction carriers, and elementary particle multiplets.

Each irreducible representation of $L_1 \times \cdots \times L_r \times \mathbb{T}^p$ is a tensor product of irreducible representations of the individual factors. The individual $L_i$-representations will be representations of special unitary groups, spin groups, symplectic groups, or one of the exceptional groups. The irreducible representations of each family of simple, simply connected, compact Lie groups, can be classified in much the same way that the irreducible representations of $SU(n)$ can be classified. Hence, with the combination of the structural decomposition theorem for a compact, connected Lie group, and the classification of the irreducible representations of any simple, simply connected, compact Lie group, one can classify the particle multiplets of any universe in which the gauge group is assumed to be a compact, connected Lie group.

Even with the gauge group $G$ fixed, the finite-dimensional, irreducible representations of this group only determine the set of possible particle multiplets in a universe. The set of actual particle multiplets instantiated in a universe appears to be contingent. In our universe, only a finite number of particle multiplets have been selected from the countably infinite number of possible finite-dimensional, irreducible representations of $SU(3) \times SU(2) \times U(1)$. Thus, universes with the same gauge group, and the same set of *possible* particle multiplets, can be further sub-classified by the particular collection of *actual* particle multiplets instantiated.

## 5.3. The standard model gauge group

Whilst it is often written that the gauge group of the standard model is $SU(3) \times SU(2) \times U(1)$, in fact, as stated in Section 5.1, the strict standard model gauge group is actually a quotient $(SU(3) \times SU(2) \times U(1))/D$ with respect to a finite central subgroup $D$. To reiterate, it is the collection of finite-dimensional irreducible representations of $SU(3) \times SU(2) \times U(1)$ which happen to be instantiated by the particle multiplets in a universe, which select the finite central subgroup $D$. The latter is the subgroup which acts trivially in all those representations. None of the fermion representations specified in Section 5.2, nor the gauge boson representations, are faithful, one-to-one representations. In each such representation, elements of $SU(3) \times SU(2) \times U(1)$ other than the identity element, are mapped to the identity transformation upon the target vector space. One says that such elements act trivially in the representation. We shall now demonstrate that, in our universe, the subgroup which acts trivially is isomorphic to $\mathbb{Z}_6$ (Baez, 1998, 1999, 2003c; Tsou, 2000, p. 31, 2003, p. 18). $G = (SU(3) \times SU(2) \times U(1))/\mathbb{Z}_6$ is the strict gauge group of our universe, the group which is faithfully represented in all the multiplet representations.

Under a choice of gauge, a gauge field with gauge group $G$ is represented by cross-sections of a vector bundle which possesses the adjoint representation of $G$ upon its typical fibre. This vector bundle is the interaction carrier bundle $T^*\mathcal{M} \otimes \mathfrak{g}(\delta)$. Each gauge group $G$ is a subgroup of the gauge group of the standard model $SU(3) \times SU(2) \times U(1)$. Hence, each interaction carrier bundle possesses, upon its typical fibre, the adjoint representation of some subgroup of $SU(3) \times SU(2) \times U(1)$. Recall that in the adjoint representation of a group $G$, each $g \in G$ is mapped to $f_*^g|_e$, the differential map at the identity $e$, of the group automorphism

$$f^g(h) = ghg^{-1}.$$

The elements of $G$ which act trivially in the adjoint representation are those in the centre of $G$. Each $g$ in the centre of $G$ maps an arbitrary $h \in G$ to

$$ghg^{-1} = gg^{-1}h = h.$$

Hence, each $g$ in the centre of $G$ is mapped to the identity automorphism upon $G$, and the differential of such an automorphism at the identity element is therefore the identity transformation upon the Lie algebra $\mathfrak{g}$. The subgroup of $SU(3) \times SU(2) \times U(1)$ which acts trivially in all of the representations, is therefore a subgroup of the centre of $SU(3) \times SU(2) \times U(1)$.

The centre of $SU(3) \times SU(2) \times U(1)$ is the product of the centre of each factor. $U(1)$ is Abelian, so its centre is the whole of $U(1)$. The centre of $SU(2)$ is generated by the matrix

$$e^{2\pi i\frac{1}{2}} \begin{pmatrix} 1 & 0 \\ 0 & 1 \end{pmatrix}.$$

In this case, there is only one power of the generator, the square power:

$$e^{2\pi i}\begin{pmatrix}1 & 0\\ 0 & 1\end{pmatrix}.$$

This two-element subgroup of $SU(2)$ is isomorphic to $\mathbb{Z}_2 \cong \{-1, 1\}$. To confirm this, note that

$$e^{2\pi i\frac{1}{2}}\begin{pmatrix}1 & 0\\ 0 & 1\end{pmatrix} = e^{\pi i}\begin{pmatrix}1 & 0\\ 0 & 1\end{pmatrix} = -1\begin{pmatrix}1 & 0\\ 0 & 1\end{pmatrix},$$

and

$$e^{2\pi i}\begin{pmatrix}1 & 0\\ 0 & 1\end{pmatrix} = \begin{pmatrix}1 & 0\\ 0 & 1\end{pmatrix}.$$

The centre of $SU(3)$ is generated by the matrix

$$e^{2\pi i\frac{1}{3}}\begin{pmatrix}1 & 0 & 0\\ 0 & 1 & 0\\ 0 & 0 & 1\end{pmatrix}.$$

In this case, there are two powers of the generator, the square power and the cube power:

$$e^{2\pi i\frac{2}{3}}\begin{pmatrix}1 & 0 & 0\\ 0 & 1 & 0\\ 0 & 0 & 1\end{pmatrix} \quad \text{and} \quad e^{2\pi i}\begin{pmatrix}1 & 0 & 0\\ 0 & 1 & 0\\ 0 & 0 & 1\end{pmatrix}.$$

This three-element subgroup of $SU(3)$ is isomorphic to $\mathbb{Z}_3$. Hence, the centre of $SU(3) \times SU(2) \times U(1)$ is a subgroup isomorphic to $\mathbb{Z}_3 \times \mathbb{Z}_2 \times U(1)$.

The elements of $\mathbb{Z}_3 \times \mathbb{Z}_2 \times U(1)$ that act trivially in all the fermion representations are generated by the triple

$$a = \left(e^{2\pi i\frac{1}{3}}\begin{pmatrix}1 & 0 & 0\\ 0 & 1 & 0\\ 0 & 0 & 1\end{pmatrix}, e^{2\pi i\frac{1}{2}}\begin{pmatrix}1 & 0\\ 0 & 1\end{pmatrix}, e^{2\pi i\frac{1}{6}}\right) \in \mathbb{Z}_3 \times \mathbb{Z}_2 \times U(1),$$

which can be more economically denoted as

$$a = \left(e^{2\pi i\frac{1}{3}}, e^{2\pi i\frac{1}{2}}, e^{2\pi i\frac{1}{6}}\right).$$

The powers of this generating element are:

$$a^2 = \left(e^{2\pi i\frac{2}{3}}, e^{2\pi i}, e^{2\pi i\frac{2}{6}}\right),$$
$$a^3 = \left(e^{2\pi i}, e^{2\pi i\frac{1}{2}}, e^{2\pi i\frac{3}{6}}\right),$$
$$a^4 = \left(e^{2\pi i\frac{1}{3}}, e^{2\pi i}, e^{2\pi i\frac{4}{6}}\right),$$
$$a^5 = \left(e^{2\pi i\frac{2}{3}}, e^{2\pi i\frac{1}{2}}, e^{2\pi i\frac{5}{6}}\right),$$
$$a^6 = \left(e^{2\pi i}, e^{2\pi i}, e^{2\pi i}\right) = Id.$$

Hence, the subgroup of $SU(3) \times SU(2) \times U(1)$ which acts trivially in all the representations is isomorphic to $\mathbb{Z}_6$.

To demonstrate that the elements of this $\mathbb{Z}_6$ subgroup do act trivially in all the representations, consider the action of the generating element in each of the representations:

- The neutrino and the 'left-handed' electron $(\nu_L, e_L)$ transform according to the $(1, 2, -1)$ irreducible representation of $SU(3) \times SU(2) \times U(1)$. The representation space[7] is $\mathbb{C}^1 \otimes \mathbb{C}^2 \otimes \mathbb{C}^1 \cong \mathbb{C}^2$. $e^{2\pi i \frac{1}{3}} \in SU(3)$ is mapped to the identity transformation on $\mathbb{C}^1$, and $e^{2\pi i \frac{1}{2}} \in SU(2)$ is mapped to itself, the negative of the identity, acting on $\mathbb{C}^2$. In the hypercharge $y$ representation of $U(1)$, $e^{i\theta}$ is mapped to $e^{i3y\theta}$, hence in the hypercharge $y = -1$ representation of $U(1)$, $e^{2\pi i \frac{1}{6}}$ is mapped to $e^{(-3)2\pi i \frac{1}{6}} = e^{(-1)\pi i} = e^{\pi i}$, the negative of the identity transformation. Hence, in the $(1, 2, -1)$ representation, the generating element of the $\mathbb{Z}_6$ subgroup is mapped to $Id \otimes -Id \otimes -Id = Id$.
- The left-handed up quark and down quark $(u_L, d_L)$ transform according to the $(3, 2, 1/3)$ representation. The representation space is $\mathbb{C}^3 \otimes \mathbb{C}^2$. In this case $e^{2\pi i \frac{1}{3}} \in SU(3)$ maps to itself acting on $\mathbb{C}^3$, $e^{2\pi i \frac{1}{2}} \in SU(2)$ maps to itself acting on $\mathbb{C}^2$, and $e^{2\pi i \frac{1}{6}} \in U(1)$ is mapped to itself. Hence, in the $(3, 2, 1/3)$ representation, the generating element of the $\mathbb{Z}_6$ subgroup is mapped to

$$\begin{aligned} &e^{2\pi i \frac{1}{3}} Id \otimes e^{2\pi i \frac{1}{2}} Id \otimes e^{2\pi i \frac{1}{6}} Id \\ &\quad = e^{2\pi i \frac{1}{3}} e^{2\pi i \frac{1}{6}} (-1)(Id \otimes Id \otimes Id) = e^{\pi i}(-1)\, Id = Id. \end{aligned}$$

- The right-handed electron $e_R$ transforms according to the $(1, 1, -2)$ representation. The representation space is therefore $\mathbb{C}^1$. $e^{2\pi i \frac{1}{6}}$ is mapped to $e^{(-6)2\pi i/6} = e^{2\pi i}$, the identity transformation.
- The right-handed up quark $u_R$ transforms according to the $(3, 1, 4/3)$ representation. The representation space is therefore $\mathbb{C}^3$. $e^{2\pi i \frac{1}{3}} \in SU(3)$ maps to itself acting on $\mathbb{C}^3$, and $e^{2\pi i \frac{1}{6}} \in U(1)$ is mapped to $e^{(4)2\pi i/6} = e^{2\pi i \frac{2}{3}}$. Now, $e^{2\pi i \frac{1}{3}} e^{2\pi i \frac{2}{3}} = e^{2\pi i}$, hence the generating element of the $\mathbb{Z}_6$ subgroup is mapped to the identity again.
- The right-handed down quark $d_R$ transforms according to the $(3, 1, -2/3)$ representation. The representation space is therefore $\mathbb{C}^3$. $e^{2\pi i \frac{1}{3}} \in SU(3)$ maps to itself acting on $\mathbb{C}^3$, and $e^{2\pi i \frac{1}{6}} \in U(1)$ is mapped to $e^{(-2)2\pi i/6} = e^{2\pi i \frac{-1}{3}}$. Now, $e^{2\pi i \frac{1}{3}} e^{2\pi i \frac{-1}{3}} = e^{2\pi i}$, hence the generating element of the $\mathbb{Z}_6$ subgroup is mapped to the identity again.

Having demonstrated that the generator of the $\mathbb{Z}_6$ subgroup does indeed act trivially in all the representations, it follows that any power of the generator will also act trivially, because any power of the identity is also the identity.

[7] See Appendix M for a closer look at representation spaces which include a tensor factor with $\mathbb{C}^1$.

Having established that the gauge group of the standard model is $(SU(3)\times SU(2)\times U(1))/\mathbb{Z}_6$, a couple of potentially confusing group isomorphisms need to be declared. Firstly, the finite group $\mathbb{Z}_6$ is isomorphic to $\mathbb{Z}_3\times\mathbb{Z}_2$, hence a number of authors assert that the gauge group of the standard model can be taken to be

$$\big(SU(3)\times SU(2)\times U(1)\big)/\mathbb{Z}_3\times\mathbb{Z}_2.$$

However, care should be taken to distinguish the $\mathbb{Z}_6$-subgroup of $SU(3)\times SU(2)\times U(1)$ from the $\mathbb{Z}_3\times\mathbb{Z}_2$-subgroup of $SU(3)\times SU(2)$. The centre of $SU(3)\times SU(2)$ is $\mathbb{Z}_3\times\mathbb{Z}_2$, but the subgroup of $SU(3)\times SU(2)\times U(1)$ which acts trivially in all the particle representations is not $\mathbb{Z}_3\times\mathbb{Z}_2\times Id$. Furthermore, although there is a $\mathbb{Z}_6$-subgroup of $U(1)$,

$$\big\{e^{2\pi i\frac{k}{6}}\colon k=1,2,3,4,5,6\big\},$$

it is not $\mathbb{Z}_3\times\mathbb{Z}_2\times\mathbb{Z}_6$, a finite group of 36 elements, which acts trivially in all the particle representations. Rather, the $\mathbb{Z}_6$-subgroup of relevance here is a subgroup of $\mathbb{Z}_3\times\mathbb{Z}_2\times\mathbb{Z}_6$. The isomorphism between $\mathbb{Z}_3\times\mathbb{Z}_2$ and this particular $\mathbb{Z}_6\subset\mathbb{Z}_3\times\mathbb{Z}_2\times\mathbb{Z}_6$ is as follows:

$$\begin{aligned}
&\big(e^{2\pi i\frac{1}{3}},e^{2\pi i\frac{1}{2}}\big)\mapsto\big(e^{2\pi i\frac{1}{3}},e^{2\pi i\frac{1}{2}},e^{2\pi i\frac{1}{6}}\big),\\
&\big(e^{2\pi i\frac{2}{3}},e^{2\pi i}\big)\mapsto\big(e^{2\pi i\frac{2}{3}},e^{2\pi i},e^{2\pi i\frac{2}{6}}\big),\\
&\big(e^{2\pi i},e^{2\pi i\frac{1}{2}}\big)\mapsto\big(e^{2\pi i},e^{2\pi i\frac{1}{2}},e^{2\pi i\frac{3}{6}}\big),\\
&\big(e^{2\pi i\frac{1}{3}},e^{2\pi i}\big)\mapsto\big(e^{2\pi i\frac{1}{3}},e^{2\pi i},e^{2\pi i\frac{4}{6}}\big),\\
&\big(e^{2\pi i\frac{2}{3}},e^{2\pi i\frac{1}{2}}\big)\mapsto\big(e^{2\pi i\frac{2}{3}},e^{2\pi i\frac{1}{2}},e^{2\pi i\frac{5}{6}}\big),\\
&\big(e^{2\pi i},e^{2\pi i}\big)\mapsto\big(e^{2\pi i},e^{2\pi i},e^{2\pi i}\big).
\end{aligned}$$

One can verify that the group product is preserved under this mapping. The group product is defined component-wise in $\mathbb{Z}_3\times\mathbb{Z}_2$, hence, for example:

$$\begin{aligned}
&\big(e^{2\pi i\frac{1}{3}},e^{2\pi i\frac{1}{2}}\big)\cdot\big(e^{2\pi i\frac{1}{3}},e^{2\pi i}\big)\\
&\quad=\big(e^{2\pi i\frac{1}{3}}e^{2\pi i\frac{1}{3}},e^{2\pi i\frac{1}{2}}e^{2\pi i}\big)\\
&\quad=\big(e^{2\pi i\frac{2}{3}},e^{2\pi i\frac{1}{2}}\big).
\end{aligned}$$

The group product is also defined component-wise in the $\mathbb{Z}_6$-subgroup of $\mathbb{Z}_3\times\mathbb{Z}_2\times\mathbb{Z}_6$, hence:

$$\begin{aligned}
&\big(e^{2\pi i\frac{1}{3}},e^{2\pi i\frac{1}{2}},e^{2\pi i\frac{1}{6}}\big)\cdot\big(e^{2\pi i\frac{1}{3}},e^{2\pi i},e^{2\pi i\frac{4}{6}}\big)\\
&\quad=\big(e^{2\pi i\frac{2}{3}},e^{2\pi i\frac{1}{2}},e^{2\pi i\frac{5}{6}}\big).
\end{aligned}$$

Given that

$$\big(e^{2\pi i\frac{2}{3}},e^{2\pi i\frac{1}{2}}\big)\mapsto\big(e^{2\pi i\frac{2}{3}},e^{2\pi i\frac{1}{2}},e^{2\pi i\frac{5}{6}}\big),$$

it is clear from this that the group product is preserved. The mapping is one-to-one, hence it is an isomorphism between a pair of six-element finite groups.

One should also bear in mind that the quotient group $(SU(3) \times SU(2) \times U(1))/\mathbb{Z}_6$ is isomorphic to $S(U(3) \times U(2))$. The latter is the subgroup of $U(3) \times U(2)$ consisting of pairs $(u, v) \in U(3) \times U(2)$ which are such that

$$\det u \cdot \det v = 1.$$

To establish the isomorphism, we need to demonstrate that there is a covering map from $SU(3) \times SU(2) \times U(1)$ onto $S(U(3) \times U(2))$, whose kernel is $\mathbb{Z}_6$. For the covering map, one can choose

$$\left(A, B, e^{i\theta}\right) \mapsto \left(e^{-i\theta 2} A, e^{i\theta 3} B\right),$$

with $A \in SU(3)$, $B \in SU(2)$, $e^{i\theta} \in U(1)$.[8] Alternatively, one can choose

$$\left(A, B, e^{i\theta}\right) \mapsto \left(e^{i\theta 2} A, e^{-i\theta 3} B\right).$$

In general, if $A \in U(3)$ is such that $\det A = 1$, then $\det e^{i\theta} A = e^{i3\theta}$, and if $B \in U(2)$ is such that $\det B = 1$, then $\det e^{i\theta} B = e^{i2\theta}$. Hence, in the first case above,

$$\det e^{-i\theta 2} A = e^{-i6\theta},$$

and

$$\det e^{i\theta 3} B = e^{i6\theta},$$

therefore,

$$\begin{aligned}\det\left(e^{-i\theta 2} A, e^{i\theta 3} B\right) &= \det e^{-i\theta 2} A \cdot \det e^{i\theta 3} B\\ &= e^{-i6\theta} \cdot e^{i6\theta}\\ &= 1.\end{aligned}$$

The kernel of the covering map from $SU(3) \times SU(2) \times U(1)$ onto $S(U(3) \times U(2))$ is the $\mathbb{Z}_6$ subgroup already defined. One can verify this by checking that the generator of the $\mathbb{Z}_6$ subgroup is mapped to the identity element of $S(U(3) \times U(2))$[9]:

$$\left(e^{2\pi i \frac{1}{3}} Id_3, e^{2\pi i \frac{1}{2}} Id_2, e^{2\pi i \frac{1}{6}}\right) \mapsto \left(e^{-2\pi i \frac{2}{6}} e^{2\pi i \frac{1}{3}} Id_3, e^{2\pi i \frac{3}{6}} e^{2\pi i \frac{1}{2}} Id_2\right) = (Id_3, Id_2).$$

Given an arbitrary $(A, B, e^{i\theta}) \in SU(3) \times SU(2) \times U(1)$, one can define an equivalence relationship

$$\left(A, B, e^{i\theta}\right) \sim \left(e^{2\pi i \frac{k}{3}} Id_3 \cdot A, e^{2\pi i \frac{k}{2}} Id_2 \cdot B, e^{2\pi i \frac{k}{6}} e^{i\theta}\right),$$

for any $k \in \{1, 2, 3, 4, 5, 6\}$. The set of equivalence classes under this equivalence relationship is the quotient group $(SU(3) \times SU(2) \times U(1))/\mathbb{Z}_6$. The covering map

[8] This map was pointed out to me by John C. Baez.

[9] $Id_n$ denotes the $n \times n$ identity matrix.

defined above is constant upon each element in such an equivalence class. Hence, by this means, the homomorphic covering map from $SU(3) \times SU(2) \times U(1)$ onto $S(U(3) \times U(2))$ induces the isomorphic map from $(SU(3) \times SU(2) \times U(1))/\mathbb{Z}_6$ onto $S(U(3) \times U(2))$.

## 5.4. Duality and triality

An irreducible representation of $SU(n)$ can be characterised by something called the '$n$-ality' of the representation (also variously called the 'class' or the 'rank index'). In the case of $SU(2)$ it is called the duality of the representation, and in the case of $SU(3)$ it is called the triality. Given the generator $e^{2\pi i/n} Id_n$ of the $\mathbb{Z}_n$-centre of $SU(n)$, the $n$-ality of the irreducible representation with spin $(s_1, \ldots, s_{n-1})$ is defined to be

$$t = \left( \sum_{j=1}^{n-1} j s_j \right) n \mod n.$$

The $n$-ality will be generically denoted as '$t$,' but in the special case of representations of $SU(2)$ and $SU(3)$, it is convenient to denote duality and triality as '$d$' and '$t$' respectively.

Under the irreducible representation with spin $(s_1, \ldots, s_{n-1})$, the generator of the centre of $SU(n)$ undergoes the mapping

$$e^{2\pi i \frac{1}{n}} Id_n \mapsto e^{2\pi i \frac{t}{n}} Id_n = e^{2\pi i (\sum_{j=1}^{n-1} j s_j)} Id_n \,.$$

The $n$-ality is therefore the power by which the generator of the centre of $SU(n)$ is represented. In the special case of $SU(2)$, the generator of the centre of $SU(2)$ undergoes the mapping

$$e^{2\pi i \frac{1}{2}} Id_2 \mapsto e^{2\pi i \frac{d}{2}} Id_2,$$

where $d \in \mathbb{Z}_2$. In the special case of $SU(3)$, the generator of the centre of $SU(3)$ undergoes the mapping

$$e^{2\pi i \frac{1}{3}} Id_3 \mapsto e^{2\pi i \frac{t}{3}} Id_3,$$

where $t \in \mathbb{Z}_3$.

## 5.5. The electroweak gauge group

It is often asserted in the physics literature that the gauge group of the electroweak interaction is $SU(2) \times U(1)$. However, there is a $\mathbb{Z}_2$-subgroup of $SU(2) \times U(1)$ which acts trivially in all the multiplet representations found in our universe. This entails that the gauge group of the electroweak interaction is $(SU(2) \times U(1))/\mathbb{Z}_2 \cong U(2)$.

We have already established that the centre of $SU(2)$ is $\mathbb{Z}_2$ and the centre of $U(1)$ is $U(1)$ itself, hence the centre of $SU(2) \times U(1)$ is $\mathbb{Z}_2 \times U(1)$. We need to find the elements of this subgroup which act trivially in all the fermion representations.

As before, the $\mathbb{Z}_2$-subgroup of $SU(2)$ is generated by the matrix:

$$e^{2\pi i \frac{1}{2}} \begin{pmatrix} 1 & 0 \\ 0 & 1 \end{pmatrix}.$$

If we take $e^{2\pi i \frac{1}{2}} \in U(1)$, and then form the pair

$$a = \left( e^{2\pi i \frac{1}{2}} \begin{pmatrix} 1 & 0 \\ 0 & 1 \end{pmatrix}, e^{2\pi i \frac{1}{2}} \right) \in \mathbb{Z}_2 \times U(1),$$

this element is the generator of a subgroup in $SU(2) \times U(1)$ which is isomorphic to $\mathbb{Z}_2$. In more economical notation $a = (e^{2\pi i \frac{1}{2}}, e^{2\pi i \frac{1}{2}}) \in SU(2) \times U(1)$ is the generator, equivalent to $-1$ in $\mathbb{Z}_2$, and $a^2 = (e^{2\pi i}, e^{2\pi i})$ is the identity element. The reader can verify for himself that the generator of this subgroup acts trivially in all the fermion representations listed in the Section 5.2.

The only representations of $SU(2) \times U(1)$ that are used to model physical particles in our universe, are representations of $U(2)$ which are composed with the covering homomorphism $\phi : SU(2) \times U(1) \to U(2)$ to form representations of $SU(2) \times U(1)$. The degree of redundancy in the representations of $SU(2) \times U(1)$ corresponds to the fact that all the representations of $SU(2) \times U(1)$ are actually representations of $U(2)$ which have been lifted to the covering group $SU(2) \times U(1)$.

A universe in which $SU(2) \times U(1)$, rather than $U(2)$, is the true electroweak symmetry group, must possess a multiplet representation of $SU(2) \times U(1)$ which does not have a $\mathbb{Z}_2$-kernel. In terms of the duality $d$ of the $SU(2)$ representation and the index of the $U(1)$ representation, an irreducible representation of $SU(2) \times U(1)$ has a $\mathbb{Z}_2$-kernel if and only if $d = m \mod 2$. In other words, an irreducible representation of $SU(2) \times U(1)$ has a $\mathbb{Z}_2$-kernel if and only if

$$e^{2\pi i \frac{1}{2}} \begin{pmatrix} 1 & 0 \\ 0 & 1 \end{pmatrix} \in SU(2) \mapsto e^{2\pi i \frac{d}{2}} \begin{pmatrix} 1 & 0 \\ 0 & 1 \end{pmatrix},$$

and

$$e^{2\pi i \frac{1}{2}} \in U(1) \mapsto e^{2\pi i \frac{m}{2}}$$

with $d = m \mod 2$.

If there is an irreducible representation of $SU(2) \times U(1)$ in which either (i) the representation of $SU(2)$ has odd duality and the representation of $U(1)$ has an even index, or (ii) the representation of $SU(2)$ has even duality and the representation of $U(1)$ has an odd index, then this representation will be a faithful representation of $SU(2) \times U(1)$, and $SU(2) \times U(1)$ will be the true symmetry group of the electroweak force. Either of these two conditions would ensure that the generator,

$$\left(e^{2\pi i \frac{1}{2}} \begin{pmatrix} 1 & 0 \\ 0 & 1 \end{pmatrix}, e^{2\pi i \frac{1}{2}}\right),$$

of the $\mathbb{Z}_2$-subgroup of $SU(2) \times U(1)$ is not mapped to $-Id_2 \otimes -Id_1$ or $Id_2 \otimes Id_1$, both of which constitute the identity on $\mathbb{C}^2 \otimes \mathbb{C}^1$.

An example of such a faithful representation is the (2, 2) representation of $SU(2) \times U(1)$ i.e., the tensor product of the standard representation of $SU(2)$ with the weak hypercharge $y = 2$ representation of $U(1)$. Recall that the standard representation of $SU(2)$ has duality $d = 1$. The index $m$ of the $y = 2$ representation of $U(1)$ is even, hence the (2, 2) representation of $SU(2) \times U(1)$ has odd $SU(2)$-duality and an even $U(1)$-index.

## 5.6. The electromagnetic subgroup

The purpose of this section is to explain how the electromagnetic gauge group $U(1)_Q$ occurs as a subgroup of the electroweak gauge group $U(2)$, and is distinct from the weak hypercharge subgroup $U(1)_Y \subset U(2)$.

In the physics literature, the $SU(2) \times U(1)$ group which provides a double cover of the electroweak gauge group $U(2)$, is often denoted as $SU(2)_L \times U(1)_Y$ to indicate that it is the group product of the weak isospin group $SU(2)_L$ and the weak hypercharge group $U(1)_Y$. Let us take a look first at the Lie algebras of these groups. The Lie algebra $\mathfrak{su}(2)$ is the real vector space of trace-free, skew-adjoint (a.k.a. 'anti-Hermitian'), complex $2 \times 2$ matrices. As remarked in Section 2.8, a basis for this vector space can be obtained from the Pauli matrices $\sigma^i$, $i = 1, 2, 3$. The Pauli matrices themselves

$$\sigma^1 = \begin{pmatrix} 0 & 1 \\ 1 & 0 \end{pmatrix}, \qquad \sigma^2 = \begin{pmatrix} 0 & -i \\ i & 0 \end{pmatrix}, \qquad \sigma^3 = \begin{pmatrix} 1 & 0 \\ 0 & -1 \end{pmatrix},$$

are self-adjoint, not skew-adjoint. However, if one takes convenient multiples of the Pauli matrices, such as $T^i = 1/2i\sigma^i$, $i = 1, 2, 3$, then one obtains three skew-adjoint matrices which span the Lie algebra $\mathfrak{su}(2)$. In the case of these particular multiples, one has a basis of the Lie algebra $\mathfrak{su}(2)$ which satisfies the (Lie bracket) commutation relations

$$\left[T^i, T^j\right] = (-1)\epsilon_{ijk} T^k,$$

where $\epsilon_{ijk}$ is the completely antisymmetric Levi-Civita tensor. As a basis of the Lie algebra $\mathfrak{su}(2)$, the skew-adjoint matrices $T^i = 1/2i\sigma^i$, $i = 1, 2, 3$, are often referred to as the 'generators' of the Lie group $SU(2)_L$, and represent the three components of weak isospin.

The Lie algebra $\mathfrak{u}(1)$ can be defined to be the set of imaginary numbers,

$$\mathfrak{u}(1) = \left\{iy\colon\ y \in \mathbb{R}^1\right\},$$

and the Lie group $U(1)$ can be defined to be the image of $\mathfrak{u}(1)$ under the Lie exponential map, the set of complex numbers of unit modulus,

$$U(1) = \{e^{i\theta}: \theta = y \mod 2\pi\}.$$

As a Lie algebra, the set of imaginary numbers $i\mathbb{R}^1$, is isomorphic to the set of real numbers $\mathbb{R}^1$. Hence, $\mathfrak{u}(1) \cong \mathbb{R}^1$. Furthermore, there is an isomorphic image of $\mathfrak{u}(1)$ as a set of diagonal matrices in $\mathfrak{u}(N)$ for any integer $N$:

$$\mathfrak{u}(1) \cong i\mathbb{R}^1 \begin{pmatrix} 1 & 0 & \cdot & 0 \\ 0 & 1 & & \\ \cdot & & \cdot & \\ 0 & & & 1 \end{pmatrix} \subset u(N).$$

Thus, whilst the weak hypercharge Lie algebra $\mathfrak{u}(1)_Y$ can be expressed as the set of imaginary numbers,

$$\mathfrak{u}(1)_Y = \{iy: y \in \mathbb{R}^1\},$$

and whilst the weak hypercharge group $U(1)_Y$ can be expressed as the set of complex numbers of unit modulus,

$$U(1)_Y = \{e^{i\theta}: \theta = y \mod 2\pi\},$$

the weak hypercharge Lie algebra can also be expressed as the subset of $\mathfrak{u}(2)$ consisting of all real multiples of the following matrix,

$$\begin{pmatrix} i & 0 \\ 0 & i \end{pmatrix},$$

and the weak hypercharge Lie group can duly be expressed as the following subset of $U(2)$,

$$U(1)_Y = \left\{\begin{pmatrix} e^{i\theta} & 0 \\ 0 & e^{i\theta} \end{pmatrix}: \theta \in [0, 2\pi)\right\} \subset U(2).$$

The Lie algebra $\mathfrak{u}(2)$ of the Lie Group $U(2)$ is the set of all skew-adjoint matrices, regardless of whether the trace is zero or non-zero. Because $SU(2) \times U(1)$ covers $U(2)$, $SU(2) \times U(1)$ and $U(2)$ have isomorphic Lie algebras,

$$\mathfrak{u}(2) \cong \mathfrak{su}(2) \oplus \mathfrak{u}(1).$$

This also follows from the fact that, in general, $\mathfrak{u}(N) \cong \mathfrak{su}(N) \oplus \mathbb{R}^1$, and from the isomorphism $\mathfrak{u}(1) \cong \mathbb{R}^1$. The consequence is that every element of $\mathfrak{u}(2)$ can be expressed as the sum of a trace-free skew-adjoint matrix and a skew-adjoint scalar matrix. Letting $I$ denote the $2 \times 2$ identity matrix, the four matrices $(iI, 1/2i\sigma^1, 1/2i\sigma^2, 1/2i\sigma^3)$ provide a basis for the Lie algebra $\mathfrak{su}(2) \oplus \mathfrak{u}(1) \cong \mathfrak{u}(2)$.

The weak hypercharge group $U(1)_Y$ is a distinct $U(1)$ subgroup of $SU(2) \times U(1)$ and $U(2)$ from the electromagnetic group, $U(1)_Q$. The electromagnetic subgroup

$U(1)_Q$ is generated by a linear combination of $T_3$ and $iI$. In fact, the default generator of the electromagnetic group is $Q = T_3 + 1/2iI$. If $Q \in \mathfrak{u}(2) \cong \mathfrak{su}(2) \oplus \mathfrak{u}(1)$, then $Q$ must be a $2 \times 2$ skew-adjoint matrix:

$$\begin{aligned} Q &= 1/2i\sigma_3 + 1/2iI \\ &= 1/2i(\sigma_3 + I) \\ &= 1/2i\left(\begin{pmatrix} 1 & 0 \\ 0 & -1 \end{pmatrix} + \begin{pmatrix} 1 & 0 \\ 0 & 1 \end{pmatrix}\right) \\ &= 1/2i\begin{pmatrix} 2 & 0 \\ 0 & 0 \end{pmatrix} \\ &= \begin{pmatrix} i & 0 \\ 0 & 0 \end{pmatrix}. \end{aligned}$$

Taking all the real multiples of this skew-adjoint matrix generates a Lie subalgebra isomorphic to $\mathfrak{u}(1)$:

$$\mathfrak{u}(1)_Q = \{Qt\colon t \in \mathbb{R}^1\}.$$

This electromagnetic Lie algebra $\mathfrak{u}(1)_Q \subset \mathfrak{u}(2) \cong \mathfrak{su}(2) \oplus \mathfrak{u}(1)$ is a distinct $\mathfrak{u}(1)$ subalgebra from the weak hypercharge $\mathfrak{u}(1)_Y$ subalgebra.

The electromagnetic subgroup $U(1)_Q \subset U(2)$ is the image of $\mathfrak{u}(1)_Q$ under the Lie exponential map:

$$U(1)_Q = \{e^{Q\theta}\colon \theta = t \mod 2\pi\}.$$

Note that $Qt$ here is not an imaginary number, but a skew-adjoint matrix, and $e^{Q\theta} = e^{Qt}$, $\theta = t \mod 2\pi$, is not a complex number, but a unitary matrix, expressible as a power series in $Qt$.

Conveniently, with the choice of the $Q$ matrix given above, and using the power series definition of the Lie exponential map,

$$e^X = I + X + \frac{X^2}{2!} + \frac{X^3}{3!} + \cdots,$$

it follows that

$$\begin{aligned} e^{Qt} &= \exp\begin{pmatrix} it & 0 \\ 0 & 0 \end{pmatrix} \\ &= \begin{pmatrix} 1 & 0 \\ 0 & 1 \end{pmatrix} + \begin{pmatrix} it & 0 \\ 0 & 0 \end{pmatrix} + \begin{pmatrix} \frac{(it)^2}{2!} & 0 \\ 0 & 0 \end{pmatrix} + \begin{pmatrix} \frac{(it)^3}{3!} & 0 \\ 0 & 0 \end{pmatrix} + \cdots \\ &= \begin{pmatrix} 1 + it + \frac{(it)^2}{2!} + \frac{(it)^3}{3!} + \cdots & 0 \\ 0 & 1 \end{pmatrix} \end{aligned}$$

$$= \begin{pmatrix} e^{it} & 0 \\ 0 & 1 \end{pmatrix}$$

$$= \begin{pmatrix} e^{i\theta} & 0 \\ 0 & 1 \end{pmatrix}, \qquad \theta = t \mod 2\pi.$$

Contrast the form of the electromagnetic subgroup $U(1)_Q \subset U(2)$,

$$U(1)_Q = \left\{ \begin{pmatrix} e^{i\theta} & 0 \\ 0 & 1 \end{pmatrix} : \theta \in [0, 2\pi) \right\},$$

with the form of the hypercharge subgroup $U(1)_Y \subset U(2)$

$$U(1)_Y = \left\{ \begin{pmatrix} e^{i\theta} & 0 \\ 0 & e^{i\theta} \end{pmatrix} : \theta \in [0, 2\pi) \right\}.$$

Chapter 6

# The Standard Model Interacting-Particle Bundle

The purpose of this chapter is to elucidate the relationship between the standard model interacting-particle bundles and the finite-dimensional irreducible representations of $SU(3) \times SU(2) \times U(1)$.

Recall from Chapter 5, that in the standard model of the particle world in our universe, a select collection of finite-dimensional irreducible representations of $SL(2, \mathbb{C}) \times SU(3) \times SU(2) \times U(1)$ define the set of interacting elementary particles, boson or fermion, consistent with the electroweak-unified standard model. These irreducible representations are said to define the elementary particle *multiplets*, when interpreted in terms of the spectrum of interacting elementary particles consistent with the electroweak-*broken* standard model. These latter particles correspond to either *reducible or irreducible* representations of $SL(2, \mathbb{C})$, tensored with irreducible representations of the broken gauge group, $SU(3) \times U(1)_Q$.

Each interacting elementary particle consistent with the electroweak-unified standard model corresponds to an interacting-particle bundle $\alpha$, or an interaction carrier bundle $T^*\mathcal{M} \otimes \mathfrak{g}(\delta)$, possessing a finite-dimensional irreducible representation of $SL(2, \mathbb{C}) \times SU(3) \times SU(2) \times U(1)$ upon its typical fibre. In contrast, the interacting elementary fermions with which we are most familiar correspond to interacting-particle bundles which possess a representation of $SL(2, \mathbb{C}) \times SU(3) \times U(1)_Q$ upon their typical fibre. The representation of $SL(2, \mathbb{C})$ upon the typical fibre of such a bundle is often a reducible direct sum representation, corresponding to the Dirac spinor bundle $\sigma = \sigma_L \oplus \sigma_R$.

To examine the electroweak-unified interacting-particle bundles of the standard model, let us begin by listing again the finite-dimensional irreducible representations of $SU(3) \times SU(2) \times U(1)$ which are conventionally associated with the particles in the first fermion generation[1]:

- The neutrino and the 'left-handed' part of the state-space of the electron $(\nu_L, e_L)$, transform according to the $(1, 2, -1)$ irreducible representation of $SU(3) \times SU(2) \times U(1)$. I.e., the tensor product of the trivial representation of $SU(3)$ with the 2-dimensional standard representation of $SU(2)$ with the 1-dimensional representation of $U(1)$ with hypercharge $-1$.

[1] Again, for pedagogical purposes, we make the assumption here that the neutrino is massless.

- The left-handed part of the state-spaces of the up quark and down quark $(u_L, d_L)$ transform according to the $(3, 2, 1/3)$ representation. I.e., the tensor product of the standard representation of $SU(3)$ with the 2-dimensional standard representation of $SU(2)$ with the 1-dimensional representation of $U(1)$ with hypercharge $1/3$.
- The right-handed part of the state-space of the electron $e_R$ transforms according to the $(1, 1, -2)$ representation.
- The right-handed part of the state-space of the up quark $u_R$ transforms according to the $(3, 1, 4/3)$ representation.
- The right-handed part of the state-space of the down quark $d_R$ transforms according to the $(3, 1, -2/3)$ representation.

Let us take the direct sum of these irreducible representations of $SU(3) \times SU(2) \times U(1)$:

$$(1, 2, -1) \oplus (3, 2, 1/3) \oplus (1, 1, -2) \oplus (3, 1, 4/3) \oplus (3, 1, -2/3).$$

Each irreducible representation here is unitary with respect to a standard inner product. The inner product on each irreducible representation is then used to determine an inner product on the direct sum, which obviously renders the reducible direct sum representation as a unitary representation itself.

In fact, any finite-dimensional compact group representation can be rendered as a unitary representation, and it can be proven that a finite-dimensional unitary compact group representation has a unique decomposition into a direct sum of orthogonal irreducible subrepresentations (up to equivalence).

To elaborate, let $G$ denote a compact group and let $V$ be a Hilbert space on which we have a unitary representation of $G$. Let $\mathcal{T}$ denote the set of irreducible representation equivalence classes. Then $V$ can be *uniquely* decomposed into an orthogonal direct sum of invariant subspaces,

$$V = \bigoplus_{\tau \in \mathcal{T}} V_\tau,$$

where each $V_\tau$ is an orthogonal direct sum of type $\tau$ subrepresentations:

$$V_\tau = \bigoplus_{j \in J_\tau} E_j.$$

The $E_j$ are all unitarily equivalent to a finite-dimensional irreducible representation space. The further decomposition of the $V_\tau$ into irreducible subrepresentations is not unique, but the cardinality of the index set $J_\tau$ is unique, and is called the multiplicity of $\tau$.

Thus if one has a decomposition of $V_\tau$ into a Hilbert space direct sum $\bigoplus_{i \in I} E_i$ of irreducible subrepresentations, and another decomposition into a Hilbert space direct sum $\bigoplus_{i' \in I'} E'_{i'}$, then there is a bijection $b : I \to I'$ and a unitary isomorphism $\phi_i : E_i \to E'_{b(i)}$.[2]

[2] Private communication with Karl H. Hofmann.

The reducible direct sum representation of $SU(3) \times SU(2) \times U(1)$ conventionally associated with the first fermion generation, does not contain more than one copy of any irreducible subrepresentation, hence this representation cannot be given any other direct sum decomposition. We shall, however, have cause later in this chapter to associate an alternative reducible direct sum representation with the first fermion generation:

$$(1, 2, -1) \oplus (1, 1, -2) \oplus (3, 2, 1/3) \oplus (3, 2, 1/3).$$

This representation is inequivalent to the conventional representation, but equally consistent with the spectrum of particles we observe to exist after symmetry breaking.

Let us proceed for the moment with the conventional representation. Each irreducible subrepresentation of $SU(3) \times SU(2) \times U(1)$ corresponds to at least one interaction bundle $\delta$ which possesses that representation upon its typical fibre. If an interaction bundle $\delta$ possesses a finite-dimensional representation of $SU(3) \times SU(2) \times U(1)$ upon its typical fibre, then given a free-particle bundle $\eta$ equipped with a finite-dimensional representation of $SL(2, \mathbb{C})$ upon its typical fibre, the interacting-particle bundle $\alpha$ constructed from $\delta$ and $\eta$ will possess a finite-dimensional representation of $SL(2, \mathbb{C}) \times SU(3) \times SU(2) \times U(1)$ upon its typical fibre. If the representation of $SU(3) \times SU(2) \times U(1)$ is irreducible, if the representation of $SL(2, \mathbb{C})$ is irreducible, and if the interacting-particle bundle is the tensor product $\alpha = \eta \otimes \delta$, then the representation of $SL(2, \mathbb{C}) \times SU(3) \times SU(2) \times U(1)$ upon the typical fibre of $\alpha$ will also be irreducible.

The only two free-particle bundles used in the standard model multiplets are $\sigma_L$ and $\sigma_R$, the left-handed and right-handed Weyl spinor bundles, respectively. These bundles possess upon their typical fibres the $(1/2, 0)$ and $(0, 1/2)$ complex, finite-dimensional, irreducible representations of $SL(2, \mathbb{C})$. Hence, the interacting-particle bundles which represent interacting elementary fermions in the electroweak-unified standard model, are obtained by tensoring a Weyl spinor bundle with an interaction bundle that possesses an irreducible finite-dimensional representation of $SU(3) \times SU(2) \times U(1)$.

## 6.1. Standard model electroweak-unified bundles

Let us examine the interaction bundles relevant to the electroweak-unified standard model. Recall that each of the irreducible representations of the gauge group of the standard model, $SU(3) \times SU(2) \times U(1)$, is (equivalent to) a tensor product of irreducible representations of the individual factors. To obtain at least one interaction bundle for every possible irreducible representation of the standard model gauge group, one can procure an interaction bundle $\rho^{s_1,s_2}$ for each possible spin-$(s_1, s_2)$ irreducible representation of $SU(3)$, an interaction bundle $\tau^s$ for each possible spin-$s$ irreducible representation of $SU(2)$, and an interaction bundle $\chi^y$ for each possible index $k = 3y$ irreducible representation of $U(1)$. $\rho^{s_1,s_2}$ is a family of strong interaction bundles, $\tau^s$ is a family of weak isospin interaction bundles, and $\chi^y$ is a family of weak hypercharge interaction bundles. The typical fibre of $\rho^{s_1,s_2}$ is the vector space $V^{s_1,s_2}$ which possesses the spin-$(s_1, s_2)$ irreducible representation of $SU(3)$; the typical fibre of $\tau^s$ is the vector space $V^s$ which possesses the spin-$s$ irreducible representation of $SU(2)$; and the typical fibre of $\chi^y$ is the vector space $V^k$ which possesses the index $k = 3y$ irreducible representation of $U(1)$. By taking the tensor products of these interaction bundles, one obtains an interaction bundle for every irreducible representation of $SU(3) \times SU(2) \times U(1)$.

If one were only dealing with the leptons, then $\chi$, and its standard representation of $U(1)$, would correspond to a particle of unit weak hypercharge. Tensor products of this line bundle, and its conjugate, would correspond to all the possible integral values of weak hypercharge. However, we are dealing with leptons and quarks, for which, in the latter case, weak hypercharge comes in integer multiples of $1/3$. Hence, we need to use the family of weak hypercharge complex line bundles $\chi^y$, where $y$ is any integer multiple of $1/3$; where each $\chi^y$ is equipped with a $U(1)$ structure in each fibre; and where $\chi = \chi^{1/3}$. Note carefully that it is the bundle denoted as $\chi^{1/3}$ which possesses the standard representation of $U(1)$, not $\chi^1$. Unconventionally, $\chi \neq \chi^1$ in this notation.

One can treat this family of weak hypercharge bundles as tensor products of the $\chi^{1/3}$ bundle, or its conjugate $\bar{\chi}^{1/3}$. For any positive value of hypercharge, $y = k(1/3)$, $k \in \mathbb{Z}_+$, one can form the $k$-fold tensor product of $\chi^{1/3}$. The typical fibre of $\chi^{1/3}$ possesses the $e^{i\theta} \mapsto e^{i1\theta}$ representation of $U(1)$, and the typical fibre of the $k$-fold tensor product $\bigotimes^k \chi^{1/3}$ possesses the $k$-fold tensor product of this representation on $\mathbb{C}^1$. So, for example, the typical fibre of the three-fold tensor product of $\chi^{1/3}$ possesses the three-fold tensor product representation

$$e^{i\theta} \mapsto e^{i\theta} e^{i\theta} e^{i\theta} = e^{i3\theta}.$$

In the event that $y$ is negative, one forms the $k$-fold tensor product of the conjugate bundle $\bar{\chi}^{1/3}$. The typical fibre of $\bar{\chi}^{1/3}$ possesses the $e^{i\theta} \mapsto e^{-i\theta}$ representation of $U(1)$.

We shall use the notation $\tau = \tau^{1/2}$ to denote the weak isospin bundle which possesses the standard representation of $SU(2)$. Hence, the tensor product bundle $\tau \otimes \chi^y$ possesses the (*dimension*, *index*) $= (2, 3y)$ representation of $SU(2) \times U(1)$. Denoting $\tau^0$ as the bundle whose typical fibre possesses the trivial representation of $SU(2)$, the tensor product bundle $\tau^0 \otimes \chi^y$ possesses the $(1, 3y)$-representation of $SU(2) \times U(1)$. Given that $\tau^0 \cong \mathcal{M} \times \mathbb{C}^1$, it follows that

$$\tau^0 \otimes \chi^y \cong (\mathcal{M} \times \mathbb{C}^1) \otimes \chi^y \cong \chi^y.$$

Hence, the typical fibre of $\chi^y$ can be considered to possess either the index-$3y$ representation of $U(1)$, or the $(1, 3y)$ representation of $SU(2) \times U(1)$.

Note that the strong interaction bundle $\rho$ referred to in previous chapters is taken to possess the standard representation of $SU(3)$. Hence, $\rho^{1/3,0} = \rho$ in the notation of this chapter.

Given that the fermionic multiplet representations in the standard model only use the standard representations of $SU(3)$ and $SU(2)$, one can obtain an interaction bundle for each multiplet representation using only the vector bundles $\rho, \tau$ and $\chi^y$. The $(1, 2, -1)$ representation corresponds to the interaction bundle $\tau \otimes \chi^{-1}$, the $(3, 2, 1/3)$ representation corresponds to the interaction bundle $\rho \otimes \tau \otimes \chi^{1/3}$, the $(1, 1, -2)$ representation corresponds to the interaction bundle $\chi^{-2}$, the $(3, 1, 4/3)$ representation corresponds to the interaction bundle $\rho \otimes \chi^{4/3}$, and the $(3, 1, -2/3)$ representation corresponds to the interaction bundle $\rho \otimes \chi^{-2/3}$.

Given the handedness ('chirality') of the coupling between fermions and the weak force, the interacting-particle bundles are obtained by tensoring the interaction bundles with either left-handed or right-handed Weyl spinor bundles as follows:

The $(1, 2, -1)$ representation for the left-handed electron and neutrino $(\nu_L, e_L)$, corresponds to the vector bundle

$$\tau \otimes \chi^{-1} \otimes \sigma_L.$$

The state space of the interacting electron contains a subset of left-handed electron states which are composed of mass-$m_e$ solutions to a $\nabla$-dependent differential equation imposed upon the space of cross-sections $\Gamma(\tau \otimes \chi^{-1} \otimes \sigma_L)$. The state space of an interacting zero mass neutrino is composed of solutions to a $\nabla$-dependent differential equation imposed upon the space of cross-sections $\Gamma(\tau \otimes \chi^{-1} \otimes \sigma_L)$.

The $(3, 2, 1/3)$ representation for the left-handed up-quark and down-quark $(u_L, d_L)$, corresponds to the vector bundle

$$\rho \otimes \tau \otimes \chi^{1/3} \otimes \sigma_L.$$

The state space of the interacting up-quark contains a subset of left-handed up-quark states which are composed of mass-$m_u$ solutions to a $\nabla$-dependent differential equation imposed upon the space of cross-sections $\Gamma(\rho \otimes \tau \otimes \chi^{1/3} \otimes \sigma_L)$. The state space of the interacting down-quark contains a subset of left-handed down-quark states which are composed of mass-$m_d$ solutions to a $\nabla$-dependent differential equation imposed upon the space of cross-sections $\Gamma(\rho \otimes \tau \otimes \chi^{1/3} \otimes \sigma_L)$.

The $(1, 1, -2)$ representation for the right-handed electron $(e_R)$, corresponds to the vector bundle

$$\chi^{-2} \otimes \sigma_R.$$

The state space of the interacting electron contains a subset of right-handed electron states which are composed of mass-$m_e$ solutions to a $\nabla$-dependent differential equation imposed upon the space of cross-sections $\Gamma(\chi^{-2} \otimes \sigma_R)$.

The $(3, 1, 4/3)$ representation for the right-handed up-quark $(u_R)$, corresponds to the vector bundle

$$\rho \otimes \chi^{4/3} \otimes \sigma_R.$$

The state space of the interacting up-quark contains a subset of right-handed up-quark states, which are composed of mass-$m_u$ solutions to a $\nabla$-dependent differential equation imposed upon the space of cross-sections $\Gamma(\rho \otimes \chi^{4/3} \otimes \sigma_R)$.

The $(3, 1, -2/3)$ representation for the right-handed down-quark ($d_R$), corresponds to the vector bundle

$$\rho \otimes \chi^{-2/3} \otimes \sigma_R.$$

The state space of the interacting down-quark contains a subset of right-handed down-quark states, which are composed of mass-$m_d$ solutions to a $\nabla$-dependent differential equation imposed upon the space of cross-sections $\Gamma(\rho \otimes \chi^{-2/3} \otimes \sigma_R)$.

The interaction carrier bundles which correspond to weak isospin, $T^*\mathcal{M} \otimes \mathfrak{g}(\tau^s)$, represent the $W$ bosons of the weak force, and the interaction carrier bundles which correspond to weak hypercharge, $T^*\mathcal{M} \otimes \mathfrak{g}(\chi^y)$, represent the $Z$ bosons of the weak force. The magnitude of the weak hypercharge in the list of bundles above indicates the strength of the coupling between a fermion and the $Z$ boson. The absence of a $\tau$-factor in the last three bundles indicates that the right-handed electrons and quarks do not couple to the $W$ boson of the weak force.

Now comes the rub: An irreducible representation of the gauge group of the standard model does not uniquely determine a vector bundle. For each irreducible representation, there is more than one vector bundle which possesses that representation upon its typical fibre. As a special case of this, one can replace the weak isospin and weak hypercharge interaction bundles used above, with electroweak interaction bundles. To understand the bundle relationships, we first need to understand some of the relationships between representations of $U(2)$ and representations of $SU(2) \times U(1)$.

The following discussion will draw upon two families of representations of $U(2)$. These will be denoted as the $s_n$ and the $t_k$ families. An arbitrary element $u \in U(2)$ can be expressed as $u = e^{i\theta}A$, where $e^{i\theta} \in U(1)$ and $A \in SU(2)$. This arbitrary element of $U(2)$ is covered by a pair of elements in $SU(2) \times U(1)$, namely $(A, e^{i\theta})$ and $(-A, -e^{i\theta})$, with $A = (\det u)^{-1/2}u$ and $e^{i\theta} = (\det u)^{1/2}$. For each integer $n$, the $s_n$ representation maps $e^{i\theta}A$ to $e^{in\theta}A$ acting upon $\mathbb{C}^2$. For odd values of $n$, the $s_n$ representation lifts to the (*dimension*, *index*) $= (2, n)$ representation of $SU(2) \times U(1)$. For each integer $k$, the $t_k$ representation maps $e^{i\theta}A$ to $e^{ik\theta}$ acting on $\mathbb{C}^1$. For even values of $k$, the $t_k$ representation lifts to the $(1, k)$ representation of $SU(2) \times U(1)$. Again, $k$ here refers to the index, not the hypercharge, of the $U(1)$ representation.

Every $s_n$-representation of $U(2)$ on $\mathbb{C}^2$ can be lifted to a representation of $SU(2) \times U(1)$ on $\mathbb{C}^2$. However, it is only in the case of the odd-valued integers $n$ that the lifted representation is equivalent to the $(2, n)$ irreducible representation of $SU(2) \times U(1)$ on $\mathbb{C}^2 \otimes \mathbb{C}^1 \cong \mathbb{C}^2$. For example, take the representation

$$s_2 : e^{i\theta}A \mapsto e^{i2\theta}A.$$

$e^{i\theta}A$ is covered by $(A, e^{i\theta})$ and $(-A, -e^{i\theta})$ in $SU(2) \times U(1)$, hence the lift of the $s_2$ representation maps both $(A, e^{i\theta})$ and $(-A, -e^{i\theta})$ to $e^{i2\theta}A$. This is clearly different from the (2, 2)-representation of $SU(2) \times U(1)$, which maps $(A, e^{i\theta})$ to $e^{i2\theta}A$, and $(-A, -e^{i\theta})$ to the distinct operator $e^{i2\theta}(-A) = -e^{i2\theta}A$.

The $t_k$-representation of $U(2)$ maps $e^{i\theta}A$ to $e^{ik\theta}$, hence the lift of the $t_k$-representation maps $(A, e^{i\theta})$ and $(-A, -e^{i\theta})$ to the same element, $e^{ik\theta}$. The $(1, k)$ representations of $SU(2) \times U(1)$ only do this for even integers. To verify this, note that the $(1, k)$ representation, for arbitrary integer $k$, provides the mapping

$$\left(-A, -e^{i\theta}\right) = \left(-A, e^{i\pi} \cdot e^{i\theta}\right) \mapsto e^{ik\pi} \cdot e^{ik\theta}.$$

For the $(1, k)$ representation to be the lift of the $t_k$-representation, it is required that $e^{ik\pi} \cdot e^{ik\theta} = e^{ik\theta}$, and $e^{ik\pi}$ equals $e^{i2\pi} = 1$ if and only if $k$ is an even integer.

Each $s_n$ representation of $U(2)$ on $\mathbb{C}^2$ induces the representation $e^{i\theta}A \mapsto e^{i2n\theta}$ of $U(2)$ on the two-fold antisymmetric tensor product $\bigwedge^2 \mathbb{C}^2$. To see this, first note that the standard representation of $SU(2)$ on $\mathbb{C}^2$ induces the trivial representation of $SU(2)$ on $\bigwedge^2 \mathbb{C}^2$. To express this fact, let $r$ denote the representation of $SU(2)$ on $\bigwedge^2 \mathbb{C}^2$, let $v_1 \wedge v_2$ denote a simple tensor element of $\bigwedge^2 \mathbb{C}^2$, and we can then assert that, for any $A \in SU(2)$,

$$\begin{aligned} r(A)(v_1 \wedge v_2) = Av_1 \wedge Av_2 &= \frac{1}{\sqrt{2}}(Av_1 \otimes Av_2 - Av_2 \otimes Av_1) \\ &= v_1 \wedge v_2. \end{aligned}$$

Now, letting $e^{i\theta}A$ denote an arbitrary element of $U(2)$, given that $s_n(e^{i\theta}A) = e^{in\theta}A$, the induced representation of $U(2)$ on $\bigwedge^2 \mathbb{C}^2$ is as follows:

$$\begin{aligned} \left(e^{in\theta}A\right)(v_1 \wedge v_2) &= \left(e^{in\theta}A\right)v_1 \wedge \left(e^{in\theta}A\right)v_2 \\ &= \frac{1}{\sqrt{2}}\left(\left(e^{in\theta}A\right)v_1 \otimes \left(e^{in\theta}A\right)v_2 - \left(e^{in\theta}A\right)v_2 \otimes \left(e^{in\theta}A\right)v_1\right) \\ &= \frac{1}{\sqrt{2}}\left(e^{i2n\theta}(Av_1 \otimes Av_2) - e^{i2n\theta}(Av_2 \otimes Av_1)\right) \\ &= e^{i2n\theta}\left[\frac{1}{\sqrt{2}}(Av_1 \otimes Av_2 - Av_2 \otimes Av_1)\right] \\ &= e^{i2n\theta}\left(r(A)(v_1 \wedge v_2)\right) \\ &= e^{i2n\theta}(v_1 \wedge v_2). \end{aligned}$$

This demonstrates that each $s_n$ representation of $U(2)$ on $\mathbb{C}^2$ induces the representation $e^{i\theta}A \mapsto e^{i2n\theta}$ of $U(2)$ on the two-fold antisymmetric tensor product $\bigwedge^2 \mathbb{C}^2$. Given that $\mathbb{C}^1 \cong \bigwedge^2 \mathbb{C}^2$, these induced representations are the even numbered members of the $t_k$ family of representations, i.e., those with $k = 2n$, for all integers $n \in \mathbb{Z}$. Given that the $t_k$ representation lifts to the $(1, k)$ representation of $SU(2) \times U(1)$ for even values of $k$, this entails that the representation of $U(2)$ on $\bigwedge^2 \mathbb{C}^2$, induced by the $s_n$ representation of $U(2)$, lifts to the $(1, 2n)$ representation of $SU(2) \times U(1)$.

Consider now the electroweak interaction bundle $\iota$, a complex vector bundle of fibre dimension 2, equipped with a $U(2)$-structure in each fibre.[3] The typical fibre of this bundle possesses the standard representation of $U(2)$ on $\mathbb{C}^2$. Now introduce a new family of electroweak interaction bundles $\iota^y$, where each $\iota^y$ is equipped with a $U(2)$ structure in each fibre. Assuming that we are dealing with quarks and leptons, $y$ is any integer multiple of $1/3$. The complex plane bundle $\iota^y$ is distinguished by the fact that its typical fibre $\mathbb{C}^2$ possesses the $s_{3y}$ representation of $U(2)$.[4] The $s_{3y}$ representation maps $e^{i\theta}A$ to $e^{i3y\theta}A$. So long as $3y$ is an odd-valued integer, this representation of $U(2)$ lifts to the $(dimension, index) = (2, 3y)$ representation of $SU(2) \times U(1)$.

Now, given that the typical fibre of $\tau \otimes \chi^y$ possesses the $(2, 3y)$-representation of $SU(2) \times U(1)$, with the constraint that $3y$ is an odd-valued integer, the typical fibre of the electroweak interaction bundle $\iota^y$ provides the same irreducible representation of $SU(2) \times U(1)$ as the typical fibre of $\tau \otimes \chi^y$. If $3y$ is odd-valued, the irreducible representation of $U(2)$ on the typical fibre of $\iota^y$, when lifted to a representation of $SU(2) \times U(1)$, is equivalent to the representation on the typical fibre of $\tau \otimes \chi^y$. In particular, the typical fibre of $\iota^{-1}$ provides the same representation as the typical fibre of $\tau \otimes \chi^{-1}$, and the typical fibre of $\iota^{1/3}$ provides the same representation as the typical fibre of $\tau \otimes \chi^{1/3}$. Hence, in the interacting-particle bundles listed above, one can substitute these electroweak bundles in the place of the weak isospin-weak hypercharge tensor product bundles. The $\iota^y$ and $\tau \otimes \chi^y$ bundles are only inter-substitutable when $3y$ is an odd integer.

Note also that a $U(2)$-bundle $\iota^y$ is only adequate as an electroweak interaction bundle because the standard model does not use faithful irreducible representations of $SU(2) \times U(1)$. In general, with no restriction upon the irreducible representations used by the standard model, one would need to use tensor product bundles $\tau^s \otimes \chi^y$ as electroweak interaction bundles.

There is a further relationship between the electroweak bundles and the weak isospin and weak hypercharge bundles. Recall that the $s_n$ representation of $U(2)$ on $\mathbb{C}^2$ induces the $t_{2n}$-representation of $U(2)$ on the two-fold exterior power $\bigwedge^2 \mathbb{C}^2$, in which $e^{i\theta}A$ acts as $e^{i2n\theta}$ upon $\bigwedge^2 \mathbb{C}^2$. Hence, the typical fibre of the two-fold exterior power $\bigwedge^2 \iota^y$ possesses the representation of $U(2)$ in which $e^{i\theta}A$ acts as $e^{i2\cdot 3y\theta} = e^{i6y\theta}$ upon $\bigwedge^2 \mathbb{C}^2$. The typical fibre of $\bigwedge^2 \iota^y$ possesses the $t_{6y}$-representation of $U(2)$. Hence, the family of vector bundles $\{\bigwedge^2 \iota^y \colon y \in \frac{1}{3}\mathbb{Z}\}$ possess, upon their typical fibres, the $t_k$-family of representations of $U(2)$ on $\bigwedge^2 \mathbb{C}^2$, with $k = 6y$ running through all the even integers. Thus, $\bigwedge^2 \iota^{1/3}$ possesses the $t_2$-representation, $\bigwedge^2 \iota^{2/3}$ possesses the $t_4$-representation, $\bigwedge^2 \iota^{3/3}$ possesses the $t_6$-representation, etc.

The typical fibre of $\chi^y$ possesses the $(1, 3y)$ representation of $SU(2) \times U(1)$. The $t_{6y}$-representation of $U(2)$ upon the typical fibre of $\bigwedge^2 \iota^y$, lifts to the $(1, 6y)$-

[3] Appendix N provides a technical discussion of electroweak interaction bundles.

[4] In this notation, $\iota \neq \iota^1$; the bundle equipped with the standard representation of $U(2)$ upon its typical fibre is $\iota = \iota^{1/3}$, not $\iota^1$.

representation of $SU(2) \times U(1)$. The $(1, 6y)$-representation is the representation possessed by the typical fibre of $\chi^{2y}$. Thus, $\chi^{2y}$ must possess the same representation as $\bigwedge^2 \iota^y$. For example, then, one can substitute $\bigwedge^2 \iota^{-1}$ in place of $\chi^{-2}$ for the right-handed electron ($e_R$). The typical fibre of $\bigwedge^2 \iota^{-1}$ possesses the $t_{-6}$-representation of $U(2)$, which lifts to the $(1, -6)$-representation of $SU(2) \times U(1)$, the same representation possessed by the typical fibre of $\chi^{-2}$.

Given that the typical fibre of the electroweak interaction bundle $\iota^y$ provides the same irreducible representation of $SU(2) \times U(1)$ as the typical fibre of $\tau \otimes \chi^y$, for odd values of $3y$, and given that the typical fibre of the bundle $\bigwedge^2 \iota^y$ provides the same irreducible representation of $SU(2) \times U(1)$ as $\chi^{2y}$, we have the following alternative list of interacting-particle bundles (see Table 6.1)[5]:

The $(1, 2, -1)$ representation for the left-handed electron and neutrino $(\nu_L, e_L)$, corresponds to the vector bundle

$$\iota^{-1} \otimes \sigma_L.$$

The state space of the interacting electron contains a subset of left-handed electron states which are composed of mass-$m_e$ solutions to a $\nabla$-dependent differential equation imposed upon the space of cross-sections $\Gamma(\iota^{-1} \otimes \sigma_L)$. The state space of a zero-mass interacting neutrino is composed of solutions to a $\nabla$-dependent differential equation imposed upon the space of cross-sections $\Gamma(\iota^{-1} \otimes \sigma_L)$.

The $(3, 2, 1/3)$ representation for the left-handed up-quark and down-quark $(u_L, d_L)$, corresponds to the vector bundle

$$\rho \otimes \iota^{1/3} \otimes \sigma_L.$$

The state space of the interacting up-quark contains a subset of left-handed up-quark states which are composed of mass-$m_u$ solutions to a $\nabla$-dependent differential equation imposed upon the space of cross-sections $\Gamma(\rho \otimes \iota^{1/3} \otimes \sigma_L)$. The state space of the interacting down-quark contains a subset of left-handed down-quark states which are composed of mass-$m_d$ solutions to a $\nabla$-dependent differential equation imposed upon the space of cross-sections $\Gamma(\rho \otimes \iota^{1/3} \otimes \sigma_L)$.

The $(1, 1, -2)$ representation for the right-handed electron $(e_R)$, corresponds to the vector bundle

$$\bigwedge^2 \iota^{-1} \otimes \sigma_R.$$

The state space of the interacting electron contains a subset of right-handed electron states which are composed of mass-$m_e$ solutions to a $\nabla$-dependent differential equation imposed upon the space of cross-sections $\Gamma(\bigwedge^2 \iota^{-1} \otimes \sigma_R)$.

[5] Note that we revert to using the hypercharge $y$, rather than the index $n = 3y$, to denote the $U(1)$ representations here.

Table 6.1
Standard model interacting-particle bundles

| Multiplet | Bundle 1 | Bundle 2 | Representation |
|---|---|---|---|
| $(\nu_L, e_L)$ | $\tau \otimes \chi^{-1} \otimes \sigma_L$ | $\iota^{-1} \otimes \sigma_L$ | $(1, 2, -1)$ |
| $(u_L, d_L)$ | $\rho \otimes \tau \otimes \chi^{1/3} \otimes \sigma_L$ | $\rho \otimes \iota^{1/3} \otimes \sigma_L$ | $(3, 2, 1/3)$ |
| $(e_R)$ | $\chi^{-2} \otimes \sigma_R$ | $\bigwedge^2 \iota^{-1} \otimes \sigma_R$ | $(1, 1, -2)$ |
| $(u_R)$ | $\rho \otimes \chi^{4/3} \otimes \sigma_R$ | $\rho \otimes \bigwedge^2 \iota^{2/3} \otimes \sigma_R$ | $(3, 1, 4/3)$ |
| $(d_R)$ | $\rho \otimes \chi^{-2/3} \otimes \sigma_R$ | $\rho \otimes \bigwedge^2 \iota^{-1/3} \otimes \sigma_R$ | $(3, 1, -2/3)$ |

The $(3, 1, 4/3)$ representation for the right-handed up-quark $(u_R)$, corresponds to the vector bundle

$$\rho \otimes \bigwedge^2 \iota^{2/3} \otimes \sigma_R.$$

The state space of the interacting up-quark contains a subset of right-handed up-quark states, which are composed of mass-$m_u$ solutions to a $\nabla$-dependent differential equation imposed upon the space of cross-sections $\Gamma(\rho \otimes \bigwedge^2 \iota^{2/3} \otimes \sigma_R)$.

The $(3, 1, -2/3)$ representation for the right-handed down-quark $(d_R)$, corresponds to the vector bundle

$$\rho \otimes \bigwedge^2 \iota^{-1/3} \otimes \sigma_R.$$

The state space of the interacting down-quark contains a subset of right-handed down-quark states, which are composed of mass-$m_d$ solutions to a $\nabla$-dependent differential equation imposed upon the space of cross-sections $\Gamma(\rho \otimes \bigwedge^2 \iota^{-1/3} \otimes \sigma_R)$.

If one takes the direct sum of these bundles, one obtains the following standard model interacting-particle bundle $\alpha_1^{sm}$ for the first fermion generation:

$$\alpha_1^{sm} = (\iota^{-1} \otimes \sigma_L) \oplus \left(\bigwedge^2 \iota^{-1} \otimes \sigma_R\right) \oplus (\rho \otimes \iota^{1/3} \otimes \sigma_L) \oplus \left(\rho \otimes \bigwedge^2 \iota^{2/3} \otimes \sigma_R\right) \oplus \left(\rho \otimes \bigwedge^2 \iota^{-1/3} \otimes \sigma_R\right).$$

## 6.2. Standard model electroweak-broken bundles

Before we can compare the standard model interacting-particle bundle with the physical world, we must incorporate the fact that the electroweak force has undergone spontaneous symmetry breaking (SSB). Under electroweak SSB the electroweak interaction bundles $\iota^y$ decompose into direct sums of electromagnetic interaction bundles. These electromagnetic bundles form a family of complex line bundles $\lambda^q$, where $q$ is any integer multiple of $1/3$, and where each $\lambda^q$ is equipped with a $U(1)$ structure in each fibre. The typical fibre of $\lambda^q$ possesses the index $m = 3q$-irreducible representation of $U(1)_Q$. Thus, in this notation, it is the bundle denoted as $\lambda^{1/3}$ which possesses the standard representation of $U(1)_Q$, not $\lambda^1$.

One can treat the family of electromagnetic bundles as tensor products of the $\lambda^{1/3}$ bundle or its conjugate. For any value of electric charge, $q = m(1/3)$, $m \in \mathbb{Z}$, one can obtain the electromagnetic bundle $\lambda^q$ as the $m$-fold tensor product of $\lambda^{1/3}$ or its conjugate. The typical fibre of $\lambda^{1/3}$ possesses the $e^{i\theta} \mapsto e^{i1\theta}$ representation of $U(1)_Q$, and the typical fibre of the $m$-fold tensor product $\bigotimes^m \lambda^{1/3}$ possesses the $m$-fold tensor product of this $q = 1/3$ representation on $\mathbb{C}^1$.

The restriction of the standard representation of $U(2)$ to $U(1)_Q$ decomposes as a direct sum of two irreducible representations of $U(1)_Q$. These two representations are possessed by the electromagnetic bundles $\lambda^{2/3}$ and $\lambda^{-1/3}$, respectively. Hence, under spontaneous symmetry breaking the bundle $\iota^{1/3}$ reduces to $\lambda^{2/3} \oplus \lambda^{-1/3}$. Thus, $\rho \otimes \iota^{1/3} \otimes \sigma_L$ reduces to $\rho \otimes (\lambda^{2/3} \oplus \lambda^{-1/3}) \otimes \sigma_L$.

The interaction bundle $\iota^{-1}$ possesses the $s_{3y}$ representation of $U(2)$ for $y = -1$. This representation induces upon $U(1)_Q$ the direct sum of the trivial representation of $U(1)_Q$ with the representation possessed by the typical fibre of $\lambda^{-1}$. Hence, under spontaneous symmetry breaking, the bundle $\iota^{-1}$ reduces to $1 \oplus \lambda^{-1}$, and $\iota^{-1} \otimes \sigma_L$ reduces to $(1 \oplus \lambda^{-1}) \otimes \sigma_L$.

The two-fold antisymmetric tensor product bundle $\bigwedge^2 \iota^y$ possesses the $t_{6y}$ representation of $U(2)$ upon its typical fibre $\mathbb{C}^1$, and the $t_{6y}$-representation of $U(2)$ induces the representation of the electromagnetic subgroup $U(1)_Q \subset U(2)$ possessed by the typical fibre of $\lambda^y$. Thus, under spontaneous symmetry breaking, $\bigwedge^2 \iota^y$ reduces to $\lambda^y$.

Taking the above facts into account, under spontaneous symmetry breaking, the standard model interacting-particle bundle for the first fermion generation,

$$\begin{aligned}\alpha_1^{sm} &= \left(\iota^{-1} \otimes \sigma_L\right) \oplus \left(\bigwedge^2 \iota^{-1} \otimes \sigma_R\right) \oplus \left(\rho \otimes \iota^{1/3} \otimes \sigma_L\right) \\ &\quad \oplus \left(\rho \otimes \bigwedge^2 \iota^{2/3} \otimes \sigma_R\right) \oplus \left(\rho \otimes \bigwedge^2 \iota^{-1/3} \otimes \sigma_R\right),\end{aligned}$$

reduces to

$$\begin{aligned}\alpha_1^{sm} &= \left[\left(1 \oplus \lambda^{-1}\right) \otimes \sigma_L\right] \oplus \left(\lambda^{-1} \otimes \sigma_R\right) \oplus \left(\rho \otimes \left(\lambda^{2/3} + \lambda^{-1/3}\right) \otimes \sigma_L\right) \\ &\quad \oplus \left(\rho \otimes \lambda^{2/3} \otimes \sigma_R\right) \oplus \left(\rho \otimes \lambda^{-1/3} \otimes \sigma_R\right) \\ &= \sigma_L \oplus \left(\lambda^{-1} \otimes \sigma\right) \oplus \left(\rho \otimes \lambda^{2/3} \otimes \sigma\right) \oplus \left(\rho \otimes \lambda^{-1/3} \otimes \sigma\right).\end{aligned}$$

This is the correct generalised particle model for the first fermion generation $(\nu_e, e, u, d)$ when only the electromagnetic and strong forces are 'turned on.' $\sigma_L$ represents the charge-less electron-neutrino $\nu_e$, $\lambda^{-1} \otimes \sigma$ represents the electron coupled to the electromagnetic field with a charge of $-1$, $\rho \otimes \lambda^{2/3} \otimes \sigma$ represents the up-quark coupled to the electromagnetic and strong forces with an electromagnetic charge of $2/3$, and $\rho \otimes \lambda^{-1/3} \otimes \sigma$ represents the down-quark coupled to the electromagnetic and strong forces with an electromagnetic charge of $-1/3$.

The discussion above assumes that the neutrino is a massless particle, represented by cross-sections of the Weyl spinor bundle $\sigma_L$, and satisfying the Weyl equation. However, current evidence indicates that the neutrino does possess mass, and although there is no consensus on how this should be accommodated within the standard model, one could represent a massive neutrino by the cross-sections of a Dirac spinor bundle $\sigma = \sigma_L \oplus \sigma_R$ which satisfy the Dirac equation for a specific value of mass. In terms of the standard model multiplet structure, the right-handed part of such a massive neutrino forms an additional singlet corresponding to the trivial $(1, 1, 0)$ representation of $SU(3) \times SU(2) \times U(1)$.

If one wishes the summands of the standard model interacting-particle bundle to correspond to the irreducible representations determining the multiplet structure, then the existence of such a massive neutrino changes the leptonic terms of the bundle from $(\iota^{-1} \otimes \sigma_L) \oplus (\bigwedge^2 \iota^{-1} \otimes \sigma_R)$ to

$$\left(\iota^{-1} \otimes \sigma_L\right) \oplus \left(\bigwedge^2 \iota^{-1} \otimes \sigma_R\right) \oplus \sigma_R.$$

Needless to say, $\sigma_R$ represents the right-handed neutrino. Under spontaneous symmetry breaking, this bundle reduces to

$$\begin{aligned}
&\left[\left(1 \oplus \lambda^{-1}\right) \otimes \sigma_L\right] \oplus \left(\lambda^{-1} \otimes \sigma_R\right) \oplus \sigma_R \\
&\quad = \sigma_L \oplus \sigma_R \oplus \left(\lambda^{-1} \otimes \sigma\right) \\
&\quad = \sigma \oplus \left(\lambda^{-1} \otimes \sigma\right).
\end{aligned}$$

$\sigma$ represents the charge-less electron-neutrino $\nu_e$, and $\lambda^{-1} \otimes \sigma$ represents the electron coupled to the electromagnetic field with a charge of $-1$.

Derdzinski suggests that in the presence of a massive neutrino, $\iota^{-1} \otimes \sigma_L \oplus \bigwedge^2 \iota^{-1} \otimes \sigma_R$ should be replaced by $\iota^{-1} \otimes \sigma$ (1992, p. 118, where I have changed Derdzinski's sign convention). This also gives an interacting-particle bundle which reduces to $\sigma \oplus \lambda^{-1} \otimes \sigma$. Noting that

$$\iota^{-1} \otimes \sigma = \left(\iota^{-1} \otimes \sigma_L\right) \oplus \left(\iota^{-1} \otimes \sigma_R\right),$$

it is clear that this bundle possesses the $(1, 2, -1) \oplus (1, 2, -1)$ representation. Hence, the move from $\iota^{-1} \otimes \sigma_L \oplus \bigwedge^2 \iota^{-1} \otimes \sigma_R$ to $\iota^{-1} \otimes \sigma$ corresponds to a move from the $(1, 2, -1) \oplus (1, 1, -2)$ representation to the $(1, 2, -1) \oplus (1, 2, -1)$ representation.

We have emphasised in this chapter that more than one bundle can be the recipient of a fixed representation. Here, however, we find that there is more than one representation of the unified symmetry group which reduces to the representation consistent with post symmetry-breaking observations. This is a type of under-determination of theory by data, and it is not confined to the representation of leptons. One can define an alternative standard model interacting-particle bundle to be the recipient of the alternative representation of the unified symmetry group. The standard model interacting-particle bundle introduced above,

$$\alpha_1^{sm} = (\iota^{-1} \otimes \sigma_L) \oplus \left(\bigwedge^2 \iota^{-1} \otimes \sigma_R\right) \oplus (\rho \otimes \iota^{1/3} \otimes \sigma_L)$$
$$\oplus \left(\rho \otimes \bigwedge^2 \iota^{2/3} \otimes \sigma_R\right) \oplus \left(\rho \otimes \bigwedge^2 \iota^{-1/3} \otimes \sigma_R\right),$$

can duly be replaced with the following alternative:

$$\alpha_1^{\prime sm} = (\iota^{-1} \otimes \sigma_L) \oplus \left(\bigwedge^2 \iota^{-1} \otimes \sigma_R\right) \oplus (\rho \otimes \iota^{1/3} \otimes \sigma)$$
$$= (\iota^{-1} \otimes \sigma_L) \oplus \left(\bigwedge^2 \iota^{-1} \otimes \sigma_R\right) \oplus \left[\rho \otimes \iota^{1/3} \otimes (\sigma_L \oplus \sigma_R)\right]$$
$$= (\iota^{-1} \otimes \sigma_L) \oplus \left(\bigwedge^2 \iota^{-1} \otimes \sigma_R\right) \oplus (\rho \otimes \iota^{1/3} \otimes \sigma_L) \oplus (\rho \otimes \iota^{1/3} \otimes \sigma_R).$$

As clear from the last line of this equation, the summands of this alternative bundle do not correspond to the irreducible multiplet representations of $SU(3) \times SU(2) \times U(1)$ hitherto considered. Instead, the summands correspond to the following representation:

$$(1, 2, -1) \oplus (1, 1, -2) \oplus (3, 2, 1/3) \oplus (3, 2, 1/3).$$

Under electroweak SSB, $\rho \otimes \iota^{1/3} \otimes \sigma_L$ reduces to $(\rho \otimes \lambda^{2/3} \otimes \sigma_L) \oplus (\rho \otimes \lambda^{-1/3} \otimes \sigma_L)$, as before, and $\rho \otimes \iota^{1/3} \otimes \sigma_R$ reduces to $(\rho \otimes \lambda^{2/3} \otimes \sigma_R) \oplus (\rho \otimes \lambda^{-1/3} \otimes \sigma_R)$. Hence, under electroweak SSB, $\alpha_1^{\prime sm}$ also reduces to

$$\alpha_1^{\prime sm} = \sigma_L \oplus (\lambda^{-1} \otimes \sigma) \oplus (\rho \otimes \lambda^{2/3} \otimes \sigma) \oplus (\rho \otimes \lambda^{-1/3} \otimes \sigma).$$

$\alpha_1^{sm}$ and $\alpha_1^{\prime sm}$ are two distinct vector bundles, but they are related by the fact that they both reduce to a common bundle under electroweak symmetry breaking. The $(\rho \otimes \bigwedge^2 \iota^{2/3} \otimes \sigma_R) \oplus (\rho \otimes \bigwedge^2 \iota^{-1/3} \otimes \sigma_R)$ summands in the original bundle can be replaced by $\rho \otimes \iota^{1/3} \otimes \sigma_R$, because $\bigwedge^2 \iota^{2/3} \oplus \bigwedge^2 \iota^{-1/3}$ and $\iota^{1/3}$ both break into $\lambda^{2/3} \oplus \lambda^{-1/3}$. The $(3, 1, 4/3) \oplus (3, 1, -2/3)$ and $(3, 2, 1/3)$ representations of $SU(3) \times SU(2) \times U(1)$ both break into the $(3, 2/3) \oplus (3, -1/3)$ representation of $SU(3) \times U(1)_Q$.

Given that each irreducible subrepresentation of the unified symmetry group $SU(3) \times SU(2) \times U(1)$ corresponds, pre-symmetry breaking, to an interacting elementary particle, the under-determination of the overall representation of the unified symmetry group entails an under-determination of the pre-symmetry breaking spectrum of interacting elementary particles. Knowledge of the unified symmetry group, and knowledge of the spectrum of interacting elementary particles which exist post-symmetry breaking, is insufficient to infer the spectrum of interacting elementary particles which existed when the electroweak force was unified. If one takes the latter set of particles to be the spectrum of *intrinsically* distinct elementary particle types, consistent with the electroweak-unified theory, and the properties defining them to be intrinsic properties, then the standard model appears to under-determine the spectrum of intrinsically distinct elementary particle types and the corresponding intrinsic properties.

This type of under-determination is a problem for structural realism in general, and is closely related to so-called 'Jones under-determination.' This occurs when a theory has more than one formulation, each of which is empirically equivalent (Pooley, 2005). (This should not to be confused with the 'under-determination of theory by data,' which occurs when there are distinct theories, each of which is empirically equivalent. The different formulations of a fixed theory have mathematical relationships which extend beyond those required for mere empirical equivalence; this makes them more closely related than different theories.) The problem for structural realism occurs when the different formulations of a theory involve different mathematical structures. For example, in the traditional formulation of canonical classical general relativity, the configuration space is the set of Riemannian metric tensor fields on a 3-manifold $\Sigma$, but in Ashtekar's 'new variables' formulation, the configuration space is the space of connections upon an $SU(2)$-principal fibre bundle over $\Sigma$. These are non-isomorphic structures. There are cases, such as the Schrodinger and Heisenberg formulations of quantum mechanics, in which the structures associated with different formulations transpire to be isomorphic. These, however, are rather special cases.

Clearly, if there are non-isomorphic formulations of a theory, then a structural realist cannot say which of these structures the theory is committed to the instantiation of. In the case of the standard model, we wish to understand what type of thing a particle is represented to be, but we cannot do so if there are non-isomorphic structures on offer. The case under consideration, involving alternative bundles $\alpha^{sm}$ and $\alpha'^{sm}$, differs from Jones under-determination in that we have an under-determination of structure within a fixed formulation. The fibre bundle formulation of the standard model has been employed throughout this text, and the Lagrangian formulation is really just an economical way of specifying the fibre bundle information relevant for the calculations performed by physicists. One can diagnose the existence of multiple formulations and alternative structures as a symptom of an incomplete theory, and, indeed, most physicists clearly consider the standard model to be merely a staging post *en route* to their final destination. However, given that we are currently bereft of a way to complete the standard model, we have to accept that it strictly fails to provide a unique structure.

For the sake of argument, however, let us chose a particular standard model bundle for each fermion generation. Recall that there are three fermion generations, $(\nu_e, e, u, d)$, $(\nu_\mu, \mu, c, s)$, and $(\nu_\tau, \tau, t, b)$, and three generations of anti-fermions, $(\bar{\nu}_e, e^+, \bar{u}, \bar{d})$, $(\bar{\nu}_\mu, \mu^+, \bar{c}, \bar{s})$, and $(\bar{\nu}_\tau, \tau^+, \bar{t}, \bar{b})$. Each fermion generation has an isomorphic standard model interacting-particle bundle, and each anti-fermion generation has a conjugate standard model interacting-particle bundle. One can take the direct sum of these six bundles to obtain the overall standard model interacting-particle bundle

$$\alpha^{sm} = \alpha_1^{sm} \oplus \alpha_2^{sm} \oplus \alpha_3^{sm} \oplus \bar{\alpha}_1^{sm} \oplus \bar{\alpha}_2^{sm} \oplus \bar{\alpha}_3^{sm}.$$

In the chosen formulation of the standard model, any interacting elementary particle in our universe, with the possible exception of those that lie in regions of very strong gravitational field, is represented by a cross-section $\psi$ of, and a connection $\nabla$ upon, the bundle $\alpha^{sm}$ over space–time; together, $\psi$ and $\nabla$ satisfy an interacting matter field equation and the coupled Yang–Mills equation. This is what the chosen formulation of the standard model represents an interacting elementary particle to be.

# Appendices

Appendix A

# Topology

A topological space $(X, \mathscr{T})$ is a non-empty set $X$ equipped with a class of subsets $\mathscr{T}$ which satisfies the following conditions:

- $\emptyset \in \mathscr{T}$ and $X \in \mathscr{T}$.
- $\bigcup_i \mathscr{O}_i \in \mathscr{T}$ for any collection of members $\mathscr{O}_i \in \mathscr{T}$.
- $\bigcap_i \mathscr{O}_i \in \mathscr{T}$, for any *finite* collection of members $\mathscr{O}_i \in \mathscr{T}$.

$\mathscr{T}$ is called a topology, and the members $\mathscr{O} \in \mathscr{T}$ are called the open subsets of $X$. For a topological space $(X, \mathscr{T})$, a subset $\mathscr{C}$ is defined to be closed if and only if its complement $X-\mathscr{C}$ is an open subset. A set will typically possess many different topologies, and the topologies on a set are equipped with a partial ordering relation. A topology $\mathscr{T}_2$ is said to be 'finer' or 'stronger' than a topology $\mathscr{T}_1$ if and only if all the open subsets of $\mathscr{T}_1$ are also open subsets of $\mathscr{T}_2$, and there is at least one open subset of $\mathscr{T}_2$ which is not an open subset of $\mathscr{T}_1$. Conversely, one says that the topology $\mathscr{T}_1$ is 'weaker' or 'coarser' than the topology $\mathscr{T}_2$. The class of subsets $\{\emptyset, X\}$ is called the 'non-discrete' topology, and the set of all subsets, the power set $\mathscr{P}(X)$, is called the discrete topology.

A continuous mapping $f : X \to Y$ between a pair of topological spaces, $(X, \mathscr{T}_1)$ and $(Y, \mathscr{T}_2)$, is one for which the inverse image $f^{-1}(V)$ of every $\mathscr{T}_2$-open subset of $Y$, is a $\mathscr{T}_1$-open subset of $X$. If the mapping $f$ is one-to-one, then the inverse function $f^{-1}$ is defined. If a continuous one-to-one mapping $f$ is such that the inverse is also continuous, then the mapping is said to be a homeomorphism, an isomorphism between topological spaces which maps the open subsets of one topological space to the open subsets of another in a one-to-one fashion.

A neighbourhood of a point $x$ in a topological space can be defined to be a subset $U$ which contains an open subset containing $x$. This definition enables both open sets and closed sets to count as neighbourhoods.

An open covering of a topological space $(X, \mathscr{T})$ is a collection of open subsets which is such that the union of the subsets contains the entire space. A topological space is defined to be compact if and only if every open covering admits a finite subcovering. Obviously, a topological space which doesn't satisfy this condition is said to be non-compact.

A topological space is connected if and only if it is not the union of two (or more) non-empty disjoint subsets. If a topological space is disconnected, then the disjoint open subsets it decomposes into are termed the 'components' of the disconnected space.

A topological manifold is a special type of topological space which is:

1. Locally Euclidean. I.e., each point $x$ possesses an open neighbourhood $U$ which is homeomorphic with an open subset of some $n$-dimensional Euclidean space $\mathbb{R}^n$. The homeomorphism $\phi : U \rightarrow \mathbb{R}^n$ is called a coordinate chart, and $n$ is the local dimension.
2. A Hausdorff topological space. I.e., each pair of points $x_1, x_2$ possess neighbourhoods, $x_1 \in U$ and $x_2 \in V$, which are disjoint from each other, $U \cap V = \emptyset$.

A connected locally Euclidean topological space must be of constant local dimension.

A $C^\infty$-differentiable $n$-manifold is a topological manifold in which there is a maximal collection of coordinate charts which cover the manifold, and which is such that wherever there is an overlap between the domains of any two charts, $(U, \phi)$ and $(V, \psi)$, then the coordinate transformation $\phi \circ \psi^{-1} : \psi(U \cap V) \rightarrow \phi(U \cap V)$ is infinitely differentiable. A collection of such charts is called an atlas of charts, and the manifold is said to be a $C^\infty$-manifold.

Mappings between $C^\infty$-manifolds $f : N \rightarrow M$ are said to be differentiable if they are differentiable at each point $x$ with respect to a coordinate chart $(U, \phi)$ of $x$ and a coordinate chart $(V, \psi)$ of $f(x)$, in the sense that $\psi \circ f \circ \phi^{-1} : \phi(U) \rightarrow \psi(V)$ is differentiable at $\phi(x)$. A diffeomorphism $f : N \rightarrow M$ between a pair of $C^\infty$-manifolds, $N$ and $M$, is an infinitely differentiable ('smooth') mapping, which is bijective, and which is such that the inverse $f^{-1}$ is also infinitely differentiable. A diffeomorphism is automatically a homeomorphism, and diffeomorphic manifolds must have the same dimension.

Given a $C^\infty$-manifold $M$, if there exists a nowhere-vanishing $C^\infty$ $n$-form field $\mu$, then $M$ is said to be orientable, and the $n$-form field is said to provide an orientation. $\mu$ can be referred to as a 'volume form' on the manifold.

Let $f, g$ be continuous maps from a topological space $X$ into a topological space $Y$, and let $I$ denote the unit interval $[0, 1]$. The maps $f, g$ are said to be homotopic if there is a continuous map

$$H : X \times I \rightarrow Y,$$

such that $f(x) = H(x, 0)$ and $g(x) = H(x, 1)$ for all $x \in X$ (Boothby, 1986, p. 267). In effect, there is a one-parameter deformation of one map into the other.

A loop through a point $x$ in a manifold $M$ is a continuous map $f : I \rightarrow M$ such that $f(0) = x = f(1)$. At each point $x$ of a manifold $M$, the set of all homotopic classes of loops at $x$ can be imbued with the structure of a group, and dubbed the fundamental group of $M$ at $x$. If $M$ is a connected manifold, the fundamental groups at the various points are isomorphic, and one can speak of the fundamental group of the manifold. A manifold is said to be simply connected if it is connected and the fundamental group is trivial. I.e., all the loops through any point are homotopic.

If a manifold $M$ is not simply connected, then one can construct from it a manifold $\tilde{M}$, called the universal covering manifold, which *is* simply connected. Basically, one fixes an arbitrary point $x \in M$, and by taking the homotopy equivalence classes of all curves between $x$ and $y$, as $y$ ranges over $M$, one obtains a simply connected manifold $\tilde{M}$, and a natural surjective mapping $f : \tilde{M} \to M$, called the covering map. Different choices of the point $x$ simply result in diffeomorphic covering manifolds (see Wald, 1984, p. 345).

Appendix B

# Lie Groups and Lie Algebras

A group $G$ is defined to be a set which is equipped with a binary operation $\circ$ called the product, a unary operation $(\cdot)^{-1}$ called the inverse, and a special element $e$ called the identity, such that, for any $a, b, c \in G$, the following conditions are satisfied:

1. $(a \circ b) \circ c = a \circ (b \circ c)$.
2. $a \circ e = e = e \circ a$.
3. $a \circ a^{-1} = e = a^{-1} \circ a$.

A Lie group $G$ is a manifold which is also a group, such that the group operations are smooth with respect to the manifold structure. I.e., the group product $G \times G \to G$ sending $(a, b)$ to $a \circ b$, and the inverse operation $G \to G$ sending $a$ to $a^{-1}$, are both infinitely differentiable mappings.

A real Lie algebra is a real vector space $\mathfrak{g}$ equipped with a bilinear mapping $[\, ,\, ] : \mathfrak{g} \times \mathfrak{g} \to \mathfrak{g}$ called the Lie bracket, which satisfies the following conditions:

1. $[X, Y] = -[Y, X]$ (skew-symmetry).
2. $\big[[X, Y], Z\big] + \big[[Z, X], Y\big] + \big[[Y, Z], X\big] = 0$ (Jacobi identity).

There is a Lie algebra canonically associated with every Lie group, namely the set of vector fields on the Lie group which are invariant under the left action of the group upon itself. This Lie algebra of 'left-invariant' vector fields can be identified with the tangent vector space at the identity element $T_e(G)$.

Let $(X_1, \ldots, X_n)$ be a basis of the $n$-dimensional Lie algebra $\mathfrak{g}$. The Lie bracket $[X_j, X_k]$ of any two elements from the basis is a linear combination $C^i_{jk} X_i$, and the numbers $C^i_{jk}$ are called the structure constants of the Lie algebra.

Given a Lie group $G$ which is not simply connected, one can construct from it a Lie group $\tilde{G}$, called the universal covering group, which *is* simply connected. $\tilde{G}$ is the universal covering manifold[1] of $G$, with the additional condition that $\tilde{G}$ is equipped with a group product which renders the covering map $f : \tilde{G} \to G$ a group homomorphism (see Wald (1984, p. 345)).

[1] See Appendix A.

Appendix C

# Fibre Bundles

A bundle $(E, M, \pi)$ consists of two topological spaces $E$ and $M$, and a continuous surjective mapping $\pi : E \to M$ called the projection mapping. $E$ is called the total space and $M$ is called the base space. For each $x \in M$, the inverse image $\pi^{-1}(x)$ is called the fibre over $x$. A fibre bundle $(E, M, \pi, F)$ is a bundle for which each fibre is homeomorphic with a topological space $F$ called the typical fibre. A vector bundle is a fibre bundle for which the typical fibre is a vector space. A global cross-section of a fibre bundle is a continuous mapping from the base space into the total space, $\psi : M \to E$, which selects an element from the fibre $\pi^{-1}(x)$ over each point $x \in M$.

A $C^\infty$-fibre bundle is such that $E$, $M$ and $F$ are differentiable manifolds, the projection mapping $\pi$ is a smooth mapping, and each fibre is diffeomorphic to the typical fibre $F$. Global cross-sections of $C^\infty$-fibre bundles are smooth mappings $\psi : M \to E$. Hereafter, it is assumed that we are dealing with $C^\infty$-fibre bundles.

In the event that the total space $E$ can be expressed as product $M \times F$ of the base space and the typical fibre, the fibre bundle is said to be trivial. Whilst a trivial fibre bundle is a very special type of fibre bundle, all fibre bundles are locally trivial. It is always possible to find a covering of the base space by open subsets $U_\alpha$ which is such that each inverse image $\pi^{-1}(U_\alpha)$ is diffeomorphic to $U_\alpha \times F$. The maps

$$\phi_\alpha : \pi^{-1}(U_\alpha) \to U_\alpha \times F,$$

are referred to as local trivialization maps. Each $p \in \pi^{-1}(U_\alpha)$ is mapped to $(\pi(p), f) \in U_\alpha \times F$ by a local trivialization map. The restriction of $\phi_\alpha$ to the fibre $\pi^{-1}(x)$ over any $x \in U_\alpha$ provides a diffeomorphism between $\pi^{-1}(x)$ and the typical fibre $F$.

Wherever two of the open subsets intersect, $U_\alpha \cap U_\beta$, one has a pair of distinct local trivializations,

$$\phi_\alpha : \pi^{-1}(U_\alpha \cap U_\beta) \to (U_\alpha \cap U_\beta) \times F,$$

and

$$\phi_\beta : \pi^{-1}(U_\alpha \cap U_\beta) \to (U_\alpha \cap U_\beta) \times F.$$

Each provides a different mapping from the fibre $\pi^{-1}(x)$ over any $x \in (U_\alpha \cap U_\beta)$ to the typical fibre $F$.

A transformation between two local trivializations is given by a mapping

$$\phi_\beta \circ \phi_\alpha^{-1} : (U_\alpha \cap U_\beta) \times F \to (U_\alpha \cap U_\beta) \times F,$$

which can be expressed as

$$\phi_\beta \circ \phi_\alpha^{-1}(x, f) = \big(x, g_{\beta\alpha}(x) f\big),$$

where $g_{\beta\alpha}(x)$ is a diffeomorphism of $F$ for each $x \in (U_\alpha \cap U_\beta)$:

$$g_{\beta\alpha}(x) : F \to F.$$

The mappings $g_{\beta\alpha}(x)$ from points of the base space into the group of automorphisms of the typical fibre, are called the transition functions of the fibre bundle. They satisfy the conditions

$$\big(g_{\beta\alpha}(x)\big)^{-1} = g_{\alpha\beta}(x), \quad x \in U_\alpha \cap U_\beta,$$

and

$$g_{\gamma\beta}(x) \circ g_{\beta\alpha}(x) = g_{\gamma\alpha}(x), \quad x \in U_\alpha \cap U_\beta \cap U_\gamma.$$

In the event that the fibre bundle is a vector bundle, with typical fibre $V$, the transition functions are valued in the general linear group $GL(V)$, or some subgroup thereof. If $V = \mathbb{C}^n$, $GL(V) = GL(n, \mathbb{C})$. For the transition functions to be valued in the automorphism group $GL(V)$ of a vector space they can either be directly valued in $GL(V)$, or they can be valued in an abstract group if specified in combination with a representation of the abstract group upon the vector space. In the event that a vector bundle over $M$ is globally trivial, the covering of $M$ and the transition functions can be chosen so that they are all valued in $G = \{Id\}$.

A principal fibre bundle $(P, \pi, M, G)$ is a fibre bundle in which the typical fibre is a Lie group $G$, which acts from the right upon the total space $P$ in such a way that the orbits of the $G$-action coincide with the fibres of $P$, and in such a way that $G$ acts simply transitively upon each fibre $\pi^{-1}(x)$. In other words, for any pair of elements in any fibre, $p, \ q \in \pi^{-1}(x)$, there is a unique $g \in G$ which is such that $pg = q$. This is equivalent to requiring that the action is transitive and free upon each fibre.

Given a principal fibre bundle $(P, \pi, M, G)$, and a representation $r : G \to GL(V)$ upon a vector space $V$, one can construct a vector bundle $P \times_{r(G)} V$ called an associated vector bundle. $P \times_{r(G)} V$ is a quotient space of the product $P \times V$. The representation of $G$ on $V$ enables one to define a right action of $G$ upon the product $P \times V$:

$$R_g(p, v) = \big(pg, r(g^{-1})v\big).$$

The orbit of $(p, v)$ under the right action of $G$ can be denoted as $[(p, v)]$, and the set of all orbits is the associated vector bundle $P \times_{r(G)} V$.

The transition functions of a principal fibre bundle $(P, \pi, M, G)$ are valued in $G$ itself,

$$g_{\beta\alpha}(x) : (U_\alpha \cap U_\beta) \to G,$$

with each $g_{\beta\alpha}(x)$ acting upon the typical fibre $G$ by left multiplication. Given a principal fibre bundle $(P, \pi, M, G)$, each inequivalent representation $r$ of $G$ on a vector space $V$, defines an associated vector bundle $P \times_{r(G)} V$, with typical fibre $V$. A vector bundle $(E, \Pi, M, V)$ is associated with a principal fibre bundle $(P, \pi, M, G)$ by a representation $r$ of $G$ on $V$ if its transition functions $g'_{\beta\alpha}(x)$ are the images under $r$ of the transition functions $g_{\beta\alpha}(x)$ of $P$. I.e.,

$$g'_{\beta\alpha}(x) = r\big(g_{\beta\alpha}(x)\big).$$

Given a principal fibre bundle $(P, \pi, M, G)$, each inequivalent representation of $G$ defines an inequivalent associated vector bundle, but each such bundle possesses the same topological 'twists,' defined by the transition functions of $P$.

Given any vector bundle $E$ over a number field $\mathbb{F}$, one can construct the general linear frame bundle $P$, the collection of all the bases in each fibre of $E$. The general linear frame bundle is a principal bundle with structure group $GL(n, \mathbb{F})$. The standard representation of $GL(n, \mathbb{F})$ on $\mathbb{F}^n$ then enables one to obtain an associated vector bundle which is isomorphic to the original vector bundle $E$. In fact, for any principal $G$-subbundle of the general linear frame bundle, where $G$ is any Lie subgroup of the matrix group $GL(n, \mathbb{F})$, the standard representation of $G$ on $\mathbb{F}^n$ enables one to obtain an associated vector bundle which is also isomorphic to the original vector bundle $E$.

Appendix D

# Representations of $SL(2, \mathbb{C}) \ltimes \mathbb{R}^{3,1}$ on $\mathcal{F}(\mathcal{M}, \mathbb{C}^n)$

Section 2.2 defined the canonical representation of $SL(2, \mathbb{C}) \ltimes \mathbb{R}^{3,1}$ upon the space of vector bundle cross-sections $\Gamma(\eta)$ as

$$\psi(x) \to \psi'(x) = \mathscr{D}^{s_1,s_2}(A) \cdot \psi\big(\Lambda^{-1}(x-a)\big),$$

where it is understood that $A \in SL(2, \mathbb{C})$, $a \in \mathbb{R}^{3,1}$, $\Lambda$ is shorthand for $\Lambda(A)$, and $\Lambda$ is the covering homomorphism $\Lambda : SL(2, \mathbb{C}) \to SO_0(3, 1)$. This appendix provides an alternative perspective upon this representation in the special case where the vector bundle is the product bundle $\mathcal{M} \times \mathbb{C}^n$, and the space of cross-sections is the space of functions $\mathcal{F}(\mathcal{M}, \mathbb{C}^n)$.

Let us begin by noting that the tensor product of two vector spaces, $\mathscr{H} \otimes \mathscr{K}$, is isomorphic to the $n$-fold direct sum of $\mathscr{H}$ with itself, where $n = \dim \mathscr{K}$ (Kadison and Ringrose, 1983, p. 140). Either $\mathscr{H}$ or $\mathscr{K}$ can be infinite-dimensional vector spaces. In the case of the tensor product $\mathcal{F}(\mathcal{M}) \otimes V$, where $V$ is some vector space, we have the isomorphism

$$\begin{aligned} \mathcal{F}(\mathcal{M}) \otimes V &\cong \bigoplus^n \mathcal{F}(\mathcal{M}) \\ &\cong \mathcal{F}\big(\mathcal{M}, \mathbb{C}^n\big), \quad n = \dim V. \end{aligned}$$

As a special case, when $V = \mathbb{C}^n$, then

$$\begin{aligned} \mathcal{F}(\mathcal{M}) \otimes \mathbb{C}^n &\cong \bigoplus^n \mathcal{F}(\mathcal{M}) \\ &\cong \mathcal{F}\big(\mathcal{M}, \mathbb{C}^n\big). \end{aligned}$$

Function spaces can be treated as cross-section spaces of trivial bundles, $\mathcal{F}(\mathcal{M}, \mathbb{C}^n) \cong \Gamma(\mathcal{M} \times \mathbb{C}^n)$ and $\mathcal{F}(\mathcal{M}) \cong \Gamma(\mathcal{M} \times \mathbb{C}^1)$, hence the isomorphism above can also be expressed as

$$\Gamma\big(\mathcal{M} \times \mathbb{C}^1\big) \otimes \mathbb{C}^n \cong \Gamma\big(\mathcal{M} \times \mathbb{C}^n\big).$$

The specific form of the isomorphism between $\mathcal{F}(\mathcal{M}) \otimes \mathbb{C}^n$ and $\bigoplus^n \mathcal{F}(\mathcal{M})$ depends upon the choice of a basis $\{v_i : i = 1, \dots, n\}$ in $\mathbb{C}^n$. With such a basis chosen, $\mathcal{F}(\mathcal{M}) \otimes \mathbb{C}^n$ can be decomposed into a direct sum $\bigoplus_i \mathcal{F}_i(\mathcal{M})$ of subspaces

$\mathcal{F}_i(\mathcal{M}) \subset \mathcal{F}(\mathcal{M}) \otimes \mathbb{C}^n$, each of which is isomorphic with $\mathcal{F}(\mathcal{M})$. Each direct summand $\mathcal{F}_i(\mathcal{M})$ is the span of the set of simple tensors $\{f(x) \otimes v_i\colon\ f(x) \in \mathcal{F}(\mathcal{M})\}$. I.e.,

$$\mathcal{F}_i(\mathcal{M}) = \bigvee \big\{ f(x) \otimes v_i\colon\ f(x) \in \mathcal{F}(\mathcal{M}) \big\}.$$

Given any $f(x) \in \mathcal{F}(\mathcal{M})$, and any $(c_1, \ldots, c_n) \in \mathbb{C}^n$, the tensor product

$$f(x) \otimes (c_1, \ldots, c_n) \in \mathcal{F}(\mathcal{M}) \otimes \mathbb{C}^n$$

can be mapped into an element of $\bigoplus^n \mathcal{F}(\mathcal{M}) \cong \mathcal{F}(\mathcal{M}, \mathbb{C}^n)$ under the following linear injection:

$$f(x) \otimes (c_1, \ldots, c_n) \mapsto \big(c_1 f(x), \ldots, c_n f(x)\big).$$

To see why, note that with the choice of basis $\{v_i\colon\ i = 1, \ldots, n\}$ in $\mathbb{C}^n$, it follows that $(c_1, \ldots, c_n)$ can be expressed as $c_1 v_1 + \cdots + c_n v_n$, and that

$$\begin{aligned} f(x) \otimes (c_1, \ldots, c_n) &= f(x) \otimes c_1 v_1 + \cdots + c_n v_n \\ &= f(x) \otimes c_1 v_1 + \cdots + f(x) \otimes c_n v_n \\ &= c_1 f(x) \otimes v_1 + \cdots + c_n f(x) \otimes v_n. \end{aligned}$$

Given that each direct summand $\mathcal{F}_i(\mathcal{M})$ is the span of the set of simple tensors $\{f(x) \otimes v_i\}$, it follows that we have the mapping

$$c_1 f(x) \otimes v_1 + \cdots + c_n f(x) \otimes v_n \mapsto \big(c_1 f(x), \ldots, c_n f(x)\big) \in \bigoplus^n \mathcal{F}(\mathcal{M}).$$

Conversely, given an arbitrary element $(f_1(x), \ldots, f_n(x)) \in \bigoplus^n \mathcal{F}(\mathcal{M}) \cong \mathcal{F}(\mathcal{M}, \mathbb{C}^n)$, with the choice of a basis $\{v_i\colon\ i = 1, \ldots, n\}$ in $\mathbb{C}^n$ we have the mapping

$$\big(f_1(x), \ldots, f_n(x)\big) \mapsto \big(f_1(x) \otimes v_1, \ldots, f_n(x) \otimes v_n\big) \in \bigoplus_i \mathcal{F}_i(\mathcal{M}),$$

which, in turn, leads to the mapping:

$$\big(f_1(x) \otimes v_1, \ldots, f_n(x) \otimes v_n\big) \mapsto f_1(x) \otimes v_1 + \cdots + f_n(x) \otimes v_n \in \mathcal{F}(\mathcal{M}) \otimes \mathbb{C}^n.$$

Given a representation of $SL(2, \mathbb{C}) \ltimes \mathbb{R}^{3,1}$ on the infinite-dimensional space $\mathcal{F}(\mathcal{M})$, and given a representation of $SL(2, \mathbb{C}) \ltimes \mathbb{R}^{3,1}$ on the finite-dimensional space $V_s$, one can form the tensor product of these representations on $\mathcal{F}(\mathcal{M}) \otimes V_s$. In the case where $V_s = \mathbb{C}^n$, the isomorphism between $\mathcal{F}(\mathcal{M}) \otimes \mathbb{C}^n$ and $\mathcal{F}(\mathcal{M}, \mathbb{C}^n)$ then entails that this tensor product representation can be transferred to a representation on $\mathcal{F}(\mathcal{M}, \mathbb{C}^n)$. This representation on $\mathcal{F}(\mathcal{M}, \mathbb{C}^n)$ is equivalent to the canonical representation on $\mathcal{F}(\mathcal{M}, \mathbb{C}^n)$.

For example, given the representation of $SL(2, \mathbb{C}) \ltimes \mathbb{R}^{3,1}$ on the base space $\mathcal{M}$, the natural representation of $SL(2, \mathbb{C}) \ltimes \mathbb{R}^{3,1}$ on the infinite-dimensional space of complex-valued functions $\mathcal{F}(\mathcal{M})$ is

$$f(x) \mapsto f\big(\Lambda^{-1}(x-a)\big).$$

Given some representation $\tau$ of $SL(2,\mathbb{C}) \ltimes \mathbb{R}^{3,1}$ on the finite-dimensional space $\mathbb{C}^n$, reducible or irreducible, one can form the tensor-product of these two representations on $\mathcal{F}(\mathcal{M}) \otimes \mathbb{C}^n$. This tensor product representation is such that

$$\begin{aligned} &f(x) \otimes (c_1, \ldots, c_n) \\ &\quad \mapsto f\big(\Lambda^{-1}(x-a)\big) \otimes \tau(A)(c_1, \ldots, c_n) = f\big(\Lambda^{-1}(x-a)\big) \otimes (c_1', \ldots, c_n'). \end{aligned}$$

Under the isomorphism between $\mathcal{F}(\mathcal{M}) \otimes \mathbb{C}^n$ and $\mathcal{F}(\mathcal{M}, \mathbb{C}^n)$ this becomes

$$\begin{aligned} &\big(c_1' f\big(\Lambda^{-1}(x-a)\big), \ldots, c_n' f\big(\Lambda^{-1}(x-a)\big)\big) \\ &\quad = \tau(A) \cdot \big(c_1 f\big(\Lambda^{-1}(x-a)\big), \ldots, c_n f\big(\Lambda^{-1}(x-a)\big)\big) \\ &\quad = \tau(A) \cdot \psi\big(\Lambda^{-1}(x-a)\big), \end{aligned}$$

where $\psi = (c_1 f, \ldots, c_n f)$. This is simply the canonical representation, with $\tau$ in the place of a representation from the $\mathscr{D}^{s_1,s_2}$-family.

Appendix E

# The Method of Induced Representation

Given a group $G$, a subgroup $L$, and a representation $s : L \to GL(V)$ of the subgroup $L$ on a vector space $V$, the representation of $L$ induces a representation of $G$. The method of induced representation enables one to construct a vector bundle $E$, with typical fibre $V$, and base space $G/L$. The induced representation of $G$ is a representation on the space of sections $\Gamma(E)$ (Sternberg, 1994, pp. 104–106).

To obtain the induced representation, one begins by taking the product $G \times V$. There is a right action of $L$ upon this product space, defined by

$$R_h(g, v) = \big(gh, s(h)^{-1}v\big),$$

for any $h \in L$, and any $(g, v) \in G \times V$.

One uses this right action to define the equivalence relation

$$(g, v) \sim \big(gh, s(h)^{-1}v\big),$$

and one defines $E$ to be the set of all equivalence classes,

$$E = G \times_L V.$$

Defining the base space $M$ as the coset space $G/L$, $E$ is a vector bundle over $G/L$, with typical fibre $V$. The map

$$\big[(g, v)\big] \mapsto gL,$$

provides a well-defined projection map $\pi : E \to M$. It is well-defined in the sense that it is constant on all the elements in each equivalence class. $(g, v) \mapsto gL$ and, for any $h \in L$, $(gh, v) \mapsto ghL = gL$ because $hL = L$.

Each fibre $E_p$ consists of the set of equivalence classes $[(g, v)]$ as $v$ ranges over $V$. I.e.,

$$E_p = \big\{\big[(g, v)\big]:\ v \in V\big\}.$$

It is also true that

$$E_p = \big\{\big[(gh, v)\big]:\ v \in V\big\},$$

for an arbitrary $h \in L$. $[(g, v)]$ is a point in the fibre $E_p$ over $p = gL$, and for any $h \in L$, $[(gh, v)]$ is also a point in the fibre $E_p$ over $p = gL$. In general, $[(g, v)]$ and $[(gh, v)]$ will be distinct points of the same fibre. Each fibre of $E$ is isomorphic to the typical fibre $V$, but there is no canonical isomorphism. Under one isomorphism, $[(g, v)]$ is mapped to $v \in V$, but under another isomorphism, $[(gh, v)]$, which is, in general, a distinct element of $E_p$, is mapped to $v \in V$.

$G$ acts upon the base space of the vector bundle $E$ in the obvious way, an arbitrary $a \in G$ mapping $gL$ to $agL$. There is also an action of $G$ upon the total space of the bundle,

$$a\big[(g, v)\big] = \big[(ag, v)\big],$$

which maps the fibre $E_p$ over $p = gL$ to the fibre $E_{ap}$ over $ap = agL$. The representation $r : G \to GL(\Gamma(E))$ of $G$ upon the space of sections $\Gamma(E)$ is defined by

$$r(a)\psi(p) = a\big[\psi\big(a^{-1}p\big)\big],$$

for any $\psi \in \Gamma(E)$. $\psi(a^{-1}p)$ is an element of the fibre $E_{a^{-1}p}$ over $a^{-1}p$, but $a[\psi(a^{-1}p)]$ is an element of the fibre $E_{aa^{-1}p} = E_p$. By definition, a cross-section of $E$ maps each point $p$ to an element in the fibre $E_p$. Hence, $p \mapsto \psi(a^{-1}p)$ is *not* a cross-section of $E$, but $p \mapsto a[\psi(a^{-1}p)]$ *is*. To define a representation upon the space of cross-sections $\Gamma(E)$, one therefore defines $r(a)\psi(p) = a[\psi(a^{-1}p)]$ rather than $r(a)\psi(p) = \psi(a^{-1}p)$.

The method of induced representation enables one to construct, from $G$, a vector bundle $E$ which is such that $G$ can be treated as a group of vector bundle endomorphisms of $E$. By this, we mean that there is an action of $G$ upon the base space $M$, and an action of $G$ upon the total space $E$, which is such that

$$a \cdot \pi\big(\big[(g, v)\big]\big) = \pi\big(a \cdot \big[(g, v)\big]\big),$$

and which is such that each $a : E_p \to E_{ap}$ is linear. In other words, each $a \in G$ acts upon the total space of $E$ as a fibre-preserving endomorphism. Each fibre is mapped to another fibre, and the map preserves the linear structure in each fibre. The condition $a \cdot \pi([(g, v)]) = \pi(a \cdot [(g, v)])$ means that the projection map $\pi : E \to M$ of the induced vector bundle is a $G$-morphism. This is equivalent to the requirement that each $a \in G$ maps each fibre into another fibre, $a : E_p \to E_{ap}$ (Sternberg, 1994, p. 100).

Given any group $H$, a vector space $N$, and a representation $\tau$ of $H$ on $N$, the semi-direct product $G = H \ltimes N$ is the Cartesian product of $H$ and $N$, equipped with the following group product operation

$$(h_1, n_1) \cdot (h_2, n_2) = \big(h_1 \cdot h_2, n_1 + \tau(h_1)n_2\big).$$

$N$ is a normal Abelian subgroup of the semi-direct product, and the representation of $H$ on $N$ coincides with the conjugation action of $H$ on $N$.

$H$ acts upon $N$ by conjugation, and $N$ acts upon itself by conjugation, hence the entire group $G = H \ltimes N$ acts upon $N$. Given the Abelian group $N$, let $\hat{N}$ denote the set of continuous homomorphisms from $N$ into $U(1) = \mathbb{T}$. Because $G$ acts upon $N$ it must also act upon any space of functions on $N$, hence $G$ acts upon $\hat{N}$. One can decompose $\hat{N}$ into orbits under the $G$-action,

$$\hat{N} = \hat{N}_1 \cup \cdots \cup \hat{N}_r,$$

and one can pick a representative $\chi_j \in \hat{N}_j$ from each orbit.

Let $H_j$ denote the subgroup of $H$ which leaves $\chi_j$ fixed. The conjugacy action of the Abelian group $N$ upon itself leaves every element of $N$ fixed, hence, $L_j = H_j \ltimes N$ is the isotropy group of $\chi_j$ under the action of $G$.

Given an irreducible representation $\rho_k$ of $H_j$, on a vector space $V_k$, there is an irreducible representation $\bar{\rho}_k$ of $L_j = H_j \ltimes N$ on the same vector space $V_k$, defined by:

$$\bar{\rho}_k(h, n) = \chi_j(n)\rho_k(h).$$

If $\rho_k$ is an irreducible unitary representation of $H_j$ on $V_k$, then $\bar{\rho}_k$ is an irreducible unitary representation of $H_j \ltimes N$ on $V_k$.

For each orbit of the $G$-action in $\hat{N}$, and for each irreducible representation of $H_j$, on a vector space $V_k$, one constructs a vector bundle $E$, with typical fibre $V_k$, whose base space is $G/L_j \cong \hat{N}_j$, which admits a representation of $G$ on the space of cross-sections $\Gamma(E)$.

The choice of an orbit in $\hat{N}$ by choosing a character $\chi_j$, and the choice of an irreducible representation $\rho_k$ of $H_j$, determines an irreducible unitary representation $U^{(\chi_j, \rho_k)}$ of $G$ upon the Hilbert space $\Gamma_{L^2}(E)$ of square-integrable cross-sections of $E$. If $\rho_k$ is an irreducible unitary representation of $H_j$, then $\rho_k$ induces an irreducible unitary representation $U^{(\chi_j, \rho_k)}$ of $G$ upon the Hilbert space $\Gamma_{L^2}(E)$. Up to unitary equivalence, all irreducible unitary representations of the semi-direct product $G$ can be obtained in this manner.

The representation on $\Gamma_{L^2}(E)$ is irreducible because there are no non-trivial *closed* subspaces which are invariant under the representation of $G$. There are, however, subspaces of $\Gamma_{L^2}(E)$ invariant under the representation of $G$. The subspace $\Gamma_0^\infty(E)$ of smooth cross-sections of compact support, the subspace $\Gamma_0^k(E)$ of $k$-times continuously differentiable cross-sections of compact support, and the subspace $\Gamma_0^0(E)$ of continuous cross-sections of compact support, are each subspaces of $\Gamma_{L^2}(E)$, each of which is invariant under the representation of $G$. However, in the topology induced by the norm of $\Gamma_{L^2}(E)$, each one of these subspaces has a closure which equals the whole of $\Gamma_{L^2}(E)$. There are no invariant subspaces of $\Gamma_{L^2}(E)$, other than the whole space or the zero subspace, which are closed in the topology of $\Gamma_{L^2}(E)$.

Given that $L_j$ is a closed subgroup of $G$, the quotient $G/L_j$ has exactly one $G$-invariant measure class. A measure class is an equivalence class of Borel measures on a space. A measure class is defined to be $G$-invariant if it is closed under the action $\mu \mapsto \mu_g$, with $\mu_g(B) = \mu(g^{-1}B)$ for any Borel subset $B$. If one fixes an irreducible unitary representation of $H_j$, then one has a Hilbert space $\Gamma_{L^2}(E)$ of square-integrable cross-sections with respect to each $G$-invariant measure $\mu$ on the base space $G/L_j = G/H_j \cong \hat{N}_j$ of the vector bundle $E$. However, each measure from the unique $G$-invariant measure class gives rise to an equivalent unitary representation.

Appendix F

# Canonical Field Quantization

The one-particle state spaces obtained in this text are obtained through what might be termed a 'covariant' approach to quantization: they are obtained as irreducible unitary representations of the Poincaré group, and, in this sense, the approach is independent of a choice of coordinates on Minkowski space–time; i.e., it is a coordinate-invariant, or 'covariant' approach. There are other approaches to quantization, such as canonical quantization (introduced in Section 2.3), and one can obtain the field operators and canonical commutation relations of the second-quantized theory from this approach also.

The canonical quantization of a free bosonic field on Minkowski space–time results in canonical field operators $\phi(\mathbf{x}, t)$ and canonically conjugate momentum operators $\pi(\mathbf{x}, t)$. These provide quantized versions of the canonically conjugate fields which evolve from the Cauchy data specified on a spacelike hyperplane in Minkowski space–time.

Given the heuristic expression for the field operator at a point,

$$\hat{\Psi}(x) = \frac{1}{(2\pi)^{3/2}} \int_{\mathscr{V}_m^+} e^{i(\mathbf{p}\cdot\mathbf{x}-\omega(\mathbf{p})t)}\hat{a}(\mathbf{p}) + e^{-i(\mathbf{p}\cdot\mathbf{x}-\omega(\mathbf{p})t)}\hat{a}^*(\mathbf{p})\, d^3\mathbf{p}/2\omega(\mathbf{p}),$$

one can cast this in canonical form as

$$\phi(\mathbf{x}, t) = \frac{1}{(2\pi)^{3/2}} \int_{\mathscr{V}_m^+} e^{i(\mathbf{p}\cdot\mathbf{x}-\omega(\mathbf{p})t)}\hat{a}(\mathbf{p}) + e^{-i(\mathbf{p}\cdot\mathbf{x}-\omega(\mathbf{p})t)}\hat{a}^*(\mathbf{p})\, d^3\mathbf{p}/2\omega(\mathbf{p}).$$

Then, defining the conjugate momentum field as

$$\pi(\mathbf{x}, t) = \frac{\partial\phi(\mathbf{x}, t)}{\partial t},$$

one can define the canonically conjugate operator as

$$\pi(\mathbf{x}, t) = \frac{1}{(2\pi)^{3/2}} \int_{\mathscr{V}_m^+} -i\omega(\mathbf{p})e^{i(\mathbf{p}\cdot\mathbf{x}-\omega(\mathbf{p})t)}\hat{a}(\mathbf{p}) + i\omega(\mathbf{p})e^{-i(\mathbf{p}\cdot\mathbf{x}-\omega(\mathbf{p})t)}\hat{a}^*(\mathbf{p})\, d^3\mathbf{p}/2\omega(\mathbf{p}).$$

Whilst these heuristic expressions try to define the canonical fields as time-dependent operators at each point of the $\mathbb{R}^3$ hyperplane, the integral for $\phi(\mathbf{x}, t)$ does not converge, so these expressions must be smeared with test functions $f$

from $\mathcal{S}_{\mathbb{R}}(\mathbb{R}^3)$, the Schwartz space of real-valued functions on $\mathbb{R}^3$. One thereby obtains operator-valued distributions, $\phi(f,t) := \int \phi(\mathbf{x},t) f(\mathbf{x})\, d\mathbf{x}$ and $\pi(f,t) := \int \pi(\mathbf{x},t) f(\mathbf{x})\, d\mathbf{x}$. These operator-valued distributions satisfy the following 'equal-time' commutation relations

$$\phi(f,t)\phi(g,t) - \phi(g,t)\phi(f,t) = 0,$$
$$\pi(f,t)\pi(g,t) - \pi(g,t)\pi(f,t) = 0,$$
$$\phi(f,t)\pi(g,t) - \pi(g,t)\phi(f,t) = i\langle f, g\rangle.$$

In this case $f$ and $g$ are both elements from the Schwartz space $\mathcal{S}_{\mathbb{R}}(\mathbb{R}^3)$. Defining the unitary operators $U(f,t) = e^{i\phi(f,t)}$ and $V(f,t) = e^{i\pi(f,t)}$, one has the following Weyl form of these commutation relations:

$$U(f,t)U(g,t) - U(g,t)U(f,t) = 0,$$
$$V(f,t)V(g,t) - V(g,t)V(f,t) = 0,$$
$$U(f,t)V(g,t) = e^{-i\langle f,g\rangle} V(g,t)U(f,t).$$

Appendix G

# Photons and the Gupta–Bleuler Technique

This appendix demonstrates, in outline, that, with a choice of gauge, the space of electromagnetic gauge connections satisfying the free Maxwell equations over Minkowski space–time, is the inverse Fourier transform of the single-particle space obtained for photons in the Wigner approach to first quantization.

The single-particle Hilbert space obtained as an irreducible representation of $O^{\uparrow}(3,1) \ltimes \mathbb{R}^{3,1}$ in the Wigner approach for a particle of mass zero $m = 0$ and spin $t = 1$, is the space of single photon states. As explained in Section 2.5, it is the direct sum $\Gamma_{L^2}(E^{+}_{0,1}) \oplus \Gamma_{L^2}(E^{+}_{0,-1})$ of the helicity $s = 1$ and helicity $s = -1$ irreducible representations of $SO_0(3,1) \ltimes \mathbb{R}^{3,1}$, which represent, respectively, the right-handed and left-handed states of the photon. This is the space obtained in the Dirac–Schwinger approach to the quantization of the electromagnetic field. As a Hilbert space it possesses a positive-definite inner product. In contrast, the Gupta–Bleuler approach to the quantization of the electromagnetic field uses a pseudo-Hilbert space, a space with an indefinite inner product, as the space of single-photon states. In particular, the pseudo-Hilbert space used in the Gupta–Bleuler approach is a Krein space (Prugovecki, 1992, pp. 260–266; Prugovecki, 1995, pp. 233–241).

As a classical free field, the electromagnetic field can be represented by a $\mathfrak{u}(1)$-valued one-form $\omega$ on a $U(1)$-principal fibre bundle $P$, satisfying the Maxwell equations. Under a choice of gauge[1] $\sigma : \mathcal{M} \to P$, the electromagnetic field can be represented as a *real* vector potential $A = \sigma^* \omega$, a cross-section of $T^*\mathcal{M}$. Under a Lorentz choice of gauge, each connection $\omega$ is mapped to a real vector potential $A$ satisfying $\operatorname{div} A = 0$. Given a real vector potential $A$, the electromagnetic field strength is $F = dA$. The Maxwell equations for a vector potential obtained with a Lorentz choice of gauge are simply:

$$\operatorname{div} dA = 0, \qquad \operatorname{div} A = 0.$$

These equations are equivalent to (Derdzinski, 2002, Appendix 48):

$$\Box A = 0, \quad \operatorname{div} A = 0.$$

The first equation here is the wave equation on Minkowski space–time $\mathcal{M}$,

$$\left( \frac{\partial^2}{\partial x_0^2} - \frac{\partial^2}{\partial x_1^2} - \frac{\partial^2}{\partial x_2^2} - \frac{\partial^2}{\partial x_3^2} \right) A = \Box A = (\Box A_\mu)\, dx^\mu = 0,$$

[1] See Section 3.4.

where $\Box$ is the d'Alembertian.

The Gupta–Bleuler approach to quantizing the electromagnetic field begins by taking a vector space of *complex*, 4-component electromagnetic potentials $A^c$ which solve the wave equation. In other words, the Gupta–Bleuler approach begins by taking the cross-sections of $T^*\mathcal{M} \otimes (\mathcal{M} \times \mathbb{C}^1)$ which solve the wave equation. Assuming a trivial cotangent bundle, this is equivalent to a function space $\mathcal{F}(\mathcal{M}, \mathbb{C}^4)$. In fact, one takes the space of tempered solutions $A^c \in \mathscr{S}'(\mathcal{M}, \mathbb{C}^4)$ of the wave equation, which are of positive energy, in the sense that their Fourier transforms are concentrated on the forward light cone $\mathscr{V}_0^+$ (Dautray and Lions, 1988, Chapter IX, §1, 7b).

This vector space is equipped with an indefinite inner product, albeit one dependent upon the choice of a global Lorentz chart $(x_0, x_1, x_2, x_3)$ upon Minkowski space–time. The pseudo-Hilbert space which results is equipped with a pseudo-unitary representation of the isochronous Poincaré group $O^\uparrow(3, 1) \ltimes \mathbb{R}^{3,1}$.

Under Fourier transform, one obtains a space of photon wave-functions consisting of 4-component functions on the forward light cone $\mathscr{V}_0^+$. This space is also equipped with a pseudo-unitary representation of $O^\uparrow(3, 1) \ltimes \mathbb{R}^{3,1}$. We can denote this space as $\tilde{K}$.

$\tilde{K}$ contains a subspace $\tilde{K}^L$ of momentum space photon wave-functions whose configuration space counterparts $A^c$ satisfy the Lorentz gauge condition

$$\operatorname{div} A^c = 0.$$

The subspace $\tilde{K}^L$ contains a further subspace $\tilde{K}^0 \subset \tilde{K}^L$ consisting of momentum space photon wave-functions whose configuration space counterparts $A^c$ are such that $dA^c = 0$.

The quotient space $\tilde{K}^L/\tilde{K}^0$ inherits a positive definite inner product and a unitary representation of a double cover of the isochronous Poincaré group. $\tilde{K}^L/\tilde{K}^0$ is a space of equivalence classes of complex vector potentials $A^c$, modulo gauge transformations. Given that a gauge transformation of a vector potential in electromagnetism corresponds to the addition of a $d\phi$ with $\phi \in C^\infty(\mathcal{M})$ (see Section 3.5), each equivalence class is a set

$$\{A^c + d\phi \colon \phi \in C^\infty(\mathcal{M})\}.$$

The unitary representation of $O^\uparrow(3, 1) \ltimes \mathbb{R}^{3,1}$ on the space $\tilde{K}^L/\tilde{K}^0$ of equivalence classes of 4-component momentum space Lorentz gauge photon wave-functions, is unitarily equivalent to the representation of $O^\uparrow(3, 1) \ltimes \mathbb{R}^{3,1}$ on the Wigner representation space of 2-component momentum space photon wave-functions.

Note that the Gupta–Bleuler quantization technique, when extended to non-Abelian gauge fields, leads to a non-unitary scattering matrix in the second-quantized theory. Faddeev–Popov 'ghost fields' are introduced into the Lagrangian for a non-Abelian gauge field to remedy this problem (Prugovecki, 1995, p. 256).

Appendix H

# Component Expression of the Curvature Two-Form

In Section 3.5 it was asserted that the connection one-form $A$ and curvature two-form $F$ of a gauge field are related by the equation $F = dA + \frac{1}{2}[A, A]$. Given a coordinate chart $(x_1, \ldots, x_n)$, and a Lie-algebra basis $\{E_a \in \mathfrak{g}\colon a = 1, \ldots, k\}$, one can write a Lie-algebra valued connection one-form $A$ as

$$A = A^a_\mu \, dx^\mu \otimes E_a,$$

and one can write the Lie-algebra valued curvature two-form $F$ as

$$F = F^a_{\mu\nu} \, dx^\mu \otimes dx^\nu \otimes E_a.$$

In these component terms, it transpires that one can write

$$F^a_{\mu\nu} = \partial_\mu A^a_\nu - \partial_\nu A^a_\mu + C^a_{bc} A^b_\mu A^c_\nu,$$

where $C^a_{bc}$ are the structure constants of the Lie algebra. An explanation is in order of how this expression can be obtained from the fact that $F = dA + \frac{1}{2}[A, A]$.

Let us first review some facts about second-rank antisymmetric tensors. If we start with an arbitrary second-rank covariant tensor field $T$, then in the domain of a coordinate chart $(x_1, \ldots, x_n)$ we can express it as

$$T = T_{\mu\nu} \, dx^\mu \otimes dx^\nu.$$

The anti-symmetrized tensor $\mathscr{A}T$ is a new tensor with components $T_{[\mu\nu]}$ defined by

$$T_{[\mu\nu]} = \frac{1}{2}(T_{\mu\nu} - T_{\nu\mu}).$$

Hence, the antisymmetric tensor can be written as

$$\mathscr{A}T = T_{[\mu\nu]} \, dx^\mu \otimes dx^\nu.$$

The components of this tensor are such that $T_{[\mu\nu]} = -T_{[\nu\mu]}$.

Now, the definition of the exterior tensor product $\wedge$ is such that, given an $r$-form $\phi$, and an $s$-form $\psi$ (Boothby, 1986, p. 209),

$$\phi \wedge \psi = \frac{(r+s)!}{r!s!} \mathscr{A}(\phi \otimes \psi).$$

Hence, this entails

$$\begin{aligned} dx^\mu \wedge dx^\nu &= 2\mathscr{A}\left(dx^\mu \otimes dx^\nu\right) \\ &= dx^\mu \otimes dx^\nu - dx^\nu \otimes dx^\mu . \end{aligned}$$

From the definition of anti-symmetrization and the definition of the exterior tensor product, it follows that

$$\mathscr{A}T = T_{[\mu\nu]}\, dx^\mu \otimes dx^\nu = \frac{1}{2} T_{\mu\nu}\, dx^\mu \wedge dx^\nu .$$

To see why this equality holds, note that

$$\begin{aligned} T_{[\mu\nu]}\, dx^\mu \otimes dx^\nu &= \frac{1}{2}(T_{\mu\nu} - T_{\nu\mu})\, dx^\mu \otimes dx^\nu \\ &= \frac{1}{2} T_{\mu\nu}\, dx^\mu \otimes dx^\nu - \frac{1}{2} T_{\nu\mu}\, dx^\mu \otimes dx^\nu \\ &= \frac{1}{2} T_{\mu\nu}\, dx^\mu \otimes dx^\nu - \frac{1}{2} T_{\mu\nu}\, dx^\nu \otimes dx^\mu \\ &= \frac{1}{2} T_{\mu\nu}\, dx^\mu \wedge dx^\nu . \end{aligned}$$

The penultimate step here uses the fact that

$$T_{\nu\mu}\, dx^\mu \otimes dx^\nu = T_{\mu\nu}\, dx^\nu \otimes dx^\mu .$$

The indices are dummy indices, and a tensor is invariant under a change in the name of the dummy indices. This should not be confused with a genuine transpose tensor, which is a different tensor. For example,

$$T_{\mu\nu}\, dx^\mu \otimes dx^\nu \neq T_{\nu\mu}\, dx^\mu \otimes dx^\nu .$$

With these facts in mind, we can now unravel the expression $F = dA + \frac{1}{2}[A, A]$ in component terms. Consider the first term, the exterior derivative $dA$. Given $A = A_\nu\, dx^\nu$, the definition of the exterior derivative entails that

$$dA = dA_\nu \wedge dx^\nu .$$

A differential $df$ of a function $f$ is defined to be such that its action upon a vector $v$ is $df(v) = v(f)$. In particular, $df(\partial/\partial x^\mu) = \partial f/\partial x^\mu$, and in the case of $dA_\nu$ this means that $dA_\nu(\partial/\partial x^\mu) = \partial A_\nu/\partial x^\mu$. Hence,

$$dA_\nu = dA_\nu\left(\frac{\partial}{\partial x^\mu}\right) dx^\mu = \frac{\partial A_\nu}{\partial x^\mu} dx^\mu ,$$

from which it follows that

$$\begin{aligned} dA &= dA_\nu \wedge dx^\nu \\ &= \left(\frac{\partial A_\nu}{\partial x^\mu} dx^\mu\right) \wedge dx^\nu \end{aligned}$$

$$= \frac{\partial A_\nu}{\partial x^\mu}(dx^\mu \wedge dx^\nu)$$
$$= \frac{\partial A_\nu}{\partial x^\mu}(dx^\mu \otimes dx^\nu - dx^\nu \otimes dx^\mu)$$
$$= \left(\frac{\partial A_\nu}{\partial x^\mu} - \frac{\partial A_\mu}{\partial x^\nu}\right) dx^\mu \otimes dx^\nu$$
$$= (\partial_\mu A_\nu - \partial_\nu A_\mu)\, dx^\mu \otimes dx^\nu.$$

Introducing the Lie algebra basis $\{E_a \in \mathfrak{g}\colon a = 1, \dots, k\}$, we can write $dA$ as

$$dA = \left(\partial_\mu A^a_\nu - \partial_\nu A^a_\mu\right) dx^\mu \otimes dx^\nu \otimes E_a.$$

Now consider the second term in the expression for $F$, namely $A \wedge A = \frac{1}{2}[A, A]$. Let us choose a Lie algebra basis at the outset, and decompose a Lie-algebra valued one-form as $A = A^a_\mu\, dx^\mu \otimes E_a$. The definition of the bracket operation upon the space of Lie-algebra valued exterior forms entails that

$$\frac{1}{2}[A, A] = \frac{1}{2}\left[A^b_\mu\, dx^\mu \otimes E_b, A^c_\nu\, dx^\nu \otimes E_c\right]$$
$$= \frac{1}{2}\left(A^b_\mu\, dx^\mu\right) \wedge \left(A^c_\nu\, dx^\nu\right)[E_b, E_c]$$
$$= \frac{1}{2} A^b_\mu A^c_\nu\, dx^\mu \wedge dx^\nu\, [E_b, E_c]$$
$$= \frac{1}{2} C^a_{bc} A^b_\mu A^c_\nu\, dx^\mu \wedge dx^\nu \otimes E_a$$
$$= C^a_{bc} A^b_{[\mu} A^c_{\nu]}\, dx^\mu \otimes dx^\nu \otimes E_a,$$

where $[E_b, E_c]$ is the Lie bracket upon the Lie algebra. Given the anti-symmetry in the Lie algebra components,

$$A^b_\mu A^c_\nu = -A^c_\mu A^b_\nu = -A^b_\nu A^c_\mu,$$

the space–time manifold components are already anti-symmetric, and one can remove the angular braces to obtain:

$$\frac{1}{2}[A, A] = C^a_{bc} A^b_\mu A^c_\nu\, dx^\mu \otimes dx^\nu \otimes E_a.$$

To sum-up, the pull-down of the curvature two-form can be written as

$$F = \left(\frac{\partial A^a_\nu}{\partial x^\mu} - \frac{\partial A^a_\mu}{\partial x^\nu}\right) dx^\mu \otimes dx^\nu \otimes E_a + C^a_{bc} A^b_\mu A^c_\nu\, dx^\mu \otimes dx^\nu \otimes E_a,$$

hence the expression for the real-valued component fields of $F$:

$$F^a_{\mu\nu} = \partial_\mu A^a_\nu - \partial_\nu A^a_\mu + C^a_{bc} A^b_\mu A^c_\nu.$$

Appendix I

# Elementary Particles in String Theory

Whilst this book provides an account of elementary particles within the standard model, to provide some contrast it is worthwhile considering the alternative notion of an elementary entity supplied by string theory.

String theory has a classical formulation, and a quantum formulation obtained by quantizing the classical formulation. In the classical formulation, a free particle is represented at a moment of time within a reference frame, as a one-dimensional manifold embedded in the relative space of the reference frame. Such a one-dimensional manifold is referred to as a string. Open strings are homeomorphic to [0, 1] or (0, 1), and closed strings are homeomorphic to the circle $S^1$. A string sweeps out a surface in space–time called the 'worldsheet,' which is reference-frame independent.

If space is represented by a three-dimensional manifold $\Sigma$, the configuration space of a closed string is the 'loop space'

$$L\Sigma = \{\alpha : S^1 \to \Sigma\},$$

an infinite-dimensional manifold. The simplest space of closed string histories is the set of all embeddings of $S^1 \times [0, 1]$ into the space–time manifold $\mathcal{M}$. More generally, one considers the set of embeddings of a world-sheet $\mathcal{W}$ into the 'target' space–time, $X : \mathcal{W} \to \mathcal{M}$. In string theory, the dimension of space–time $\mathcal{M}$ is, in general, much larger than 4.

Whilst an elementary particle in the standard model can be treated as a cross-section of a bundle over space–time, an elementary entity in string theory must be treated as an object whose domain is one of these infinite-dimensional embedding spaces.

A string worldsheet possesses its own Lorentzian metric tensor $q$, and a property called the string tension. Each type of string has a constant $\alpha'$ called the string constant, in terms of which the string tension is $\hbar c/\alpha'$. Modern string theory has focused upon superstrings, and the superstring theory of current interest holds that string tension is independent of string length.

Putting superstring theory aside for the moment to simplify the technicalities, the basic rationale of string theory can be understood by considering the Polyakov action for a bosonic string worldsheet (Isham, 1992, Section 3.2.1),

$$S[q, X] = \frac{1}{4\pi\alpha'} \int_{\mathcal{W}} q^{ij}(\sigma)\partial_i X^\mu(\sigma)\partial_j X^\nu(\sigma) g_{\mu\nu}\big(X(\sigma)\big)\big(\det q(\sigma)\big)^{\frac{1}{2}}\, d^2\sigma,$$

where $g_{\mu\nu}$ are the components of the metric tensor on the space–time. Classical bosonic string histories are those which constitute extrema of this action, and these are the histories in which the worldsheet metric $q$ equals the metric induced by the embedding into the space–time $(\mathcal{M}, g)$. The action is invariant under a conformal transformation of the string geometry, $q_{ij}(\sigma) \mapsto f(\sigma)q_{ij}(\sigma)$, $f(\sigma) > 0$. The embedding functions $X^\mu$ and the worldsheet metric $q_{ij}$ undergo quantization, and the requirement that conformal invariance be preserved under quantization is only satisfied if one picks certain combinations of $\mathcal{M}$-dimension, background fields on $\mathcal{M}$, and if the metric $g$ on $\mathcal{M}$ satisfies equations which, to first order, equal the vacuum Einstein equations, the equations for a Ricci-flat space–time $R_{\mu\nu} = 0$. Whilst bosonic string theory requires 26 dimensions, superstring theory requires 10 dimensions. To be consistent with our observation that there are only 4 macroscopic dimensions, it is postulated that the remaining dimensions are 'compactified.' I.e., it is postulated that the remaining dimensions are represented by a manifold of compact topology, and a very small geometrical diameter.

In terms of the background fields on space–time, a realistic superstring theory requires the existence of a massless 'dilaton,' represented classically by a scalar field $\phi$, and a massless 'vector' particle, represented classically by a tensor field $H_{\mu\nu\rho}$, in the space–time $\mathcal{M}$.

One particularly interesting claim made by the proponents of string theory is that a string has almost no intrinsic properties, beside the string tension. For example, Richard Dawid claims that "in conventional quantum physics elementary particles carry quantum numbers which determine their behaviour. A particle's characteristics like spin, charge or hypercharge, which are expressed by quantum numbers, constitute intrinsic and irreducible properties. Strings do not have quantum numbers but can differ from each other by their topological shape and their dynamics: Strings can be open, meaning that they have two endpoints, or closed like a rubber band. If they are closed, they can be wrapped around the various compactified dimensions in different ways. Finally, both open and closed strings can assume different oscillation modes... To the observer who does not have sufficient resolution to perceive the stringy structure, a string in a specific oscillation mode and topological position looks like a point-like particle with certain quantum numbers. A change of, let's say, its oscillation mode would be perceived as a transmutation into a different particle. Strings at a fundamental level do not have coupling constants either. The strength of their interaction with each other again can be reduced to some aspect of their dynamics. (The ground state of a certain mode of the string expansion, the dilaton, gives the string coupling constant.) All characteristic numbers of a quantum field theory are thus being dissolved into geometry and dynamics of an oscillating string" (2003, p. 8). Proponents of string theory will speak, for example, of the spin-1 and spin-2 oscillation ('excitation') modes of a single type of string.

However, it is not clear why the oscillation mode or topology of a string should not be considered as an intrinsic property, for there is no claim that the oscillation mode or topology of a string is a relationship between it and something else. Certainly, the oscillation mode of a string can change, hence the oscillation mode is not an *invariant* property of the string, but the invariant properties of a string do not equal its intrinsic properties. If two purportedly different types of elementary particle are merely two different oscillation modes of the same type of string, then those two particle types differ by virtue of the different intrinsic properties possessed by the underlying string type. String theory, then, reinforces the conclusion of Section 4.13, in the sense that strings are hypothetical elementary entities which can possess different intrinsic properties at different times, and which therefore do not possess a single intrinsic state.

Lee Smolin argues that the notion of unification requires the different kinds of elementary particle to be merely different states of a single underlying elementary entity, and he argues that string theory permits this, where the notion of a point particle doesn't: "If the elementary particle was something of a certain size, there would be no difficulty imagining it to exist in different states. It might be, for example, that the particle could take on different shapes. But it is hard to imagine how something that is just a point, that has no shape and takes up no space, could exist in different states or configurations... String theory resolves this paradox, because it says that the end of the process of reductionism is that the most fundamental entities are one-dimensional strings and not points... it is [the] different modes of vibration of the string that are understood in string theory as being the different elementary particles" (1997, p. 65).

Smolin's argument here relies upon a misleading pre-quantum notion of a point particle. The quantization of a point-particle provides one with an entity which is extended in space: The state of a point particle is specified by a wave-function, an object which is extended in space, and, as argued at length in Section 4.13, such objects have an infinite-dimensional state space, and possess many different intrinsic states.

Each type of string is characterised by a value for the string constant $\alpha'$, and, as befits a constant, the value of $\alpha'$, and therefore the value of the string tension, is not state-dependent. The string tension is an invariant intrinsic property of each string type. Whilst string theory might be able to reduce a property previously thought to be invariant, such as spin, to the state-dependent properties of a single underlying entity, it would do so at the cost of introducing a new invariant property, the string tension. Whilst it may be the aspiration of string theory to represent all the different types of elementary particle as merely different states of a single type of underlying entity, a single type of string with a specific tension, this aspiration has not, as yet, been realised.

It is also doubtful that string theory will be able to predict those aspects of the standard model which currently have to be fixed by experiment. Referring to a choice of $(\mathcal{M}, g, \phi, H_{\mu\nu\rho})$ as a choice of string 'background,' Smolin points out that "there are consistent string backgrounds for 4 large, uncompactified dimensions that correspond to a large range of possible values for the number of generations, for the number of Higgs fields and for the gauge group. Thus, string theory makes no prediction for these characteristics of the standard model" (2003, p. 37).

Appendix J

# The Structure Theorem for Compact, Connected Lie Groups

As stated in Section 5.1, the structure theorem for compact, connected Lie groups entails that any compact, connected Lie group $G$ is isomorphic to a quotient of a finite direct product,

$$G \cong L_1 \times L_2 \times \cdots \times L_r \times \mathbb{T}^p/D,$$

where each $L_i$ is a compact, simple, and simply connected Lie group, $\mathbb{T}^p$ is a $p$-dimensional torus, and $D$ is a finite central subgroup of $L_1 \times L_2 \times \cdots \times L_r \times \mathbb{T}^p$ (Simon, 1996, p. 155; Hofmann and Morris, 1998, pp. 204–207).

Before deriving this structural decomposition theorem, let us define some of the terms in our discourse. Recall that each Lie Group $G$ possesses a Lie algebra $\mathfrak{g}$ isomorphic to the tangent vector space at the identity element of the Lie group. A Lie algebra is a vector space equipped with an antisymmetric, bilinear product operation $[\,,\,]$ called the Lie bracket. An Abelian Lie algebra is such that $[X, Y] = \mathbf{0}$ for all $X, Y \in \mathfrak{g}$. An ideal in a Lie algebra is a Lie subalgebra $\mathfrak{h} \subset \mathfrak{g}$ which is such that $[X, Y] \in \mathfrak{h}$ for all $X \in \mathfrak{h}$, $Y \in \mathfrak{g}$. An ideal is also said to be an invariant subalgebra.

An ideal is the Lie algebra equivalent of a closed, normal subgroup of a connected Lie group. A closed subgroup $H \subset G$ of a Lie group $G$ is a Lie subgroup under the inclusion mapping $i : H \rightarrow G$. Hence, one can equivalently refer to either a closed subgroup or a Lie subgroup of a Lie group. If $H \subset G$ is a Lie subgroup of a connected Lie group $G$, with $\mathfrak{g}$ denoting the Lie algebra of $G$, and $\mathfrak{h} \subset \mathfrak{g}$ denoting the Lie algebra of $H$, then $H$ is a normal subgroup of $G$ if and only if $\mathfrak{h}$ is an ideal of $\mathfrak{g}$ (Fulton and Harris, 1991, p. 122).

A Lie algebra $\mathfrak{g}$ is defined to be simple if $\dim \mathfrak{g} > 1$ and $\mathfrak{g}$ contains no nontrivial ideals (Fulton and Harris, 1991, p. 122). A connected Lie group can be defined to be simple if its Lie algebra is simple, or equivalently, if it contains no non-trivial, closed, *connected* normal subgroups. Under this definition, a simple, connected Lie group *can* possess non-trivial, closed, normal subgroups, but if they exist they must be discrete. If a simple, connected Lie group $G$ possesses a non-trivial, discrete, closed normal subgroup $H \subset G$, then the Lie subalgebra $\mathfrak{h} \subset \mathfrak{g}$ is an ideal, but $\mathfrak{h} = \{\mathbf{0}\}$, consistent with the fact that there is no *non-trivial* ideal in $\mathfrak{g}$. Although $H$ here is a non-trivial subgroup, $H \neq \{I\}$, the identity component of $H$ is trivial,

$H_0 = \exp(\mathfrak{h}) = \{I\}$ (Hofmann and Morris, 1998, pp. 193–194). Given a Lie group $G$ and a closed, normal subgroup $H$, the quotient $G/H$ is a Lie group. A Lie group which has no non-trivial, closed, normal subgroups, has no quotient Lie groups. As defined here, a simple, connected Lie group *can* have quotient Lie groups, but they can only be quotient Lie groups with respect to a discrete subgroup.

An Abelian Lie algebra cannot be simple because any Lie subalgebra of an Abelian Lie algebra must be an ideal. For any subalgebra $\mathfrak{h}$ of an Abelian Lie algebra $\mathfrak{g}$, $[X, Y] = \mathbf{0} \in \mathfrak{h}$, for all $X \in \mathfrak{h}$, $Y \in \mathfrak{g}$. Correspondingly, every connected subgroup of a connected, Abelian Lie group must be a normal subgroup. Hence, an Abelian Lie group cannot be simple.

A semi-simple Lie algebra can be defined as a Lie algebra which has no non-trivial Abelian ideals, but it will be more useful to characterise it as a Lie algebra which is the direct sum of simple Lie algebras. The only non-trivial ideals of a semi-simple Lie algebra are the non-Abelian direct summands. Semi-simple Lie groups are the direct products of simple Lie groups. The only non-trivial normal subgroups of a semi-simple Lie group are the factors in the direct product. Needless to say, every simple Lie algebra is semi-simple, and every simple Lie group is a semi-simple Lie group.

Now, the Lie algebra of a compact, connected Lie group can be decomposed as the direct sum of a semi-simple Lie algebra and an Abelian Lie algebra. In other words, the Lie algebra of a compact, connected Lie group can be decomposed as a direct sum of simple Lie algebras and Abelian Lie algebras. Hence a compact, connected Lie group $G$ must be locally isomorphic to a direct product

$$G_1 \times G_2 \times \cdots \times G_r \times \mathbb{T}^p,$$

where each $G_i$ is a simple, compact Lie group, and $\mathbb{T}^p$ is the $p$-fold direct product of $U(1) = \mathbb{T}$, the unique compact connected 1-dimensional Abelian Lie group.

In more economical notation, a compact connected Lie group must be locally isomorphic to a direct product

$$K \times N,$$

where $K$ is a compact, semi-simple Lie group, and $N$ is a compact Abelian Lie group.

The only connected 1-dimensional Lie groups are $\mathbb{R}^1$ and $U(1) = \mathbb{T}$, both of which are Abelian. Every connected Abelian Lie group, of arbitrary dimension, is isomorphic to a direct product of these two 1-dimensional Lie groups. Any compact connected Abelian Lie group, of dimension $p$, is isomorphic to the direct product of $p$ copies of $U(1) = \mathbb{T}$. Hence the Abelian factor $N$ in the above decomposition must be $\mathbb{T}^p$ for some integer $p$.

Locally isomorphic groups share the same universal cover, hence the universal cover of $G$ must equal the universal cover of $G_1 \times G_2 \times \cdots \times G_r \times \mathbb{T}^p$. The universal cover of a direct product is given by the product of the individual universal covers, hence the universal cover of $G$ must be

$$\tilde{G} \cong \tilde{G}_1 \times \tilde{G}_2 \times \cdots \times \tilde{G}_r \times \mathbb{R}^p,$$

where each $\tilde{G}_i$ is a simple and simply connected Lie group, and $\mathbb{R}^1$ is the universal cover of $U(1) = \mathbb{T}$.

It can be proven that a compact connected Lie group which has the property of being semi-simple, must have a compact universal covering group. As a trivial consequence of being simple, each $G_i$ must also be semi-simple, hence each of the $\tilde{G}_i$ must be compact as well as simple and simply connected. $U(1) = \mathbb{T}$ is not semi-simple, by virtue of being Abelian, hence there is no inconsistency with the fact that $\mathbb{R}^1$ is non-compact.

In more economical notation, one can express the universal cover of a compact connected Lie group $G$ as

$$\tilde{G} \cong \tilde{K} \times \mathbb{R}^p,$$

where $\tilde{K}$ is a compact, semi-simple, simply connected Lie group.

Any connected Lie group $G$ must be isomorphic to a quotient $\tilde{G}/J$ of the universal cover, where $J$ is a discrete central subgroup of $\tilde{G}$, hence

$$G \cong \tilde{G}/J = \tilde{G}_1 \times \cdots \times \tilde{G}_r \times \mathbb{R}^p/J.$$

The centre $Z(\tilde{G})$ of the universal cover is given by

$$Z(\tilde{G}) = Z(\tilde{G}_1) \times \cdots \times Z(\tilde{G}_r) \times \mathbb{R}^p.$$

From the 'Finite Discrete Centre Theorem' (Hofmann and Morris, 1998, p. 180), it can be proven that the universal covering group of a compact connected semi-simple Lie group must have a finite centre, hence each $Z(\tilde{G}_i)$ is finite.[1] If at least one of the $Z(\tilde{G}_i)$ is non-trivial, this entails that the centre $Z(\tilde{G})$ of the universal covering group has multiple components. Although $\tilde{G}$ is connected, it is perfectly possible for its centre $Z(\tilde{G})$ to have multiple components. The identity component of the centre $Z_0(\tilde{G})$ is

$$Z_0(\tilde{G}) = Id \times \cdots \times Id \times \mathbb{R}^p.$$

If $G \cong \tilde{G}/J$, then the centre $Z(G)$ must be isomorphic to $Z(\tilde{G})/J$, hence

$$Z(G) \cong Z(\tilde{G}_1) \times \cdots \times Z(\tilde{G}_r) \times \mathbb{R}^p/J.$$

Once again, although $G$ is connected, it is perfectly possible for its centre $Z(G)$ to have multiple components. Because $G$ is compact, its centre $Z(G) \cong Z(\tilde{G})/J$ must also be compact. Every compact *connected* Abelian Lie group must be a product of $p$ copies of $\mathbb{T}$, hence the *identity component* of the centre, $Z_0(G)$, must be $\mathbb{T}^p$.

Whilst $J$ belongs to the centre $Z(\tilde{G})$ of the universal cover, it is not necessarily contained within the identity component of the centre. Hence, we can introduce a further

[1] Private communication with Karl H. Hofmann.

subgroup $F \subset J$ which is the subgroup of $J$ that belongs to $Z_0(\tilde{G})$.[2] Defining

$$F = Z_0(\tilde{G}) \cap J = \big(Id \times \cdots \times Id \times \mathbb{R}^p\big) \cap J,$$

we obtain

$$\tilde{G}_1 \times \cdots \times \tilde{G}_r \times \mathbb{T}^p \cong \tilde{G}_1 \times \cdots \times \tilde{G}_r \times \mathbb{R}^p / F,$$

and then

$$G \cong \tilde{G}/J \cong (\tilde{G}/F)/(J/F) \cong \tilde{G}_1 \times \cdots \times \tilde{G}_r \times \mathbb{T}^p / D,$$

where $D = J/F$ is a finite central subgroup of $\tilde{G}_1 \times \cdots \times \tilde{G}_r \times \mathbb{T}^p$. This is the structure theorem, with $\tilde{G}_i = L_i$. The quotient group $D = J/F$ is finite because it is a discrete, and therefore closed subgroup of the compact group $\tilde{G}_1 \times \cdots \times \tilde{G}_r \times \mathbb{T}^p$. A closed subgroup of a compact group must be compact itself, and a discrete compact group must be finite.

[2] Private communication with Karl H. Hofmann.

Appendix K

# Irreducible Representations of $U(1)$

As stated in Section 5.2, the irreducible representations of the unitary group $U(1)$ are parameterized by the integers $\mathbb{Z}$. For any integer $n \in \mathbb{Z}$, one has the representation

$$e^{i\theta} \mapsto e^{in\theta},$$

acting upon $\mathbb{C}^1$ by complex multiplication. A linear transformation of any 1-dimensional vector space is multiplication by a complex number.

The irreducible representations of $U(1)$ are parameterized by integers rather than rational numbers or real numbers because of the constraint that $\theta \in \mathbb{R}/2\pi\mathbb{Z}$. To see the implications of this constraint, begin by recalling that a representation $\rho$ must, by definition, be a homomorphism, hence

$$\rho(e^{i\theta_1})\rho(e^{i\theta_2}) = \rho(e^{i\theta_1}e^{i\theta_2}) = \rho(e^{i\theta_1+\theta_2}).$$

If we let $r$ denote any real number, then we can attempt to define a representation

$$e^{i\theta} \mapsto e^{ir\theta}.$$

In terms of this representation, the homomorphism condition is:

$$e^{ir\theta_1}e^{ir\theta_2} = e^{ir(\theta_1+\theta_2)}.$$

The constraint that $\theta \in \mathbb{R}/2\pi\mathbb{Z}$ means that if $\theta_2 = \theta_1 + m2\pi$, for any integer $m$, then it must be the case that $\rho(e^{i\theta_2}) = \rho(e^{i\theta_1})$. In terms of the $r$-indexed representation under consideration here, this is the requirement that $e^{ir\theta_2} = e^{ir\theta_1}$. Now

$$e^{ir\theta_2} = e^{ir(\theta_1+m2\pi)} = e^{ir\theta_1}e^{irm2\pi}.$$

Hence $e^{ir\theta_2} = e^{ir\theta_1}$ if and only if $e^{irm2\pi} = e^{i0} = 1$. This condition is satisfied if and only if $rm2\pi$ is an integral multiple of $2\pi$. I.e., if and only if $rm$ is an integer. Given that $m$ is an integer, this requires $r$ to be an integer, proving that a representation of $U(1)$ must be indexed by an integer.

For each integer $n$, the irreducible representation $e^{i\theta} \mapsto e^{in\theta}$ defines a surjective endomorphism $U(1) \to U(1)$. This map is only injective in the special cases $n = 1$ and $n = -1$. Unless $|n| = 1$, the kernel of the endomorphism is non-trivial, and the domain of the mapping, $U(1)$, provides an $n$-fold cover of the image. For any integer $n$, the set of $|n|$ elements

$$\{e^{2\pi i\frac{k}{n}}: k = 0, 1, \dots, n-1\},$$

is mapped to the identity element of $U(1)$.

The image of $U(1)$ under an irreducible representation is still $U(1)$, hence $U(1)$ is isomorphic to the group it is mapped onto under an irreducible representation. However, unless $|n| = 1$, the representation map is merely a group homomorphism, and does not express the isomorphism which exists between the domain and range of the map.[1] Unless $|n| = 1$, the representation map $e^{i\theta} \mapsto e^{in\theta}$ is a non-bijective covering map, under which $U(1)$ provides an $n$-fold cover of itself.

[1] Private communication with Gert Roepstorff.

Appendix L

# Non-Integral Charge and Hypercharge

It was explained in Section 5.2 that, although the irreducible representations of $U(1)$ are indexed by the integers, for the representation of weak hypercharge and electric charge it is particularly convenient to parameterize the representations of $U(1)$ by $\frac{1}{3}\mathbb{Z}$. To do so, one simply deems that, say, the weak hypercharge $y$ representation corresponds to the index $n = 3y$ representation of $U(1)$. There is, however, an alternative approach pursued by Roepstorff and Vehns (2001), and by Rizov and Vitev (2000), and the purpose of this appendix is to expound this approach.

Roepstorff and Vehns point out that there is an exact sequence[1]:

$$1 \rightarrow SU(3) \rightarrow S\big(U(3) \times U(2)\big) \rightarrow U(2) \rightarrow 1.$$

Restricting to the centre of each group in the sequence, one obtains another exact sequence:

$$1 \rightarrow \mathbb{Z}_3 \rightarrow \tilde{U}(1)_Y \rightarrow U(1)_Y \rightarrow 1.$$

The group $\tilde{U}(1)_Y$ is a three-fold covering of the weak hypercharge group $U(1)_Y$. Roepstorff and Vehns recommend that the integer-indexed irreducible representations of the three-fold cover $\tilde{U}(1)_Y$ be used, rather than $U(1)_Y$, to deal with the non-integral weak hypercharge of the quark representations. This is different from simply re-parameterizing the representations of $U(1)_Y$. Whilst $U(1)_Y$ is the group of complex numbers of unit modulus, or their image embedded in $U(2)$, $\tilde{U}(1)_Y$ is a subgroup of $S(U(3) \times U(2))$.

$\tilde{U}(1)_Y$, the centre of $S(U(3) \times U(2))$, consists of pairs of matrices of the form

$$\big(e^{i\beta} Id_3, e^{i\alpha} Id_2\big) \in U(3) \times U(2),$$

which satisfy the condition to be of unit determinant,

$$\det\big(e^{i\beta} Id_3, e^{i\alpha} Id_2\big) = \big(e^{i\beta 3}\big)\big(e^{i\alpha 2}\big) = 1.$$

This condition can be equivalently stated as

$$3\beta + 2\alpha = 0 \bmod 2\pi.$$

[1] In other words, there is a sequence of objects and homomorphic mappings between them, such that the image of each mapping belongs to the kernel of the next.

Using this condition, Roepstorff and Vehns point out that $\tilde{U}(1)_Y$ can be treated as a closed curve upon the surface of a 2-dimensional torus. One can treat $\alpha$ and $\beta$ as the two coordinates upon the torus, and the curve in question is the set of points for which $\beta = -2/3\alpha$. To trace out the curve, begin with $\alpha = 0$ and $\beta = 0$, and then allow $\alpha$ to increase, with $\beta$ changing according to the function $\beta = -2/3\alpha$. $\alpha$ and $\beta$ are individually mod $2\pi$, but $\alpha$ must run from 0 to $6\pi$ before the pair $(\alpha, \beta)$ returns to $(0 \mod 2\pi, 0 \mod 2\pi)$. Passing along the curve, one encounters the following sequence of reference points:

$$(\beta = 0, \alpha = 0),$$
$$\left(\beta = -\frac{2}{3}2\pi, \alpha = 2\pi\right),$$
$$\left(\beta = -\frac{2}{3}4\pi, \alpha = 4\pi\right),$$
$$\left(\beta = -\frac{2}{3}6\pi = -4\pi, \alpha = 6\pi\right).$$

Bearing in mind the mod $2\pi$ condition on each individual variable, this is the same sequence of $(e^{i\beta}, e^{i\alpha})$ pairs as

$$\left(e^{\beta=0}, e^{\alpha=0}\right), \qquad \left(e^{\beta=\frac{1}{3}2\pi}, e^{\alpha=0}\right), \qquad \left(e^{\beta=\frac{2}{3}2\pi}, e^{\alpha=0}\right), \qquad \left(e^{\beta=0}, e^{\alpha=0}\right).$$

These points along the curve correspond to a $\mathbb{Z}_3$-subgroup of $\tilde{U}(1)_Y$. Recalling that $\alpha$ and $\beta$ correspond to $(e^{i\beta}\, Id_3, e^{i\alpha}\, Id_2) \in S(U(3) \times U(2))$, the subgroup is

$$\mathbb{Z}_3 \cong \left\{\left(e^{2\pi i \frac{1}{3}}\, Id_3, Id_2\right), \left(e^{2\pi i \frac{2}{3}}\, Id_3, Id_2\right), (Id_3, Id_2)\right\} \subset \tilde{U}(1)_Y.$$

Now, there is a covering map from $S(U(3) \times U(2))$ onto $U(2)$, which induces a covering map from $\tilde{U}(1)_Y$ onto $U(1)_Y \subset U(2)$ when restricted to the centre of $S(U(3) \times U(2))$. Restricted to $\tilde{U}(1)_Y$, it is simply the map

$$\left(e^{i\beta}\, Id_3, e^{i\alpha}\, Id_2\right) \mapsto e^{i\alpha}\, Id_2.$$

The kernel of this map is the set of pairs $(e^{i\beta}\, Id_3, e^{i\alpha}\, Id_2) \in \tilde{U}(1)_Y$ which are mapped to $Id_2 = e^{i0}\, Id_2$. I.e., the kernel is the set of pairs $(e^{i\beta}\, Id_3, e^{i\alpha}\, Id_2)$ which are such that $3\beta + 2\alpha = 0 \mod 2\pi$ and $\alpha = 0$. Hence, the kernel corresponds to the set

$$\left\{\left(e^{i\beta}\, Id_3, Id_2\right) \colon 3\beta = 0 \mod 2\pi\right\}.$$

i.e., $\beta = 2\pi\frac{k}{3}$, $k = 0, 1, 2$. This is the same $\mathbb{Z}_3$-subgroup as specified above. In accordance with the exact sequence, it is the injective image in $\tilde{U}(1)_Y$ of the $\mathbb{Z}_3$-centre of $SU(3)$.

The fact that the kernel of the covering map $\tilde{U}(1)_Y \to U(1)_Y$ is a $\mathbb{Z}_3$-subgroup of $\tilde{U}(1)_Y$, entails that $\tilde{U}(1)_Y/\mathbb{Z}_3 \cong U(1)_Y$. Thinking of $\tilde{U}(1)_Y$ as the curve upon the torus, one can divide it up into three segments: (i) the points with $\alpha \in [0, 2\pi)$, (ii) the points with $\alpha \in [2\pi, 4\pi)$, and (iii) the points with $\alpha \in [4\pi, 6\pi)$. The quotient action of the $\mathbb{Z}_3$-subgroup identifies the points in the first segment with the corresponding points in the second and third segments.

Roepstorff and Vehns also suggest replacing the electromagnetic subgroup $U(1)_Q \subset U(2)$ with a three-fold cover $\tilde{U}(1)_Q \subset S(U(3) \times U(2))$. The three-fold cover $\tilde{U}(1)_Q$, is the subgroup of $S(U(3) \times U(2))$ consisting of pairs of matrices of the form

$$\left(\begin{pmatrix} e^{i\beta} & 0 & 0 \\ 0 & e^{i\beta} & 0 \\ 0 & 0 & e^{i\beta} \end{pmatrix}, \begin{pmatrix} e^{i\alpha} & 0 \\ 0 & 1 \end{pmatrix}\right) \in U(3) \times U(2),$$

which satisfy the condition to be of unit determinant,

$$\left(e^{i\beta}\right)^3\left(e^{i\alpha}\right) = 1.$$

This condition can be equivalently stated as

$$3\beta + \alpha = 0 \mod 2\pi.$$

As with the three-fold cover of the weak hypercharge group, $\tilde{U}(1)_Q$ can be treated as a closed curve upon the surface of a 2-dimensional torus. Once again, one can treat $\alpha$ and $\beta$ as the two coordinates upon the torus, and the curve in question is the set of points which satisfy the condition that $\beta = -1/3\alpha$. To trace out the curve, start with $\alpha = 0$ and $\beta = 0$, and then allow $\alpha$ to increase, with $\beta$ changing according to the function $\beta = -1/3\alpha$. $\alpha$ and $\beta$ are individually mod $2\pi$, but one must allow $\alpha$ to run from 0 to $6\pi$ before the pair $(\alpha, \beta)$ returns to $(0 \mod 2\pi, 0 \mod 2\pi)$. Passing along this toroidal curve, one encounters the following sequence of reference points:

$$(\beta = 0, \alpha = 0),$$

$$\left(\beta = -\frac{1}{3}2\pi, \alpha = 2\pi\right),$$

$$\left(\beta = -\frac{1}{3}4\pi = -\frac{2}{3}2\pi, \alpha = 4\pi\right),$$

$$\left(\beta = -\frac{1}{3}6\pi = -2\pi, \alpha = 6\pi\right).$$

These points correspond to a $\mathbb{Z}_3$-subgroup of $\tilde{U}(1)_Q$. Recalling that $\alpha$ and $\beta$ correspond to

$$\left(e^{i\beta} Id_3, \begin{pmatrix} e^{i\alpha} & 0 \\ 0 & 1 \end{pmatrix}\right) \in S\big(U(3) \times U(2)\big),$$

the subgroup is

$$\mathbb{Z}_3 \cong \left\{\left(e^{-\frac{1}{3}2\pi i} Id_3, \begin{pmatrix} e^{i\alpha} & 0 \\ 0 & 1 \end{pmatrix}\right), (e^{-\frac{2}{3}2\pi i} Id_3, \begin{pmatrix} e^{i\alpha} & 0 \\ 0 & 1 \end{pmatrix}\right), \left(e^{-2\pi i} Id_3, \begin{pmatrix} e^{i\alpha} & 0 \\ 0 & 1 \end{pmatrix}\right)\right\}.$$

The covering map from $S(U(3) \times U(2))$ onto $U(2)$ induces the following covering map from $\tilde{U}(1)_Q$ onto $U(1)_Q \subset U(2)$:

$$\left(e^{i\beta} Id_3, \begin{pmatrix} e^{i\alpha} & 0 \\ 0 & 1 \end{pmatrix}\right) \mapsto \begin{pmatrix} e^{i\alpha} & 0 \\ 0 & 1 \end{pmatrix}.$$

The kernel of this map is the $\mathbb{Z}_3$-subgroup of $\tilde{U}(1)_Q$ specified above. Hence, $\tilde{U}(1)_Q/\mathbb{Z}_3 \cong U(1)_Q$. Thinking of $\tilde{U}(1)_Q$ as the curve upon the torus, one can, once again, divide it up into three segments: (i) the points with $\alpha \in [0, 2\pi)$, (ii) the points with $\alpha \in [2\pi, 4\pi)$, and (iii) the points with $\alpha \in [4\pi, 6\pi)$. The quotient action of the $\mathbb{Z}_3$-subgroup identifies the points in the first segment with the corresponding points in the second and third segments.

Appendix M

# Representations onto Spaces Tensored with $\mathbb{C}^1$

In general, for any complex vector space $V$, the tensor product $V \otimes \mathbb{C}^1$ or $\mathbb{C}^1 \otimes V$ is isomorphic to $V$. This means that a representation on $V \otimes \mathbb{C}^1$ or $\mathbb{C}^1 \otimes V$ is equivalent to a representation on $V$. In terms of the particle multiplet representations of $SU(3) \times SU(2) \times U(1)$ introduced in Section 5.2, this means:

- The $(1, 2, -1)$ representation on $\mathbb{C}^1 \otimes \mathbb{C}^2 \otimes \mathbb{C}^1$ (left-handed electron and neutrino) is equivalent to a representation on $\mathbb{C}^2$.
- The $(3, 2, 1/3)$ representation on $\mathbb{C}^3 \otimes \mathbb{C}^2 \otimes \mathbb{C}^1$ (left-handed up quark and down quark) is equivalent to a representation on $\mathbb{C}^3 \otimes \mathbb{C}^2$.
- The $(1, 1, -2)$ representation on $\mathbb{C}^1 \otimes \mathbb{C}^1 \otimes \mathbb{C}^1$ (right-handed electron) is equivalent to a representation on $\mathbb{C}^1$.
- The $(3, 1, 4/3)$ representation on $\mathbb{C}^3 \otimes \mathbb{C}^1 \otimes \mathbb{C}^1$ (right-handed up quark) is equivalent to a representation on $\mathbb{C}^3$.
- The $(3, 1, -2/3)$ representation on $\mathbb{C}^3 \otimes \mathbb{C}^1 \otimes \mathbb{C}^1$ (right-handed down quark) is equivalent to a representation on $\mathbb{C}^3$.

Let us look a little more closely at the effect of tensoring with $\mathbb{C}^1$. As a simple example, take the tensor product $\mathbb{C}^2 \otimes \mathbb{C}^1$, isomorphic to $\mathbb{C}^2$. Let $w$ be an element of $\mathbb{C}^2$ and let $z$ be an element of $\mathbb{C}^1$. A simple tensor $w \otimes z$ equals $zw \otimes 1$ by definition of the tensor product. A linear combination of simple tensors $w_1 \otimes z_1 + w_2 \otimes z_2$ equals $z_1 w_1 \otimes 1 + z_2 w_2 \otimes 1 = (z_1 w_1 + z_2 w_2) \otimes 1$. So each element in $\mathbb{C}^2 \otimes \mathbb{C}^1$ can be mapped to an element of $\mathbb{C}^2 \otimes 1$, and $\mathbb{C}^2 \otimes 1$ is isomorphic to $\mathbb{C}^2$ in a very trivial fashion. The map which sends $w \otimes z \in \mathbb{C}^2 \otimes \mathbb{C}^1$ to $zw \in \mathbb{C}^2$ is the linear isomorphism between $\mathbb{C}^2 \otimes \mathbb{C}^1$ and $\mathbb{C}^2$; its inverse[1] is the map which sends $v \in \mathbb{C}^2$ to $v \otimes 1 \in \mathbb{C}^2 \otimes \mathbb{C}^1$.

Given that $\mathbb{C}^2 \otimes \mathbb{C}^1$ is linearly isomorphic to $\mathbb{C}^2$, a representation of, say, $SU(2) \times U(1)$ on the tensor product $\mathbb{C}^2 \otimes \mathbb{C}^1$ is equivalent to a representation of $SU(2) \times U(1)$ on $\mathbb{C}^2$, as we shall now demonstrate.

Let $A$ denote an element of $SU(2)$ and let $e^{i\theta}$ denote an element of $U(1)$. Then define $\rho_n$ to be the following tensor product representation of $SU(2) \times U(1)$ on $\mathbb{C}^2 \otimes$

[1] Private communication with John C. Baez.

$\mathbb{C}^1$:

$$\rho_n\big(A, e^{i\theta}\big) = A \otimes e^{in\theta},$$

so that

$$\big(A \otimes e^{in\theta}\big) w \otimes z = Aw \otimes e^{in\theta} z.$$

Define $\tau_n$ to be the following representation of $SU(2) \times U(1)$ on $\mathbb{C}^2$:

$$\tau_n\big(A, e^{i\theta}\big) = e^{in\theta} A.$$

Let $T$ denote the linear isomorphism

$$T : \mathbb{C}^2 \otimes \mathbb{C}^1 \to \mathbb{C}^2,$$

which is such that

$$T(w \otimes z) = zw \in \mathbb{C}^2.$$

For each $n$, the representation $\rho_n$ is equivalent to the representation $\tau_n$. This means that for each $a \in SU(2) \times U(1)$, the following condition is satisfied

$$\tau_n(a)T = T\rho_n(a).$$

To see this, let $w \otimes z$ denote an arbitrary element of $\mathbb{C}^2 \otimes \mathbb{C}^1$, and note that

$$\big(T \circ \rho_n(a)\big)(w \otimes z) = T\big(Aw \otimes e^{in\theta} z\big) = e^{in\theta} zAw,$$

and

$$\big(\tau_n(a) \circ T\big)(w \otimes z) = \tau_n(a)(zw) = e^{in\theta} Azw.$$

Appendix N

# Electroweak Interaction Bundles

The purpose of this appendix is to define three possible electroweak interaction bundles, and to explore the relationships between them. As noted in Section 5.5, whilst the fermion multiplets which exist in our own universe indicate that the gauge group of the electroweak interaction is $U(2)$, other universes could possess fermion multiplets which require an $SU(2) \times U(1)$ gauge group. Moreover, as emphasised in Section 3.2, a choice of gauge group $G$ does not, in general, determine a unique principal fibre bundle with structure group $G$, hence the choice of a gauge group does not even determine a unique vector bundle structure for the interaction bundle.

Let us consider one obvious choice of $SU(2) \times U(1)$-interaction bundle and two obvious choices of $U(2)$-interaction bundle. Let $\tau$ denote a complex vector bundle, of fibre dimension two, equipped with an $SU(2)$ structure, and let $\chi$ denote a complex vector bundle, of fibre dimension one, equipped with a $U(1)$-structure. $SU(2)$ acts freely on the bases in each fibre of $\tau$ and $U(1)$ acts freely on the bases in each fibre of $\chi$. Each fibre of $\tau$ is equipped with a Hermitian inner product and a compatible volume form, which together single out an orbit of the $SU(2)$-action, namely the set of oriented, orthonormal bases in that fibre. Each fibre of $\chi$ is equipped with a Hermitian inner product which singles out an orbit of the $U(1)$-action, namely the set of unit norm vectors in that fibre.

The three choices of electroweak interaction bundle are:

- $\iota$, a complex vector bundle of fibre dimension two, equipped with a $U(2)$-structure. The $U(2)$-structure corresponds to the existence of a Hermitian inner product in each fibre. $\iota$ is isomorphic to the associated vector bundle obtained from a $U(2)$-principal fibre bundle using the standard representation of $U(2)$ on $\mathbb{C}^2$.
- $\varsigma = \tau \oplus \chi$, a complex vector bundle of fibre dimension three, equipped with an $SU(2) \times U(1)$-structure. Each fibre decomposes into the direct sum of a 2-dimensional vector space which possesses a Hermitian inner product and compatible volume form, and a 1-dimensional vector space which possesses a Hermitian inner product. $\varsigma = \tau \oplus \chi$ is isomorphic to the associated vector bundle obtained from an $SU(2) \times U(1)$-principal fibre bundle by using the standard representation of $SU(2) \times U(1)$ on $\mathbb{C}^2 \oplus \mathbb{C}^1$.
- The tensor product bundle $\tau \otimes \chi$, a complex vector bundle of fibre dimension two, which possesses a $U(2)$-structure.[1] $\tau \otimes \chi$ is isomorphic to the associated vector

[1] Private communication with Andrzej Derdzinski.

bundle obtained from an $SU(2) \times U(1)$-principal fibre bundle by using the representation $r = \rho \circ \phi$ of $SU(2) \times U(1)$ on $\mathbb{C}^2$, which is obtained by composing the covering homomorphism $\phi : SU(2) \times U(1) \to U(2)$ with the standard representation $\rho$ of $U(2)$ on $\mathbb{C}^2$.

Take the simplest case first, the bundle $\iota$ equipped with a $U(2)$ structure. This structure corresponds to a principal fibre bundle with structure group $U(2)$, which I shall denote as $Q$. The fibre of $Q$ over a point $x$ consists of vector pairs $F = (F_1, F_2)$, where each $(F_1, F_2)$ is an orthonormal pair of vectors from the fibre of $\iota$ over $x$. Where $\rho$ denotes the standard representation of $U(2)$ on $\mathbb{C}^2$, the associated bundle $Q \times_\rho \mathbb{C}^2$ is isomorphic to $\iota$. The associated bundle consists of equivalence classes $[F, v]$, where $F \equiv (F_1, F_2) \in Q$ and $v \equiv (v^1, v^2) \in \mathbb{C}^2$, and where the equivalence relationship is defined by

$$(F, v) \sim \big(FX, \rho\big(X^{-1}\big)v\big),$$

for any $X \in U(2)$. $(FX, \rho(X^{-1})v)$ denotes the pair obtained from the right action of $X$ on the basis $F$, and the action of the inverse $X^{-1}$ upon the vector $v$ under the representation $\rho$. Under the isomorphism between $Q \times_\rho \mathbb{C}^2$ and $\iota$, each pair $(F, v) \equiv ((F_1, F_2), v) \in Q \times \mathbb{C}^2$ is mapped to $v^1 F_1 + v^2 F_2 \in \iota$, and this map is constant on the elements within an equivalence class.

Now consider the direct sum bundle $\varsigma = \tau \oplus \chi$, the bundle equipped with an $SU(2) \times U(1)$ structure. The structure in this vector bundle corresponds to a principal fibre bundle with structure group $SU(2) \times U(1)$, which I shall denote as $P$. The fibre of $P$ over a point $x$ consists of vector triples $E = (F_1, F_2, G)$, where each $(F_1, F_2)$ is an oriented orthonormal pair of vectors from the fibre of $\tau$ over $x$, and G is a unit norm vector from the fibre of $\chi$ over $x$.

Where $s$ denotes the direct sum of the standard representation of $SU(2)$ on $\mathbb{C}^2$ with the standard representation of $U(1)$ on $\mathbb{C}^1$, the associated bundle $P \times_s (\mathbb{C}^2 \oplus \mathbb{C}^1)$ is isomorphic to $\tau \oplus \chi$. The associated bundle consists of equivalence classes $[E, u]$, where $E \equiv (F_1, F_2, G) \in P$ and $u \equiv (v^1, v^2, w) \in \mathbb{C}^2 \oplus \mathbb{C}^1$, and where the equivalence relationship is defined by

$$(E, u) \sim \big(ET, s\big(T^{-1}\big)u\big),$$

for any $T \in SU(2) \times U(1)$. The isomorphism between $P \times_s (\mathbb{C}^2 \oplus \mathbb{C}^1)$ and $\tau \oplus \chi$ maps $(E, u) = ((F_1, F_2, G), (v^1, v^2, w))$ to $(v^1 F_1 + v^2 F_2) \oplus wG$. Let us demonstrate that this map is constant on the elements in an equivalence class. To do so, we will show that for any $T$, the pair $(ET, s(T^{-1})u)$ is mapped to the same element as $(E, u)$. An arbitrary $T \in SU(2) \times U(1)$ can be written as $(A, e^{i\theta})$, hence, using again the notation $F \equiv (F_1, F_2)$ and $v \equiv (v^1, v^2)$, we can write:

$$\begin{aligned}\big(ET, s\big(T^{-1}\big)u\big) &= \big((F, G)\big(A, e^{i\theta}\big), \big(A^{-1}, e^{-i\theta}\big)(v, w)\big) \\ &= \big(\big(FA, e^{i\theta}G\big), \big(A^{-1}v, e^{-i\theta}w\big)\big).\end{aligned}$$

The pair $((FA, e^{i\theta}G), (A^{-1}v, e^{-i\theta}w))$ maps to

$$\begin{aligned}(FA)\big(A^{-1}v\big) \oplus e^{-i\theta}we^{i\theta}G &= Fv \oplus wG \\ &= \big(v^1F_1 + v^2F_2\big) \oplus wG,\end{aligned}$$

the same element that $(E, u)$ is mapped to.

Now consider the tensor product bundle $\tau \otimes \chi$, the bundle equipped with a $U(2)$ structure, but obtainable from an $SU(2) \times U(1)$ principal fibre bundle $P$. Where $r$ denotes the representation $r = \rho \circ \phi$ of $SU(2) \times U(1)$ on $\mathbb{C}^2$, obtained by composing the covering homomorphism $\phi : SU(2) \times U(1) \to U(2)$ with the standard representation $\rho$ of $U(2)$ on $\mathbb{C}^2$, the associated vector bundle $P \times_r \mathbb{C}^2$ is isomorphic to the tensor product bundle $\tau \otimes \chi$. The associated bundle is the set of equivalence classes $[E, v]$ where $E \equiv (F_1, F_2, G) \in P$ and $v \in \mathbb{C}^2$, and the equivalence relationship is defined by

$$(E, v) \sim \big(ET, r\big(T^{-1}\big)v\big),$$

for any $T \in SU(2) \times U(1)$. There is an isomorphism between $P \times_r \mathbb{C}^2$ and $\tau \otimes \chi$ which maps each pair $(E, v)$ to $(v^1F_1 + v^2F_2) \otimes G$. Let us demonstrate that this map is constant on each element in an equivalence class $[E, v]$. To do so, we can write

$$\begin{aligned}\big(ET, r\big(T^{-1}\big)v\big) &= \big((F, G)\big(A, e^{i\theta}\big), r\big(A^{-1}, e^{-i\theta}\big)v\big) \\ &= \big(\big(FA, e^{i\theta}G\big), \big(e^{-i\theta}A^{-1}v\big)\big),\end{aligned}$$

and the pair $((FA, e^{i\theta}G), (e^{-i\theta}A^{-1}v)) \in P \times \mathbb{C}^2$ maps to

$$\begin{aligned}(FA)\big(e^{-i\theta}A^{-1}v\big) \otimes e^{i\theta}G &= \big(e^{-i\theta}\big)(FA)\big(A^{-1}v\big) \otimes e^{i\theta}G \\ &= e^{-i\theta}Fv \otimes e^{i\theta}G \\ &= Fv \otimes e^{-i\theta}e^{i\theta}G \\ &= Fv \otimes G \\ &= \big(v^1F_1 + v^2F_2\big) \otimes G.\end{aligned}$$

Let us now consider some of the relationships between these bundles. First note that there is no faithful (1-to-1) 2-dimensional representation of $SU(2) \times U(1)$, and there is no vector bundle with 2-dimensional fibres that possesses an $SU(2) \times U(1)$-structure. There is no isomorphism between the associated bundle $P \times_r \mathbb{C}^2$, isomorphic to the tensor product bundle $\tau \otimes \chi$, and the direct sum vector bundle $\tau \oplus \chi$. Suppose that one tries to map the equivalence class $[E, v]$ to $(v^1F_1 + v^2F_2) \oplus G \in \tau \oplus \chi$. The pair $(E, v) = ((F_1, F_2, G), v)$ can be mapped to $(v^1F_1 + v^2F_2) \oplus G$, but $(ET, r(T^{-1})v)$ maps to

$$e^{-i\theta}(FA)\big(A^{-1}v\big) \oplus e^{i\theta}G = e^{-i\theta}Fv \oplus e^{i\theta}G = e^{-i\theta}\big(v^1F_1 + v^2F_2\big) \oplus e^{i\theta}G.$$

Unlike a tensor product, the definition of a direct sum does not allow one to move the multiplicative factor $e^{-i\theta}$ to the other side of the direct sum, so the $e^{-i\theta}$ and $e^{i\theta}$ factors don't cancel out.

The curiosity here is that the bundle $\tau \otimes \chi$, isomorphic to a bundle associated with an $SU(2) \times U(1)$-principal fibre bundle, does not possess an $SU(2) \times U(1)$-structure, but a $U(2)$-structure. The $U(2)$-structure consists of vector pairs $(F_1 \otimes G, F_2 \otimes G)$, where the $(F_1, F_2)$ are oriented orthonormal pairs in $\tau$ and the $G$ are unit norm vectors in $\chi$. $U(2) \cong SU(2) \times U(1)/\mathbb{Z}_2$ acts freely and transitively upon the set of such bases. Let $X$ denote an arbitrary element in $U(2)$, which is covered by the pair of $SU(2) \times U(1)$ elements, $(A, e^{i\theta})$ and $(-A, -e^{i\theta})$. Then define the action of $X$ to be

$$(F_1 \otimes G, F_2 \otimes G)X = (F_1' \otimes G', F_2' \otimes G'),$$

where $(F_1', F_2') = (F_1, F_2)A$ and $G' = e^{i\theta}G$.

The action would be the same if defined with respect to the other covering element, $(-A, -e^{i\theta})$. To see this, first note carefully that $(-A, -e^{i\theta}) \neq (A^{-1}, e^{-i\theta})$. To be explicit, $-A = e^{i\pi}A$ and $-e^{i\theta} = e^{i\pi}e^{i\theta}$, hence

$$(F_1, F_2)(-A) = e^{i\pi}(F_1', F_2'),$$

and

$$-e^{i\theta}G = e^{i\pi}e^{i\theta}G.$$

Thus,

$$\begin{aligned}\left(e^{i\pi}F_1' \otimes e^{i\pi}e^{i\theta}G, e^{i\pi}F_2' \otimes e^{i\pi}e^{i\theta}G\right) &= \left(F_1' \otimes e^{i\theta}G, F_2' \otimes e^{i\theta}G\right)\\ &= (F_1' \otimes G', F_2' \otimes G'),\end{aligned}$$

by the properties of the tensor product operation.

The $U(2)$ action is not directly upon the set of bases in each fibre of $\tau$, otherwise there would be elements in $U(2)$ which would map oriented orthonormal pairs $(F_1, F_2)$ to oppositely oriented orthonormal pairs $(F_1', F_2')$. The extra tensor product factor $\chi$ 'absorbs' the additional mappings in $U(2)$ which would map outside the $SU(2)$-structure in $\tau$. To elaborate, consider the matrix

$$R = \begin{pmatrix} -1 & 0 \\ 0 & 1 \end{pmatrix}.$$

This is a unitary $2 \times 2$ matrix, i.e., an element of $U(2)$, which does not belong to $SU(2)$ because it has det $= -1$. This matrix reverses the orientation of orthonormal bases in a fibre of $\tau$:

$$(F_1, F_2)\begin{pmatrix} -1 & 0 \\ 0 & 1 \end{pmatrix} = (-F_1, F_2).$$

This orientation-reversing element of $U(2)$ is covered by the pair

$$\tilde{R} = \left(\begin{pmatrix} e^{i\pi/2} & 0 \\ 0 & -e^{i\pi/2} \end{pmatrix}, e^{i\pi/2}\right) \in SU(2) \times U(1).$$

The matrix

$$S = \begin{pmatrix} e^{i\pi/2} & 0 \\ 0 & -e^{i\pi/2} \end{pmatrix}$$

belongs to $SU(2)$, having unit determinant, $\det = e^{i\pi/2}(-e^{i\pi/2}) = 1$, and having an adjoint matrix,

$$S^* = \begin{pmatrix} -e^{i\pi/2} & 0 \\ 0 & e^{i\pi/2} \end{pmatrix},$$

which equals the inverse $S^{-1}$. The matrix $S$ therefore describes an orientation-preserving transformation.

The covering homomorphism $\phi : SU(2) \times U(1) \to U(2)$ maps the pair $\tilde{R} = (S, e^{i\pi/2})$ to:

$$e^{i\pi/2}\begin{pmatrix} e^{i\pi/2} & 0 \\ 0 & -e^{i\pi/2} \end{pmatrix} = \begin{pmatrix} e^{i\pi} & 0 \\ 0 & -e^{i\pi} \end{pmatrix} = \begin{pmatrix} -1 & 0 \\ 0 & 1 \end{pmatrix} = R.$$

In this sense, the $U(1)$ factors are capable of changing an orientation preserving transformation $S$ into an orientation-reversing transformation $R$. The $U(2)$ action on the bases in $\tau \otimes \chi$ is such that the $SU(2)$ factors act on the bases of $\tau$, and the $U(1)$ factors are 'absorbed' by the presence of the extra tensor product factor $\chi$.

The relationship between the direct sum $\tau \oplus \chi$ and the tensor product $\tau \otimes \chi$ is defined by a mapping which is equivariant with respect to the homomorphism $\phi : SU(2) \times U(1) \to U(2)$. The $G$-structure of the direct sum $\tau \oplus \chi$ consists of orbits of bases of the form $(F_1, F_2, G)$. The $G$-structure of the tensor product $\tau \otimes \chi$ consists of orbits of bases of the form $(F_1 \otimes G, F_2 \otimes G)$. Let $\varpi$ denote the equivariant mapping $\varpi(F_1, F_2, G) = (F_1 \otimes G, F_2 \otimes G)$. $\varpi$ is equivariant in the sense that

$$\varpi\big((F_1, F_2, G), (A, e^{i\theta})\big) = \big(\varpi(F_1, F_2, G)\big)\phi\big(A, e^{i\theta}\big).$$

To demonstrate this, let

$$(F_1, F_2, G), \big(A, e^{i\theta}\big) = \big(FA, e^{i\theta}G\big) = (F_1', F_2', G').$$

Then

$$\varpi\big((F_1, F_2, G), (A, e^{i\theta})\big) = \varpi(F_1', F_2', G') = (F_1' \otimes G', F_2' \otimes G').$$

Given that

$$\varpi(F_1, F_2, G) = (F_1 \otimes G, F_2 \otimes G),$$

it follows that

$$\begin{aligned} \big(\varpi(F_1, F_2, G)\big)\phi\big(A, e^{i\theta}\big) &= (F_1 \otimes G, F_2 \otimes G)\phi\big(A, e^{i\theta}\big) \\ &= (F_1 \otimes G, F_2 \otimes G)e^{i\theta}A \\ &= \big(e^{i\theta}F_1' \otimes G, e^{i\theta}F_2' \otimes G\big) \end{aligned}$$

$$= \left(F_1' \otimes e^{i\theta}G,\, F_2' \otimes e^{i\theta}G\right)$$
$$= (F_1' \otimes G',\, F_2' \otimes G').$$

The fact that $\iota$ and $\tau \otimes \chi$ are both vector bundles of fibre dimension two, over the same space–time $\mathcal{M}$, both equipped with a $U(2)$-structure, does not entail that $\iota$ and $\tau \otimes \chi$ are $G$-bundle isomorphic, or even vector bundle isomorphic. Recall that for a general space–time $\mathcal{M}$ there can be many non-isomorphic principal fibre bundles with structure group $U(2)$. In fact, as stated in Section 3.2, the isomorphism classes of such principal fibre bundles will be in one-to-one correspondence with the elements of $H^2(\mathcal{M};\mathbb{Z}) \oplus H^4(\mathcal{M};\mathbb{Z})$. The associated vector bundles obtained from such non-isomorphic principal fibre bundles, via the standard representation of $U(2)$ on $\mathbb{C}^2$, will be non-isomorphic as vector bundles. Hence, there is no reason why a pair of $U(2)$-bundles must be isomorphic. In particular, there is no reason why a $U(2)$-bundle $\iota$ should be isomorphic to a tensor product $U(2)$-bundle $\tau \otimes \chi$; such a tensor product bundle is not constructible out of an arbitrary $U(2)$-bundle $\iota$.[2]

There is an asymmetry in the relationship between the existence of a $U(2)$-bundle and the existence of an $SU(2) \times U(1)$ bundle. If one starts with $\tau$ and $\chi$, and the direct sum $SU(2) \times U(1)$ bundle $\tau \oplus \chi$, then one can naturally construct a $U(2)$-bundle, the tensor product $\tau \otimes \chi$. However, the opposite is not true. If one starts with a $U(2)$ bundle $\iota$, then in general $\tau$ and $\chi$ may not exist, and even if they do exist, they are not constructible out of $\iota$, hence the $U(2)$-bundle provided by their tensor product is not necessarily isomorphic to the $U(2)$-bundle $\iota$.

Let us quickly prove[3] that there are $U(2)$-bundles $\iota$ which are not isomorphic to a tensor product $U(2)$-bundle $\tau \otimes \chi$. If it were the case that $\iota \cong \tau \otimes \chi$, then it would follow that

$$\bigwedge^2 \iota^* = (\bigwedge^2 \tau^*) \otimes \chi \otimes \chi.$$

However, $\bigwedge^2 \tau^*$ is trivial in the sense that $\bigwedge^2 \tau^* \cong \mathcal{M} \times \mathbb{C}^1$ (the $SU(2)$ structure in $\tau$ is precisely a trivializing section for $\bigwedge^2 \tau^*$), hence

$$\bigwedge^2 \iota^* = \chi \otimes \chi.$$

Thus, a $U(2)$-bundle $\iota$ is isomorphic to a tensor product bundle only if $\bigwedge^2 \iota^*$ is the tensor square of some complex line bundle. Now, the first Chern class of a complex vector bundle over a real manifold $\mathcal{M}$ is an element of the second cohomology group $H^2(\mathcal{M})$. I.e., the first Chern class is an equivalence class of closed 2-forms on the real manifold $\mathcal{M}$. The integral of a representative element from the first Chern class over a 2-cycle[4] in $\mathcal{M}$ is an integer which is independent of the chosen representative.

---

[2] Private communication with Andrzej Derdzinski.

[3] This proof was suggested to me by Andrzej Derdzinski.

[4] A $p$-chain is basically a formal linear combination of $p$-dimensional domains of integration, and a $p$-chain with no boundary is called a $p$-cycle.

If a complex vector bundle $\bigwedge^2 \iota^*$ is isomorphic to the tensor square of some complex line bundle, it entails that the integral of a representative closed 2-form from the first Chern class of $\bigwedge^2 \iota^*$ over a 2-cycle in $\mathcal{M}$, is an integer divisible by 2. It also follows that the integral of a closed 2-form from the first Chern class of $\iota$, must be an integer divisible by 2. Hence, a $U(2)$-bundle $\iota$ is only isomorphic to a tensor product bundle $\tau \otimes \chi$ if this condition is satisfied.

However, there are many cases in which this condition fails to be satisfied. Take, for example, the complex projective plane $\mathcal{M} = \mathbb{C}P^2$, a 4-dimensional real manifold, and consider its tangent bundle $\iota = T\mathbb{C}P^2$, a complex vector bundle with two-dimensional fibres. Given a representative from the first Chern class of $\iota = T\mathbb{C}P^2$, a closed 2-form on $\mathbb{C}P^2$, its integral over any complex line in $\mathbb{C}P^2$ (a cycle of real dimension 2 in $\mathbb{C}P^2$), equals 3 rather than an even integer.

$\mathbb{C}P^2$ is not considered to be a physically realistic space–time because it is a compact 4-manifold of non-zero Euler characteristic, and therefore cannot be equipped with a Lorentzian metric. However, one can obtain a physically realistic space–time from $\mathbb{C}P^2$ which *does* accept a Lorentzian metric. One can choose a vector field on $\mathbb{C}P^2$ with just 3 zero points, and by removing those points, one obtains a compact manifold of Euler characteristic zero. The vector field this manifold inherits is non-zero everywhere, and can be used to define both a Lorentzian metric and a time orientation. In this manifold, there are still plenty of projective lines, and so the argument above concerning the tangent bundle remains valid.

In the special case where the space–time is Minkowski space–time, all the bundles can be trivialized as product bundles. In this case, we can define the three bundles discussed above as product bundles: $\iota = \mathcal{M} \times \mathbb{C}^2$, equipped with a $U(2)$-structure; $\tau = \mathcal{M} \times \mathbb{C}^2$, equipped with an $SU(2)$-structure; and $\chi = \mathcal{M} \times \mathbb{C}^1$, equipped with a $U(1)$-structure. The direct sum bundle $\tau \oplus \chi$ is an $SU(2) \times U(1)$ bundle defined as

$$\varsigma = \tau \oplus \chi = (\mathcal{M} \times \mathbb{C}^2) \oplus (\mathcal{M} \times \mathbb{C}^1) = \mathcal{M} \times (\mathbb{C}^2 \oplus \mathbb{C}^1),$$

and the tensor product bundle $\tau \otimes \chi$ is a $U(2)$-bundle defined as

$$\tau \otimes \chi = (\mathcal{M} \times \mathbb{C}^2) \otimes (\mathcal{M} \times \mathbb{C}^1) = \mathcal{M} \times (\mathbb{C}^2 \otimes \mathbb{C}^1) \cong \mathcal{M} \times \mathbb{C}^2.$$

Hence, in the special case of these product bundles at least, the $U(2)$-bundle $\iota = \mathcal{M} \times \mathbb{C}^2$ is vector bundle isomorphic to a tensor product bundle $U(2)$-bundle $\tau \otimes \chi$.

# Bibliography


Aitchison, I. J. R. (1991). The vacuum and unification. In S. W. Saunders, & H. R. Brown (Eds.), *The philosophy of vacuum* (pp. 159–195). Oxford: Clarendon Press.

Baez, J. C. (1996). This Week's Finds in Mathematical Physics, Week 93, October 27, 1996. http://math.ucr.edu/home/baez/week93.html.

Baez, J. C. (1998). This Week's Finds in Mathematical Physics, Week 119, April 13, 1998. http://math.ucr.edu/home/baez/week119.html.

Baez, J. C. (1999). This Week's Finds in Mathematical Physics, Week 133, April 23, 1999. http://math.ucr.edu/home/baez/week133.html.

Baez, J. C. (2003c). The True Internal Symmetry Group of the Standard Model, May 19, 2003. http://math.ucr.edu/home/baez/qg-spring2003/true.

Baez, J. C. (2005). Symmetries, Groups, and Categories, May 16, 2005. http://math.ucr.edu/home/baez/symmetries.html.

Barrow, J. D., & Tipler, F. J. (1986). *The Anthropic Cosmological Principle*. Oxford: Oxford University Press.

Berestetskii, V. B., Lifshitz, E. M., & Pitaevskii, L. P. (1982). *Quantum Electrodynamics* (2nd ed.). Oxford: Pergamon Press.

Berg, M., DeWitt-Morette, C., Gwo, S., & Kramer, E. (2001). The Pin Groups in Physics: C, P and T. *Rev. Math. Phys. 13*, 953–1034.

Bilson-Thompson, S. O., Markopoulou, F., & Smolin, L. (2006). Quantum gravity and the standard model. hep-th/0603022.

Bleecker, D. (1981). *Gauge Theory and Variational Principles*. Reading, MA: Addison–Wesley.

Bogolubov, N. N., Logunov, A. A., & Todorov, I. T. (1975). *Introduction to Axiomatic Quantum Field Theory*. Reading, MA: Benjamin.

Bogolubov, N. N., Logunov, A. A., Oksak, A. I., & Todorov, I. T. (1990). *General Principles of Quantum Field Theory*. Dordrecht: Kluwer.

Boothby, W. M. (1986). *An Introduction to Differentiable Manifolds and Riemannian Geometry* (2nd ed.). London: Academic Press.

Bueno, O. (2001). Weyl and von Neumann: symmetry, group theory, and quantum mechanics, draft paper presented. In *Symmetries in Physics, New Reflections*, Oxford Workshop, January 2001.

Callender, C. (2005). Answers in search of a question: 'proofs' of the tri-dimensionality of space. In *Studies in History and Philosophy of Modern Physics*, Vol. 36, Issue 1, March 2005, pp. 113–136.

Cao, T. Y. (1999). Introduction: Conceptual issues in quantum field theory. In T. Y. Cao (Ed.), *Conceptual Foundations of Quantum Field Theory*. Cambridge: Cambridge University Press.

Castellani, E. (1998). *Introduction to Interpreting Bodies*. Princeton, NJ: Princeton University Press.

Choquet-Bruhat, Y., DeWitt-Morette, C., & Dillard-Bleick, M. (1982). *Analysis, Manifolds and Physics, Vol. 1: Basics*. Amsterdam: North-Holland.

Dautray, R., & Lions, J.-L. (1988). *Mathematical Analysis and Numerical Methods for Science and Technology* (pp. 6–7). Berlin: Springer-Verlag (Chapter IX, §1).

Dawid, R. (2003). Realism in the age of string theory, Paper presented at *Symmetries in Physics, New Reflections*, Oxford Workshop, January 2001.

Derdzinski, A. (1992). *Geometry of the Standard Model of Elementary Particles. Texts and Monographs in Physics*. Berlin: Springer-Verlag.

Derdzinski, A. (2002). Quantum electrodynamics and gauge fields for mathematicians, unpublished manuscript, Department of Mathematics, The Ohio State University, September 2002.

DeWitt, B. S., Hart, C. F., & Isham, C. J. (1979). Topology and quantum field theory. *Physica A 96*, 197–211.

Earman, J. (2002). Laws, Symmetry, and Symmetry Breaking; Invariance, Conservation Principles, and Objectivity, Presidential address PSA 2002.

Earman, J., & Fraser, D. (2006). Haag's theorem and its implications for the foundations of quantum field theory, Erkenntnis, in press.

Emch, G. G. (1984). *Mathematical and Conceptual Foundations of 20th-Century Physics*. Amsterdam: North-Holland.

Fell, J. M. G., & Doran, R. S. (1988). *Representations of *-Algebras, Locally Compact Groups, and Banach *-Algebraic Bundles, Vol. 1*. Boston: Academic Press.

Fleming, G. N., & Butterfield, J. (1999). Strange positions. In J. Butterfield, & C. Pagonis (Eds.), *From Physics to Philosophy*. Cambridge: Cambridge University Press.

Fraser, D. (2005). The fate of 'particles' in quantum field theories with interactions, Department of History and Philosophy of Science, University of Pittsburgh, preprint.

French, S., & Rickles, D. (2003). Understanding Permutation Symmetry. In K. Brading, & E. Castellani (Eds.), *Symmetries in Physics: Philosophical Reflections* (pp. 212–238). Cambridge: Cambridge University Press.

French, S. (2006). Identity and Individuality in Quantum Theory. In E. N. Zalta (Ed.), *The Stanford Encyclopedia of Philosophy* (Spring 2006 Edition).

Fulton, W., & Harris, J. (1991). *Representation Theory*. New York: Springer-Verlag.

Geroch, R., & Horowitz, G. (1979). Global structure of spacetimes. In S. W. Hawking, & W. Israel (Eds.), *General Relativity: An Einstein Centenary Survey* (pp. 212–293). Cambridge: Cambridge University Press.

Giere, R. N. (2000). Theories. In W. H. Newton-Smith (Ed.), *A Companion to the Philosophy of Science* (pp. 515–524). Oxford: Blackwell.

Halvorson, H. (2001). *Locality, Localization, and the Particle Concept: Topics in the Foundations of Quantum Field Theory*, University of Pittsburgh Dissertation.
Hawley, K. (2004). Temporal Parts. In E. N. Zalta (Ed.). *The Stanford Encyclopedia of Philosophy* (Winter 2004 Edition).
Hofmann, K. H., & Morris, S. A. (1998). *The Structure of Compact Groups*. Berlin: Walter de Gruyter.
Huang, K. (1992). *Quarks, Leptons and Gauge Fields* (2nd ed.). Singapore: World Scientific.
Isham, C. J. (1992). Conceptual and geometrical problems in quantum gravity. In H. Mitter, & H. Gausterer (Eds.), *Recent Aspects of Quantum Fields. Lecture Notes in Physics, Vol. 396* (pp. 123–230). Berlin: Springer-Verlag.
Kadison, R. V., & Ringrose, J. R. (1983). *Fundamentals of the Theory of Operator Algebras, Vol. 1*. New York: Academic Press.
Kuhlmann, M. (2006). Quantum Field Theory. In E. N. Zalta (Ed.), *The Stanford Encyclopedia of Philosophy* (Fall 2006 Edition).
Ladyman, J. (1998). What is structural realism? In *Studies in History and Philosophy of Science, Vol. 29* (pp. 409–424).
Manin, Y. I. (1988). *Gauge Field Theory and Complex Geometry*. Berlin: Springer-Verlag.
Morse, P. M., & Feshbach, H. (1953). *Methods of Theoretical Physics*. McGraw–Hill.
Nachtmann, O. (1990). *Elementary Particle Physics: Concepts and Phenomena*. Berlin: Springer-Verlag.
Ne'eman, Y., & Kirsh, Y. (1996). *The Particle Hunters* (2nd ed.). Cambridge: Cambridge University Press.
Newton, T. D., & Wigner, E. P. (1949). Localized states for elementary systems. *Review of Modern Physics 21*, 400–406.
Okun, L. B. (1996). Fundamental role of experiments in high energy physics. hep-ph/9614428.
O'Neill, B. (1983). *Semi-Riemannian Geometry: With Applications to Relativity*. Academic Press.
O'Raifeartaigh, L. (1986). *Group Structure of Gauge Theories*. Cambridge: Cambridge University Press.
Pickering, A. (1984). *Constructing Quarks: A Sociological History of Particle Physics*. Chicago: University of Chicago Press.
Penrose, R. (2004). *The Road to Reality*. London: Jonathan Cape.
Pooley, O. (2003). Handedness, parity violation, and the reality of space. In K. Brading, & E. Castellani (Eds.), *Symmetries in Physics: Philosophical Reflections* (pp. 250–280). Cambridge: Cambridge University Press.
Pooley, O. (2005). Points, particles, and structural realism. In S. French, D. Rickles, & J. Saatsi (Eds.), *Structural Foundations of Quantum Gravity*. Oxford: Oxford University Press.
Prugovecki, E. (1992). *Quantum Geometry*. Dordrecht: Kluwer.

Prugovecki, E. (1995). *Principles of Quantum General Relativity*. Singapore: World Scientific.

Rajasekaran, G. (1997). Perspectives in high energy physics. hep-ph/9709212.

Reed, M., & Simon, B. (1975). *Methods of Modern Mathematical Physics, Vol. II: Fourier Analysis, Self-Adjointness*. New York: Academic Press.

Rizov, V., & Vitev, I. (2000). A note on the gauge group of the electroweak interactions. hep-ph/0009169.

Roepstorff, G., & Vehns, Ch. (2001). Towards a unified theory of gauge and Yukawa interactions. hep-ph/0006065.

Rugh, S. E., & Zinkernagel, H. (2002). The quantum vacuum and the cosmological constant problem. *Studies in History and Philosophy of Modern Physics 33*(4), 663–705.

Ryder, L. H. (1986). *Elementary Particles and Symmetries* (Revised Edition). London: Gordon and Breach.

Saunders, S. (1991). The negative energy sea. In S. Saunders, & H. Brown (Eds.), *Philosophy of Vacuum* (pp. 65–109). Oxford: Clarendon Press.

Schücker, T. (1997). Geometries and forces. hep-th/9712095.

Schwartz, L. (1968). *Application of Distributions to the Theory of Elementary Particles in Quantum Mechanics*. New York: Gordon and Breach.

Simon, B. (1996). *Representations of Finite and Compact Groups. Graduate Studies in Mathematics, Vol. 10*. American Mathematical Society.

Smith, Q. (1995). Change. In J. Kim, & E. Sos (Eds.), *A Companion to Metaphysics* (pp. 83–85). Oxford: Blackwell.

Smolin, L. (1997). *The Life of the Cosmos*. London: Weidenfeld and Nicolson.

Smolin, L. (2003). How far are we from the quantum theory of gravity? hep-th/0303185.

Sneed, J. D. (1971). *The Logical Structure of Mathematical Physics*. Dordrecht: Reidel.

Sternberg, S. (1994). *Group Theory and Physics*. Cambridge: Cambridge University Press.

Streater, R. F. (1988). Why should anyone want to axiomatize quantum field theory? In H. R. Brown, & R. Harre (Eds.), *Philosophical Foundations of Quantum Field Theory* (pp. 137–148). Oxford: Clarendon Press.

Suppe, F. (1989). *The Semantic Conception of Theories and Scientific Realism*. Urbana, IL: University of Illinois Press.

Suppes, P. (1969). *Studies in the Methodology and Foundation of Science: Selected Papers from 1951 to 1969*. Dordrecht: Reidel.

Swoyer, C. (2000). Properties. In E. N. Zalta (Ed.). *The Stanford Encyclopedia of Philosophy* (Winter 2000 Edition).

Tegmark, M. (1998). Is 'the Theory of Everything' merely the ultimate ensemble theory? *Annals of Physics 270*, 1–51.

Teller, P. (1988). Three problems of renormalization. In H. R. Brown, & R. Harre (Eds.), *Philosophical Foundations of Quantum Field Theory* (pp. 73–89). Oxford: Clarendon Press.

Teller, P. (1995). *An Interpretive Introduction to Quantum Field Theory*. Princeton, NJ: Princeton University Press.

Ticciati, R. (1999). *Quantum Field Theory for Mathematicians*. Cambridge: Cambridge University Press.

Torretti, R. (1983). *Relativity and Geometry*. Oxford: Pergamon Press.

Torretti, R. (1999). *The Philosophy of Physics*. Cambridge: Cambridge University Press.

Tsou, S.-T. (2000). Concepts in gauge theory leading to electric–magnetic duality. hep-ph/0006178.

Tsou, S.-T. (2003). Electric–magnetic duality and the dualized Standard Model. *International Journal of Modern Physics A 18S2*, 1–40. hep-th/0110256.

Votsis, I. (2003). Is structure not enough? *Philosophy of Science 70*(5), 879–890.

Wallace, D. (2001). In defence of naivete: The conceptual status of Lagrangian quantum field theory. *Synthese*, in press.

Wallace, D. (2005). Emergence of particles from bosonic quantum field theory. *Studies in the History and Philosophy of Modern Physics*, in preparation.

Wald, R. M. (1984). *General Relativity*. Chicago: University of Chicago Press.

Weatherson, B. (2002). Intrinsic vs. Extrinsic Properties. In E. N. Zalta (Ed.), *The Stanford Encyclopedia of Philosophy* (Fall 2004 Edition).

Weyl, H. (1952). *Symmetry*. Princeton, NJ: Princeton University Press.

Wightman, A. S. (1973). Relativistic wave equations as singular hyperbolic systems. In *Proc. Symp. in Pure Math., Vol. XXIII, Berkeley 1971*. Providence, RI: AMS.

## Further reading

Baez, J. C. (2000). This Week's Finds in Mathematical Physics, Week 162, December 17, 2000. http://math.ucr.edu/home/baez/week162.html.

Baez, J. C. (2003a). Elementary Particles, April 29, 2003. http://math.ucr.edu/home/baez/qg-spring2003/elementary.

Baez, J. C. (2003b). Hypercharge and Weak Isospin, May 12, 2003. http://math.ucr.edu/home/baez/qg-spring2003/hypercharge.

Booss, B., & Bleecker, D. (1985). *Topology and Analysis: The Atiyah–Singer Index Formula and Gauge Theoretic Physics*. Berlin: Springer-Verlag.

Derdzinski, A. (1993). Geometry of Elementary Particles. In R. E. Greene, & S.-T. Yau (Eds.), *Proceedings of Symposia in Pure Mathematics, Vol. 54* (pp. 157–171) (Part 2).

Frogatt, C. D., & Nielsen, H. D. (1999). Masses and mixing angles and going beyond the Standard Model. hep-ph/9905445.

Herrero, M. (1998). The Standard Model. hep-ph/9812242.

Hughes, R. I. G. (1989). *The Structure and Interpretation of Quantum Mechanics*. Cambridge, MA: Harvard University Press.

Kobzarev, I. Y., & Manin, Y. I. (1989). *Elementary Particles: Mathematics, Physics and Philosophy*. Dordrecht: Kluwer.

Ne'eman, Y., & Sternberg, S. (1991). Internal symmetry and superconnections. In P. Donato (Ed.), *Symplectic Geometry and Mathematical Physics, Proceedings in Honor of Jean-Marie Souriau. Progress in Mathematics, Vol. 99* (pp. 326–354). Boston: Birkhäuser.

Roepstorff, G. (2000). A class of anomaly-free gauge theories. hep-th/0005079.

Russell, B. (1927). *The Analysis of Matter*. London: Allen and Unwin.

Saller, H. (1998). The central correlations of hypercharge, isospin, colour and chirality in the standard model. hep-ph/9802112.

Streater, R. F., & Wightman, A. S. (1964). *PCT, Spin and Statistics, and All That*. New York: Benjamin Cummings.

# Subject Index